NUREG/CP-0192

# Proceedings of the Nuclear Fuels Sessions of the 2004 Nuclear Safety Research Conference

**Held at**
**Marriott Hotel at Metro Center**
**Washington, DC**
**October 22-27, 2004**

Manuscript Completed: October 2005
Date Published: October 2005

Compiled by:
M. Snell

**Prepared by**
**Division of Systems Analysis and Regulatory Effectiveness**
**Office of Nuclear Regulatory Research**
**Washington, DC 20555-0001**

# ABSTRACT

This report contains papers from the nuclear fuels sessions of the 2004 Nuclear Safety Research Conference held at the Marriott Hotel at Metro Center in Washington, DC, October 25-27, 2004.

This information describes programs and results of work sponsored by the U.S. Nuclear Regulatory Commission's Office of Nuclear Regulatory Research. Also included are invited papers from others involved in nuclear fuels research.

The summaries, presentation slides, and full papers have been compiled here to provide a basis of information that was exchanged during the course of the meeting, and these are provided in the order they were presented.

Proceedings of the Nuclear Fuels Sessions of the
2004 Nuclear Safety Research Conference
October 25-27, 2004
Washington, DC, USA

CONTENTS

# Tuesday, October 26

*SESSION 3A :SPENT FUEL RESEARCH*

Chaired by: R. Meyer (NRC), M. Hodges (NRC)

Dry storage and transportation casks impose unique temperature and pressure conditions on spent fuel rods, and the consequences of these conditions are considered in cask licensing. Research results under these conditions and their implications were presented in this session.

## SESSION 3B: HIGH BURNUP FUEL UNDER LOCA CONDITIONS

Chaired by: R. Meyer, F. Akstulewicz (NRC), and R. Yang (EPRI)

Risk-informed, performance-based rulemaking is planned to revise 10 CFR 50.46 and Appendix K for LOCA analysis. Results of recent research will be presented, and these results will provide the technical basis for changing the embrittlement criteria in the rule.

# High-Burnup Cladding Mechanical Performance during Cask Storage and Post-Storage Handling and Transportation

**Robert S. Daum, Hanchung Tsai, Yung Liu, and Michael C. Billone**

Argonne National Laboratory (ANL)
Argonne, Illinois, USA

The assessment of cladding performance during spent-nuclear-fuel (SNF) pre-storage operations, dry-cask storage, and post-storage handling and transport, including hypothetical accidents, is important in ensuring that sub-criticality is maintained, that radioactivity is contained, that cask external dose rates are limited, and that SNF assemblies can be safely retrieved, handled and transported at the end of dry-cask storage. Although cladding failure is not prohibited by federal regulations (CFR 71 and 72), the failure mode and extent of cladding failure have a significant effect on the possible reconfiguration of SNF within storage and transport casks. High-burnup, SNF Zircaloy-4 (Zry-4) cladding from pressurized-water reactors is more susceptible to failure and possible fuel dispersal than low-burnup Zry-4 due to higher hydrogen pickup, higher internal pressure and higher corrosion level. Hydride precipitation, corrosion (i.e., wall thinning), irradiation-induced defects, and possibly even post-reactor thermal creep, will reduce cladding ductility and impact resistance. Also, transfer-and-drying operations conducted from the SNF pool to the storage cask may result in thermo-mechanical conditions that promote radial-hydride-induced degradation of cladding ductility and impact resistance during severe loadings associated with hypothetical accidents. Such cladding would be more susceptible to brittle failure at the end of dry storage when temperatures are expected to be 150-250°C – below the ductile-to-brittle transition temperature for zirconium hydrides. A testing program has been developed to investigate these effects in high-burnup Zry-4 cladding, with particular emphasis on the conditions that promote radial-hydride precipitation and the effects of these radial hydrides on cladding integrity. Preliminary experimental results are presented from axial tensile and thermal creep tests, as well as conditions for radial-hydride precipitation. The test plan, which includes ring-compression-ductility and ring-crush-impact screening tests, is also described

The current regulations for storage and transportation of SNF are designed primarily to maintain sub-criticality and to ensure that doses are less than regulatory limits, that the cask provides adequate fuel confinement and containment, and that the fuel is retrievable. Interim Staff Guidance No. 11, Revision 3 limits high-burnup cladding temperatures to ≤400°C during short-term operations and normal storage conditions. Although this temperature limit is intended to minimize radial-hydride precipitation in high-burnup cladding, license applications for high-burnup storage, storage-and-transport, and transport casks are evaluated on a case-by-case basis due to the lack of pertinent data to assess high-burnup cladding behavior during the complete cycle of pre-storage operations, storage and post-storage handling and transport. The data generated within the ANL test program are to be used both by applicants for high-burnup cask licenses (e.g., nuclear industry and DOE-RW) and by cask-license evaluators (NRC-SFPO).

Experiments have concentrated on the microstructural characterization and mechanical-property testing of stress-relieved Zry-4 (15x15 design) irradiated in H.B. Robinson Unit No. 2 to a rod-average burnup of 67 GWd/MTU and a fast neutron fluence (E > 1 MeV) of $14 \times 10^{26}$ $n/m^2$. In Grid Spans 3 and 4 (fuel midplane to 0.7-m above the fuel midplane), destructive examinations of the as-irradiated cladding have been performed. Circumferentially averaged (8 locations) corrosion layers of 70±10 to 100±10 μm, with no spalling, have been measured, along with cross-section-averaged (90° segments) hydrogen contents of 550±80 to 750±90 wppm. Precipitation of circumferential hydrides shows varying distribution, density, and particle size along the axial, azimuthal, and radial directions of the cladding. In a few cladding locations, transverse metallography in an etched condition shows a significantly high hydride density localized in roughly a 90° arc directly under the outer-surface corrosion layer and to a radial depth of ≈100 μm, suggesting the presence of a hydride "lens." Such hydride microstructures are known to reduce cladding ductility under tensile loading, but little is known about the evolution of these microstructures under thermo-mechanical conditions associated with SNF drying, transfer and storage.

As compared to non-irradiated Zry-4 (15x15 design), room-temperature axial tensile properties of the high-burnup (690±40 wppm H) Zry-4 at a strain rate of 0.1%/s show an increase in yield (600 → 770 MPa) and ultimate tensile (765 → 950 MPa) strengths and a decrease in uniform (6 → 3%) and total (14 → 4%) elongations. The strength increase appears to be due mainly to radiation-induced hardening, while the ductility decrease appears to be due to both radiation- and hydride-induced embrittlement. Thermal annealing tests with high-burnup Zry-4 samples show that strength properties (based on microhardness data) appear to recover by ≈75% and circumferential hydrides tend to homogenize across the cladding radius after 72 hours at 420°C. These results suggest that SNF drying operations may partially anneal radiation-induced hardening. The degree of ductility recovery with annealing remains to be demonstrated.

Two thermal creep tests (C14 and C15) have been completed using defueled high-burnup Zry-4 cladding specimens, which are top-welded to active internal-gas-pressurization systems in order to maintain constant gas pressure inside the creep specimens. These tests were conducted at 400°C for 101 days at a pressure of 29.5 MPa and an initial hoop stress of 190 MPa. The specimens were depressurized periodically prior to cooling to room-temperature for diameter measurements. Both specimens remained intact, no localized bulging (precursor to rupture) was observed, the average hoop creep strains were ≈3.6% and the peak hoop creep strains were ≈5%. In order to induce radial-hydride precipitation, the C15 specimen was cooled at ≈2.4°C/s from 400°C under full pressure during the final shut-down. The sample depressurized at 205°C and a midplane true hoop stress of ≈205 MPa due to failure in the upper weld region. Post-test metallography at three axial locations showed significant radial-hydride and negligible circumferential-hydride precipitation. However, post-test hydrogen measurements indicated substantial loss of hydrogen from the C15 specimen (≈670 → 320 wppm at the midplane) to the Zircadyne-702 end-fittings (12 → 210 wppm at the bottom end-plug). Redesign of the thermal creep test train and furnace is in progress to minimize the hydrogen loss and the axial temperature gradient from the specimen midplane (400°C) to 30 mm above the midplane (≈390°C). However, it is interesting to note that the hydrogen solubility of non-irradiated Zry-4 is ≈210 wppm at 400°C. The absence of visible circumferential hydrides at the specimen midplane (with 320 wppm) suggests that high-burnup Zry-4 is capable of trapping about 100 wppm of hydrogen, which most likely precipitates during rapid cooling as sub-micron-size hydrides. It will be interesting to determine if this excess hydrogen precipitates as visible radial hydrides under the slow cooling rates (<4°C/day) typical of drying-transfer-storage.

The preliminary axial-tensile, thermal-creep, and hydride-reorientation results have been used to develop a test plan to better understand the mechanical behavior of high-burnup Zry-4 cladding under drying-transfer, storage, and post-storage handling-transport conditions. In additional to tensile and creep tests of pool-stored high-burnup Zry-4, sealed specimens will be annealed for ≈3 days at 380-420°C and at hoop stresses of 0, 60, 90, 120, and 150 MPa and slow-cooled at ≈3°C/day under decreasing pressure. Rings cut from these 100-mm-long samples will be subjected to ductility (diametral compression at 0.1%/s and 100%/s) and crush-impact failure-energy screening tests. These tests will be conducted at room-temperature and 150°C. The decreases in ductility and failure-impact energy will be correlated to the extent of radial hydride formation to map out cooling conditions – especially stress at 400°C – that are detrimental to high-burnup Zry-4 cladding integrity. Additional tests (e.g., fracture toughness) may be conducted on cladding subjected to these detrimental cooling conditions.

# High-Burnup Cladding Mechanical Performance during Cask Storage and Post-Storage Handling and Transportation

*Rob Daum, Hanchung Tsai, Yung Liu, and Mike Billone*

*Nuclear Safety Research Conference*
*Washington, D.C., October 25-27, 2004*

## Argonne National Laboratory

A U.S. Department of Energy
Office of Science Laboratory
Operated by The University of Chicago

- **Characterize high-burnup, spent-nuclear-fuel cladding behavior during**

  - Pre-storage operations
  - Dry-cask storage
  - Post-storage handling and transport, including postulated accidents

- **Provide cladding mechanical-properties and characterization data relevant to certification of high-burnup-fuel transport and storage casks**

  - NRC (license evaluation)
  - Industry (license application)
  - DOE-RW (license application)

Pioneering
Science and
Technology

*Nuclear Safety Research Conference*
*Washington, D.C., October 25-27, 2004*

# Storage Process - Generalized

## Impact on Cladding Microstructure and Integrity

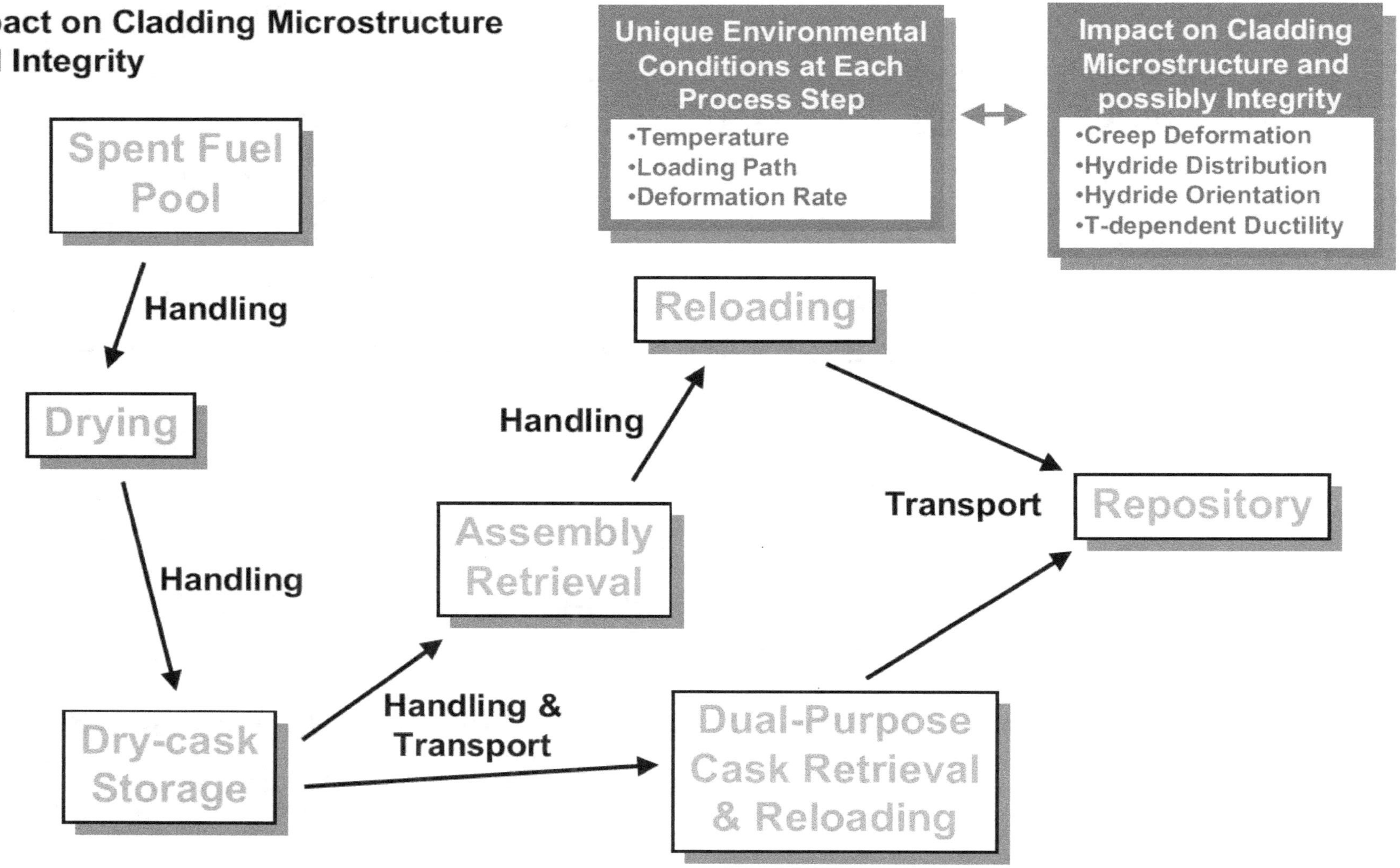

**Impact on Cladding Microstructure and possibly Integrity**
- Creep Deformation
- Hydride Distribution
- Hydride Orientation
- T-dependent Ductility

**Unique Environmental Conditions at Each Process Step**
- Temperature
- Loading Path
- Deformation Rate

Spent Fuel Pool

**Handling** →

Drying

**Handling** →

Dry-cask Storage

**Handling & Transport** →

Assembly Retrieval

**Handling** →

Reloading

**Transport** →

Repository

Dual-Purpose Cask Retrieval & Reloading

**Transport** →

Pioneering Science and Technology

*Nuclear Safety Research Conference*
*Washington, D.C., October 25-27, 2004*

5

# Data Needs based on Modeling

- ## End-Drop Accidents

  - Axial impact (mitigated somewhat by cask impact limiters)
  - Axial bending due to buckling
  - Data needs (in addition to effects of fuel on cladding response)

    Cladding AXIAL stress/plastic-strain properties (≤100%/s, RT-400°C)

    Failure limits: uniform elongation, plastic failure strain, CSED

- ## Side-Drop Accidents

  - Axial bending due to impact response (see data needs above)
  - Impact loads at grid spacers induce **HOOP** bending stresses
  - Data needs (in addition to effects of fuel on cladding response)

    Stress-temperature-cooling conditions for radial hydride formation

    Ring-compression screening tests (≤100%/s, RT, 150°C) → failure strain

    Crush-impact screening tests (RT, 150°C) → impact failure energy

    Fracture toughness data

*Nuclear Safety Research Conference*
*Washington, D.C., October 25-27, 2004*

Pioneering
Science and
Technology

# Data Needs: Radial-hydride formation

What temperature-stress-cooling conditions during pre-storage operations and storage promote radial-hydride precipitation?

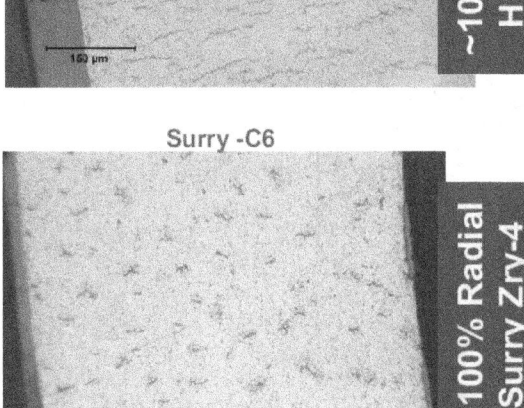

HBR -C15

~100% Radial HBR Zry-4

Surry -C6

~100% Radial Surry Zry-4

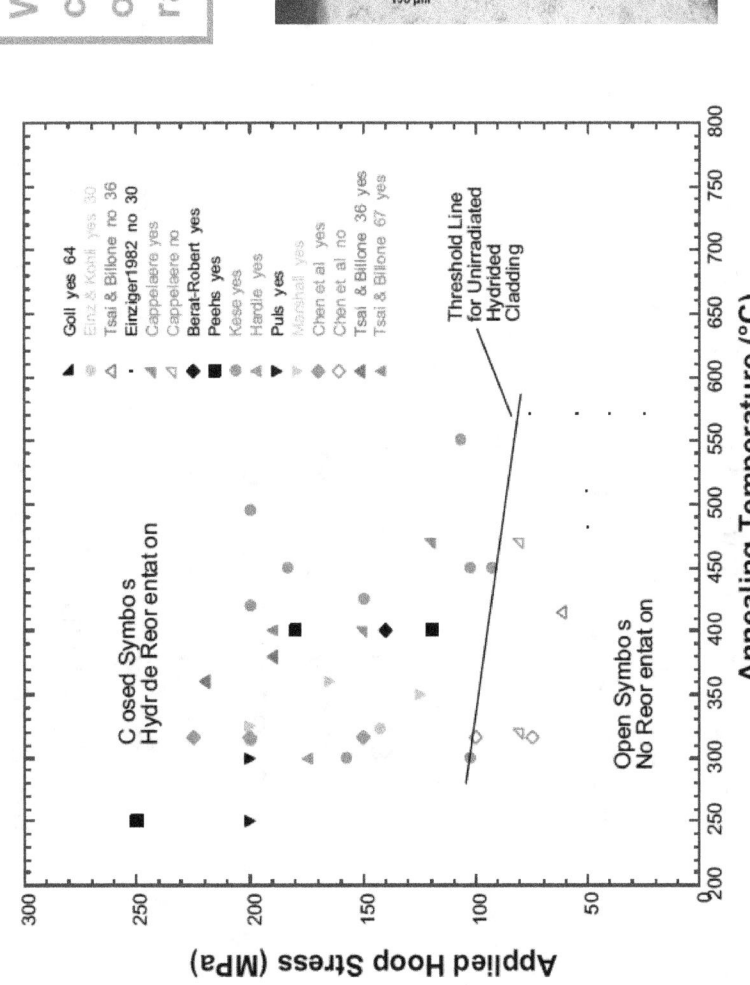

Closed Symbols
Hydride Reorientation

Open Symbols
No Reorientation

Threshold Line for Unirradiated Hydrided Cladding

Applied Hoop Stress (MPa)

Annealing Temperature (°C)

Goll yes 64
Einz & Kohi yes 30
Einziger1982 no 30
Cappelaere yes
Cappelaere no
Berat-Robert yes
Peehs yes
Kese yes
Hardie yes
Puls yes
Marshall yes
Chen et al yes
Chen et al no
Tsai & Billone 36 yes
Tsai & Billone 67 yes

H. Chung, "Understanding Hydride- and Hydrogen-Related Process in High-Burnup Cladding in Spent-Fuel Storage and Accident Situations," 2004 International Meeting on LWR Fuel Performance, Paper 1064, September 19-22, 2004.

Nuclear Safety Research Conference
Washington, D.C., October 25-27, 2004

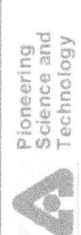

Pioneering Science and Technology

# Cladding Properties & Performance

## High Burnup → Radiation-Hardening and Hydrogen Effects

- Condition of High-Burnup Cladding after Pool Storage

  - Assumed to be the same as cladding after reactor discharge

  - Radiation-induced + hydride-induced (?) hardening → strength increase

  - Radiation-induced + hydride-induced embrittlement → ductility decrease

  - Hydrides are mainly circumferential & more dense near outer-surface

- Condition of High-Burnup Cladding after Drying-Transfer-Storage

  - Properties may improve:

    - *partial annealing of radiation damage at $\leq 400°C$*

    - *some homogenization of hydrogen → hydrides*

  - Properties may degrade:

    - *radial hydride formation during cooling from ($400°C$ to $150$-$250°C$) with $100$-$150$ MPa peak hoop stresses; hydrides are brittle below DBTT (?)*

  - Propose ductility & impact-failure-energy screening tests to map out drying-transfer conditions that result in radial-hydride-induced degradation due to hoop stress loading (e.g., slow and fast crushing loads)

*Nuclear Safety Research Conference*
*Washington, D.C., October 25-27, 2004*

Pioneering
Science and
Technology

8

# Overview

- Cladding Characterization
  - **Metallography and Fractography (SEM)**
  - **Hot Vacuum Extraction (hydrogen content)**
- Current Mechanical Testing
  - **Uniaxial Longitudinal Tension**
  - **Thermal Creep under Constant Internal Pressure**
- Proposed Future Test Plans
  - **Radial-hydride and Annealing Treatments**
  - **Screening Tests**
    - *Ring compression for ductility*
    - *Impact for failure energy*
- Summary

9

Pioneering
Science and
Technology

*Nuclear Safety Research Conference*
*Washington, D.C., October 25-27, 2004*

# H.B. Robinson (HBR) High-Burnup Zircaloy-4 (Zry-4) Cladding Characterization

(as-irradiated, assembly edge rod)

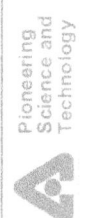

# *Cladding Materials*

## Relevant to Entire Storage Process

- High-burnup HBR Zry-4 Cladding
  - Rod average burnup = 67 GWd/MTU
  - Fast Fluence = $14 \times 10^{25}$ n(E > 1 MeV)/m$^2$
  - 15x15 Design ($D_o$ = 10.76 → 10.78 mm)
  - ≈85 µm oxide layer, ≈670 wppm H (interpolated)

- HBR Microstructural Conditions:
  - Two hydrogen contents of interest
    - <550 wppm (lower grid spans)
    - >550 wppm (higher grid spans)
  - Varying uniform hydride distribution and orientation
    - Circumferential vs. Radial
  - With and without creep deformation

- Non-irradiated, Hydrogen-charged Zry-4 (F-ANP 15x15)

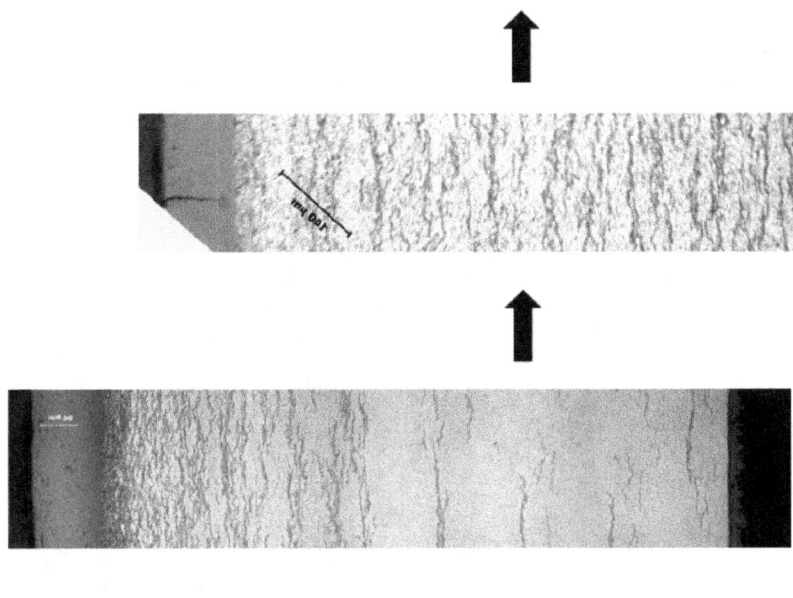

HBR C19
As-irradiated
Localized Circumferential
Hydrides

HBR C11
Annealed
Uniform Circumferential
Hydrides

HBR C15
Creep Tested
Radial Hydrides

*Nuclear Safety Research Conference*
*Washington, D.C., October 25-27, 2004*

# As-Irradiated HBR Microstructure for Edge Rod A02

### Radial/Azimuthal Hydride Profile

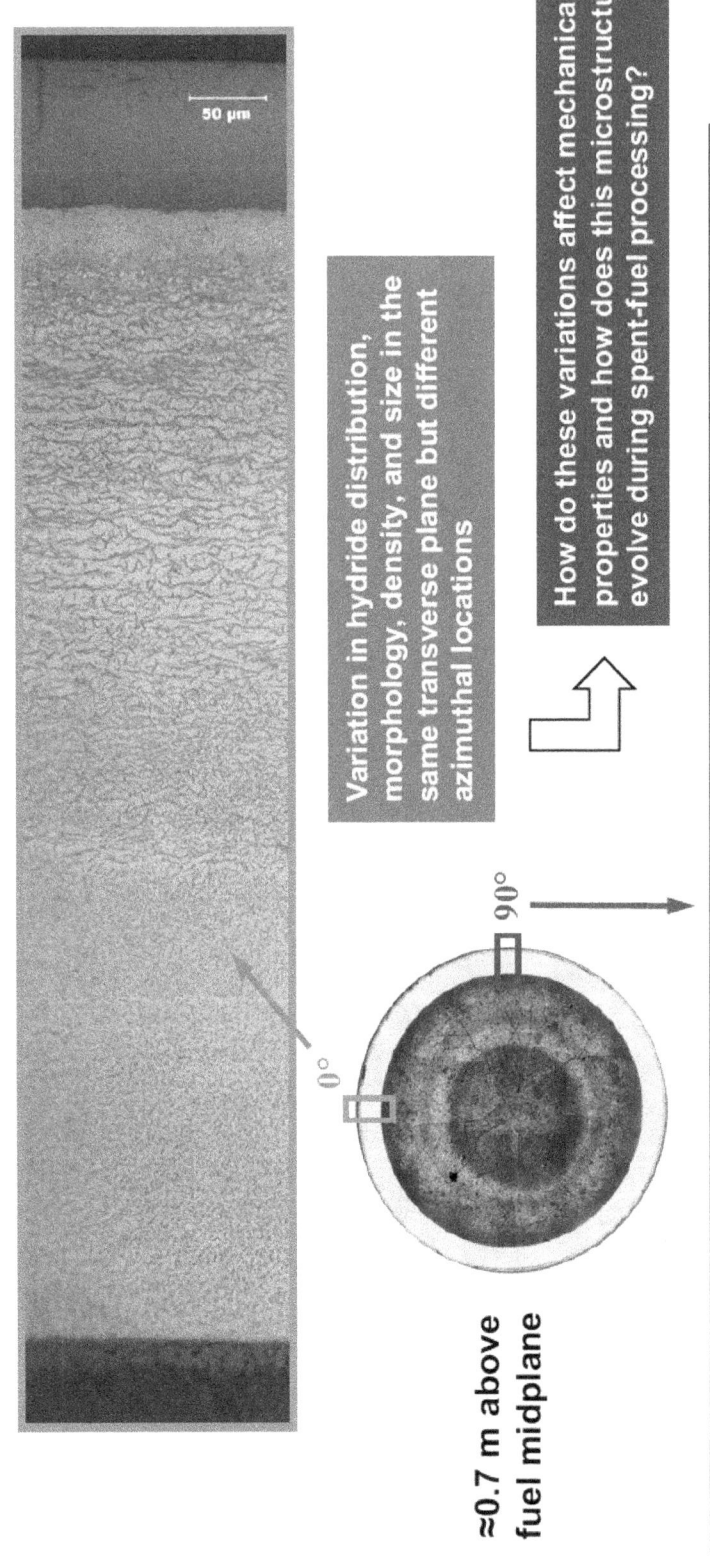

Variation in hydride distribution, morphology, density, and size in the same transverse plane but different azimuthal locations

How do these variations affect mechanical properties and how does this microstructure evolve during spent-fuel processing?

≈0.7 m above
fuel midplane

0°

90°

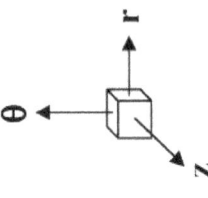

# As-Irradiated HBR Axial Profiles (Oxide and H)

## Rod A02 – Oxide/Hydrogen Axial Profile

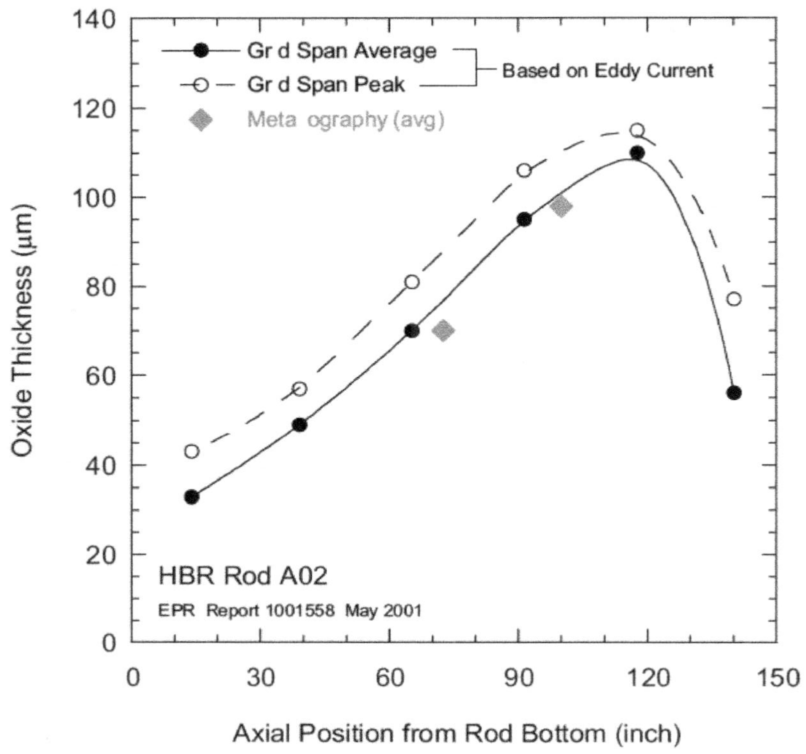

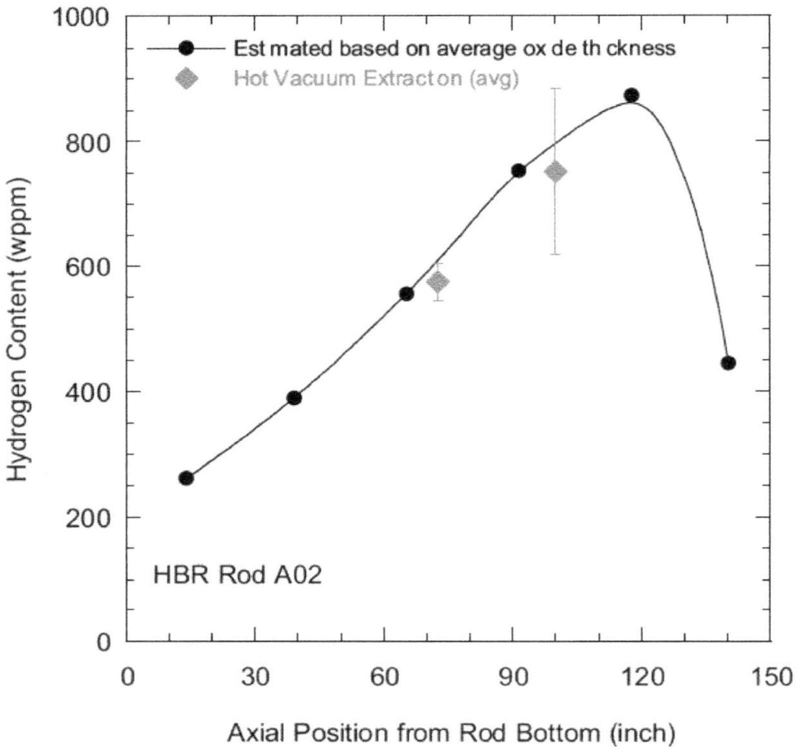

**Oxide Thickness (µm)** vs **Axial Position from Rod Bottom (inch)**

Legend:
- Gr d Span Average — ● Based on Eddy Current
- Gr d Span Peak — ○
- Meta ography (avg) — ◆

HBR Rod A02

EPR Report 1001558 May 2001

**Hydrogen Content (wppm)** vs **Axial Position from Rod Bottom (inch)**

Legend:
- ● Est mated based on average ox de th ckness
- ◆ Hot Vacuum Extract on (avg)

HBR Rod A02

Some Hot Vacuum Extraction measurements suggest significant (±150 wppm) hydrogen content variations along azimuthal positions (see bars).

*Nuclear Safety Research Conference*
*Washington, D.C., October 25-27, 2004*

Pioneering
Science and
Technology

13

# HBR Longitudinal (Axial) Tensile Testing

Nuclear Safety Research Conference
Washington, D.C., October 25-27, 2004

Pioneering
Science and
Technology

# Axial Tensile Properties

## Room Temperature Testing

- **Yield Strength (YS)** ≈600 → 800 MPa

- **Ultimate Tensile Strength (UTS)** ≈770 → 980 MPa

- **Total Elongation (TE)** ≈14% → 4%
  (Ductility Limit)

**CSED = Area under Curves**
(including elastic strain)

**Non-irrad. CSED = 106 MPa**
**High-Burnup CSED = 46 MPa**
Average CSED for 2 Tests = 42 MPa
(for 690±40 wppm H)

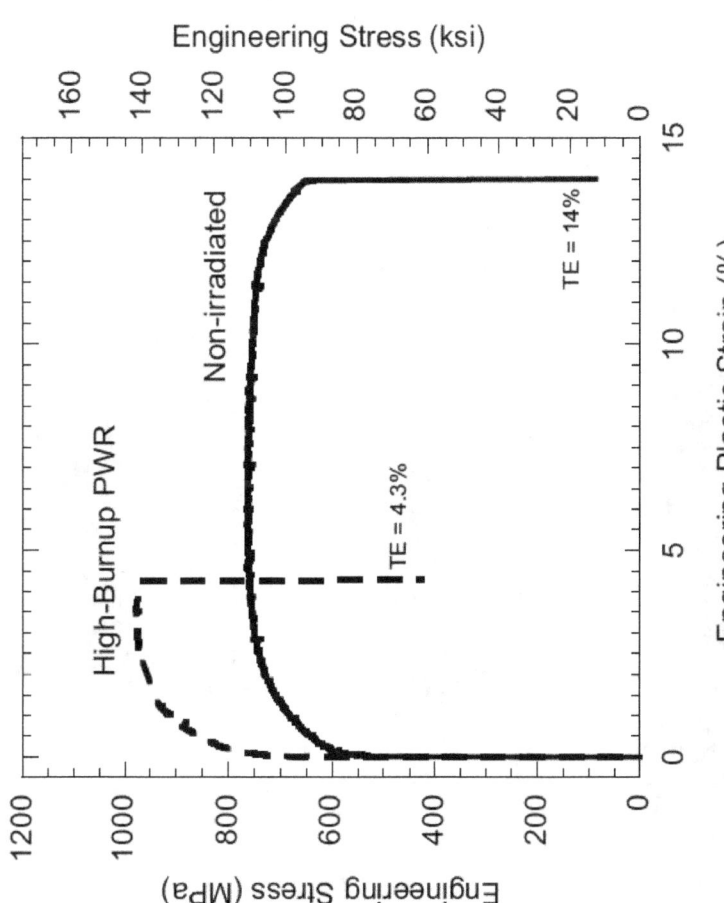

Non-irradiated

High-Burnup PWR

TE = 4.3%

TE = 14%

Engineering Stress (ksi): 160 140 120 100 80 60 40 20 0

Engineering Stress (MPa): 1200 1000 800 600 400 200 0

Engineering Plastic Strain (%): 0 5 10 15

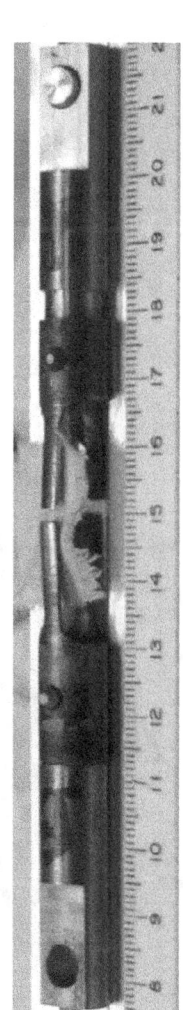

Nuclear Safety Research Conference
Washington, D.C., October 25-27, 2004

Pioneering
Science and
Technology

# Axial Tensile Properties (cont'd)

## Room Temperature Testing

HBR Axial Tensile Specimen C18

100.0 um

20.0kV

Surry Axial Tensile Specimen

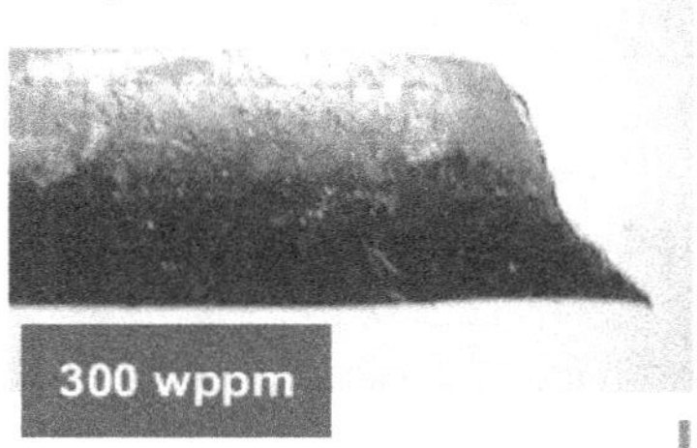

TMI-1 Ring-Stretch Specimen

450 wppm

300 wppm

Effects of localized hydrides force failure from plasticity-induced slip (across gauge width) to radial-crack initiation and propagation

HBR Axial Tensile Specimen

750 wppm

Nuclear Safety Research Conference
Washington, D.C. October 25-27, 2004

Pioneering
Science and
Technology

# Axial Tensile Properties (cont'd)

## Room Temperature Testing

HBR Metallography C19
(as-irradiated)

HBR Axial Tensile Specimen C18

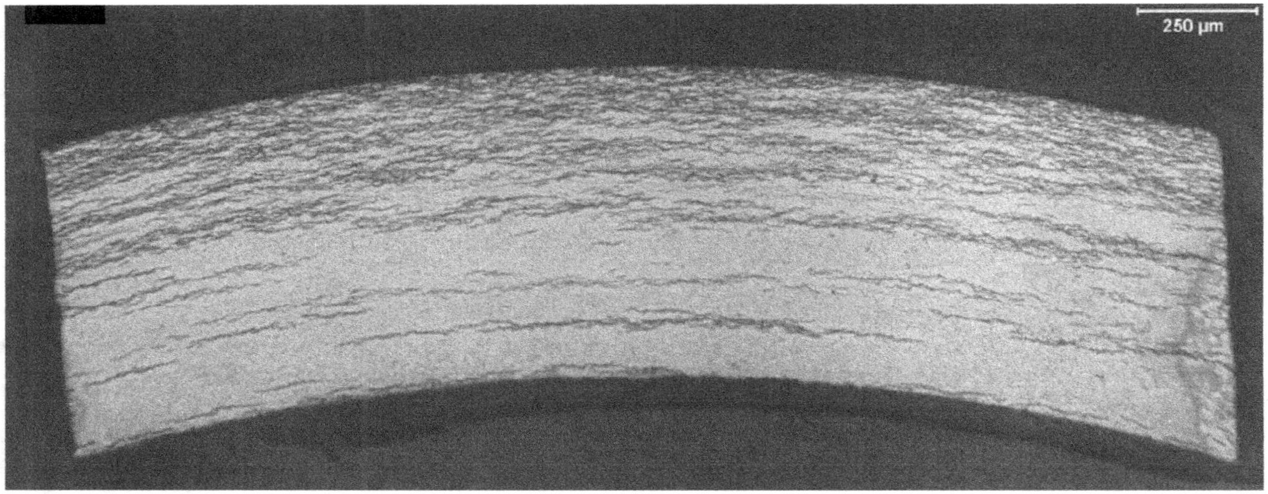

250 µm

**Post-test
Gauge Width**
(<0.5 mm from Fracture Surface)

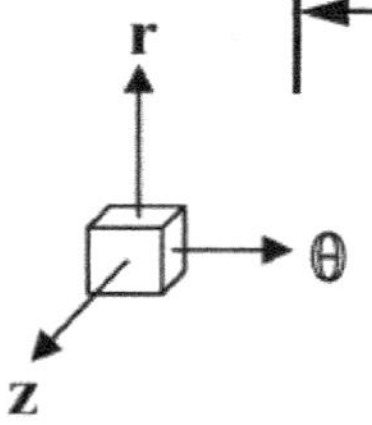

*Nuclear Safety Research Conference*
*Washington, D.C., October 25-27, 2004*

Pioneering
Science and
Technology

# HBR Thermal Creep Experiments under Constant Internal Pressure

## (Progress Update)

Nuclear Safety Research Conference
Washington, D.C., October 25-27, 2004

Pioneering
Science and
Technology

# HBR Creep Test Plan

| Temperature (°C) | Initial Applied Hoop Stress (MPa) | | | | |
|---|---|---|---|---|---|
| | 100 | 160 | 190 | 220 | 250 |
| 320 | | | | | |
| 360 | | | Planned (1) | Planned (1) | |
| 380 | | Planned (1) | In-Progress (1) (Re-run) | In-Progress (1) (Re-run) | |
| 400 | | Planned (1) | Completed (2) | Planned (1) | |
| 420 | | Planned (1) | | | |
| 450 | | | | | |

**Legend:**
- Completed (2)
- In-Progress (2) (Re-run)

C14 tested for 2427 hours at 400°C/190 MPa
(shutdown to induce circumferentially oriented hydrides) → Destructive Examination & Testing

C15 tested for 2439 hours at 400°C/190 MPa
(shutdown to induce radially oriented hydrides) → Destructive Examination & Testing

C16 tested for 404 hours at 380°C/190 MPa → Re-run Due to Hydrogen Depletion (to be discussed later)

C17 tested for 404 hours at 380°C/220 MPa → Re-run Due to Hydrogen Depletion

Pioneering Science and Technology

Nuclear Safety Research Conference
Washington, D.C., October 25-27, 2004

# C15 Creep Experiment and Shutdown

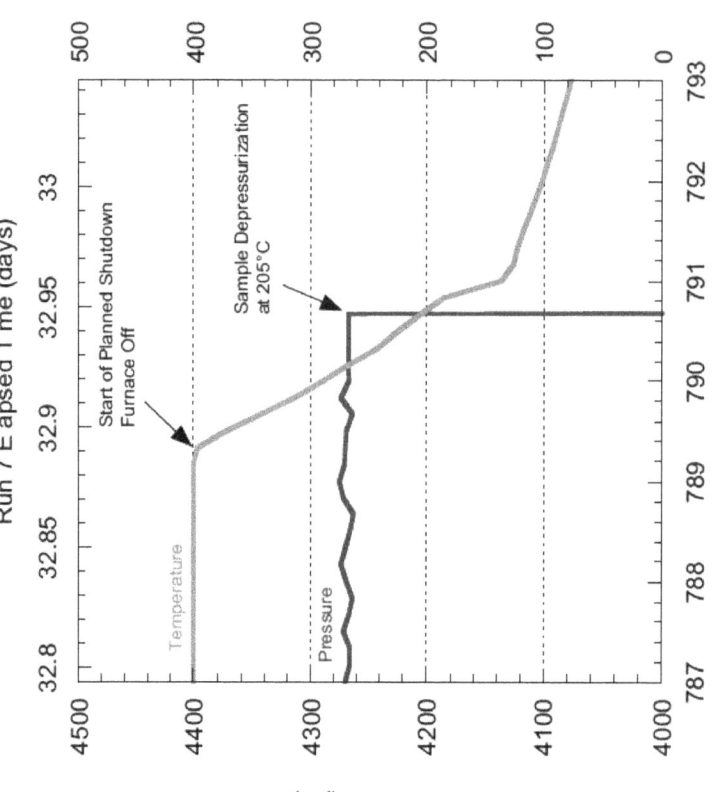

- Creep Test History

  - 7 runs at T = 400°C and $\sigma_\theta$ = 190 MPa for 2439 h

  - Average creep $\varepsilon_\theta$ = 3.5% at 2439 h

  - Same thermal & stress history as C14: $\varepsilon_\theta$ = 3.7% after 2427 h

  - Final C15 cool-down under full pressure (29.5 MPa, 4270 psig) to induce radial hydrides; C14 depressurized before cooling

- Sharp Depressurization Observed at 205°C

  - Estimated true $\sigma_\theta \approx$ 205 MPa at midplane ($\approx$20% of UTS)

  - Failures at top (pressure) and bottom (impact) welds

- Examination and Testing

Nuclear Safety Research Conference
Washington, D.C. October 25-27, 2004

Pioneering
Science and
Technology

20

# C15 Examination and Testing

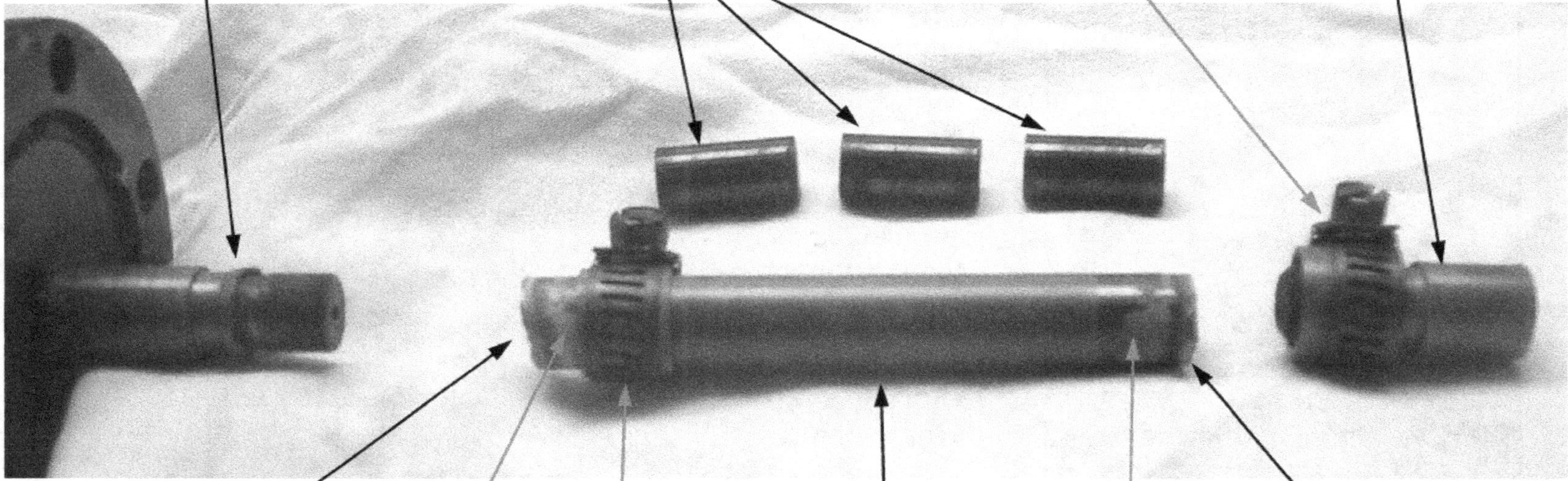

**Bottom End Fitting** (20 g)

**Bottom Weld Region**

**Bottom Hose Clamp**

**"Skid" Marks**

**Solid Zr-702 Pellets** (24 g)

**Intact Sample** (12 g Zry-4)

**Top Hose Clamp**

**Top End Fitting**

**"Skid" Marks**

**Top Weld Region**

21

*Nuclear Safety Research Conference*
*Washington, D.C., October 25-27, 2004*

Pioneering
Science and
Technology

# C15 Examination and Testing (cont'd)

## Radial Hydrides ($F_n \approx 1$) at Specimen Midplane

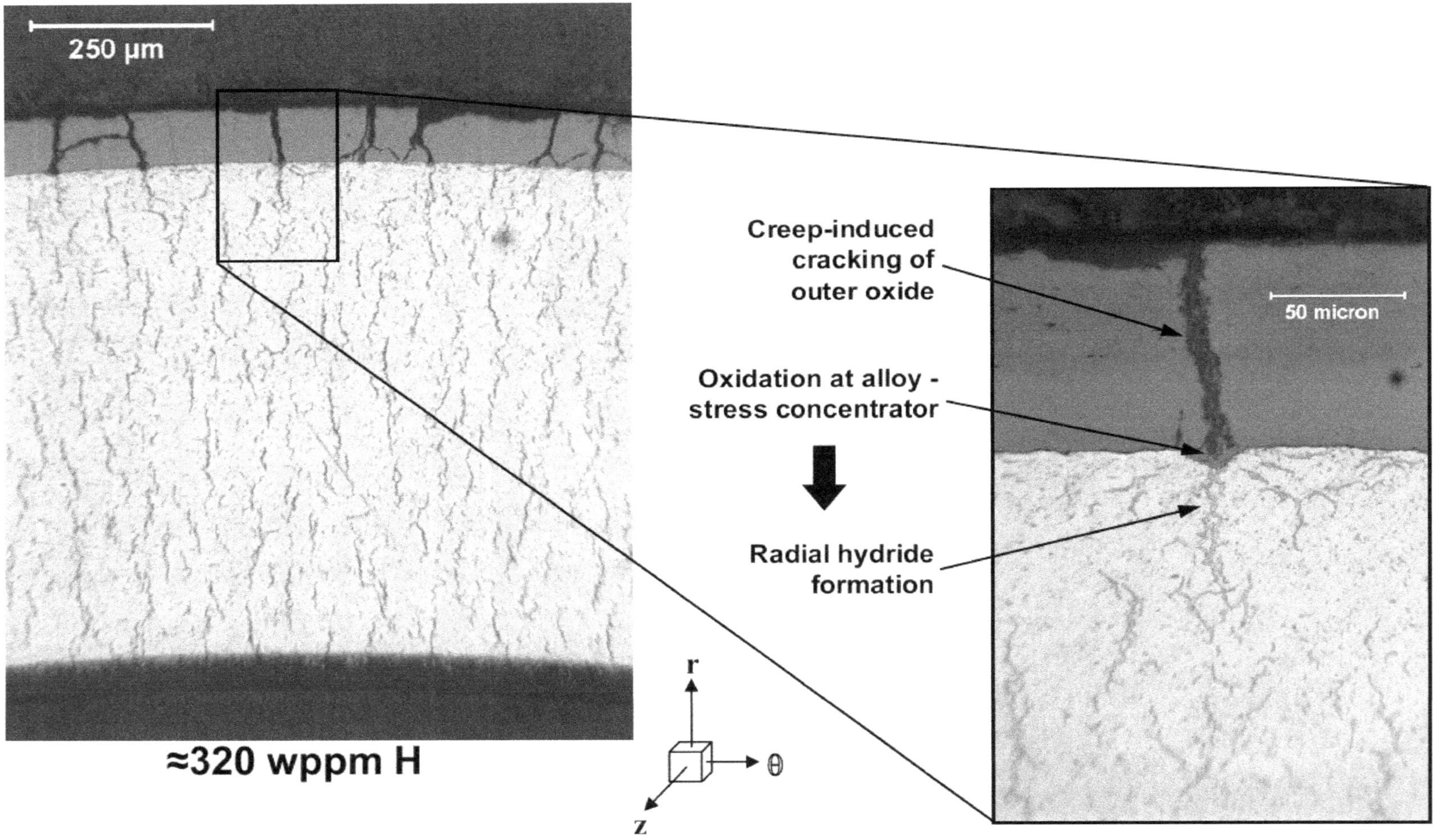

250 µm

50 micron

Creep-induced cracking of outer oxide

Oxidation at alloy - stress concentrator

Radial hydride formation

r

z

≈320 wppm H

Nuclear Safety Research Conference
Washington, D.C., October 25-27, 2004

Pioneering
Science and
Technology

# Proposed Hydride Effects Testing

## (Effect of Radial Hydrides on Cladding Ductility and Impact Resistance)

Pioneering
Science and
Technology

Nuclear Safety Research Conference
Washington. D.C., October 25-27, 2004

# Proposed Test Plans (initial screening)

- Both non-irradiated and defueled HBR cladding (≈85-mm long) with 400-700 wppm hydrogen and internally pressurized according to:

| Temperature (°C) | Initial Hoop Stress (MPa) at Temperature | | | | | |
|---|---|---|---|---|---|---|
| | 0 | 60 | 90 | 120 | 150 | |
| 380 | TBD | TBD | TBD | TBD | TBD | |
| 400 | Planned (2) | Planned (2) | Planned (2) | Planned (2) | Planned (2) | |
| 420 | TBD | TBD | TBD | TBD | TBD | |

Held at temperature for 3 days and then slow cooled at ≈2-4°C/sec

**Planned (2)** → Duplicates

→ **Initial Screening Tests (At room temperature)**

150°C to follow

**Ring Compression (developed)**

**Ring Impact (to be developed)**

24

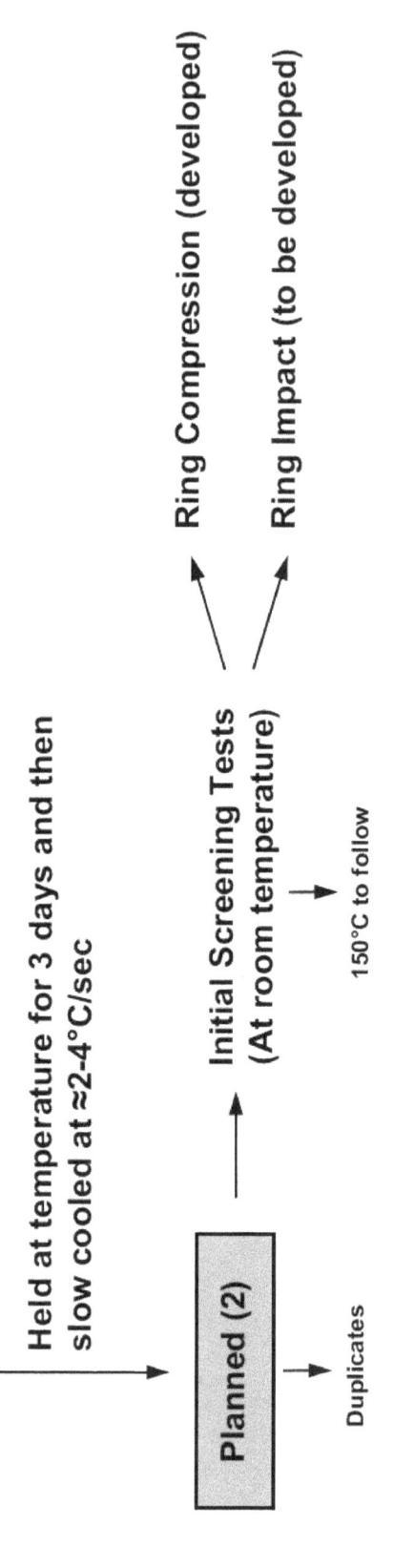

# Proposed Test Plans (cont'd)

- Ring Compression and Impact Tests

  - Currently employed to determine post-quench ductility for LOCA licensing (hydrogen + oxygen embrittlement)

  - Completed development for irradiated testing capability

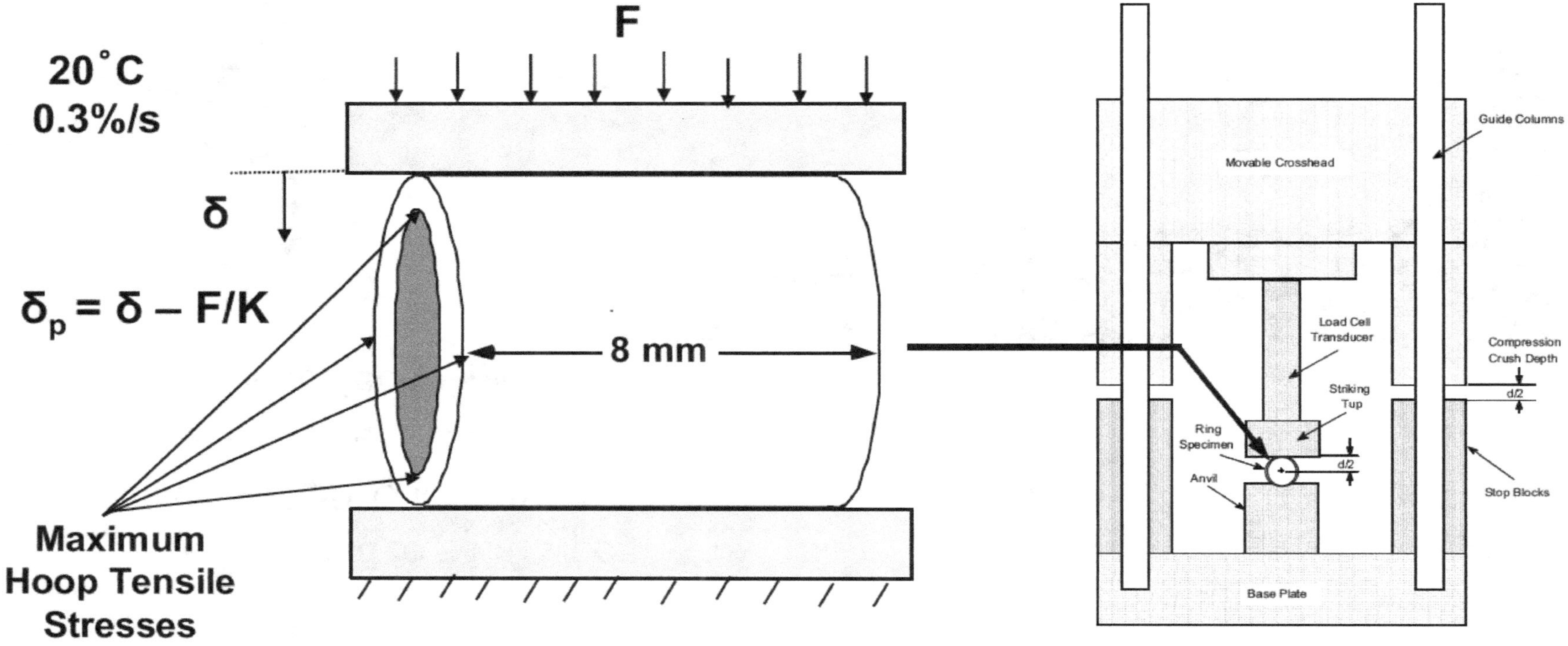

20°C
0.3%/s

$\delta$

$\delta_p = \delta - F/K$

Maximum
Hoop Tensile
Stresses

F

8 mm

Pioneering
Science and
Technology

Nuclear Safety Research Conference
Washington, D.C., October 25-27, 2004

# Radial Hydride Effects: Preliminary Results

- Non-irradiated Zry-4
  - Pre-hydrided to ≈450 wppm
  - Held at 360°C for 1 hr at 0.7kN hoop tensile load
  - Cooled under load
  - Tensile tested to failure at room temperature

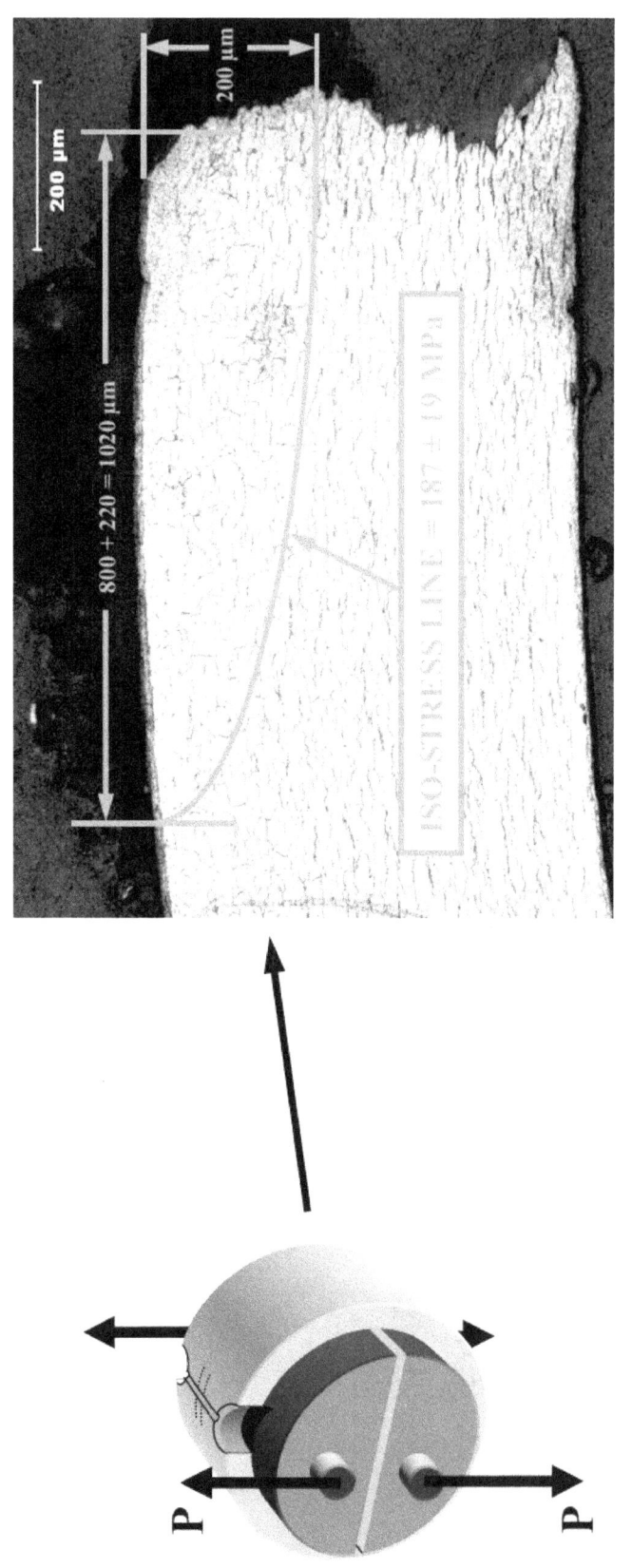

*Nuclear Safety Research Conference*
*Washington, D.C., October 25-27, 2004*

Pioneering
Science and
Technology

# Radial Hydride Effects: Prelim. Results (cont'd)

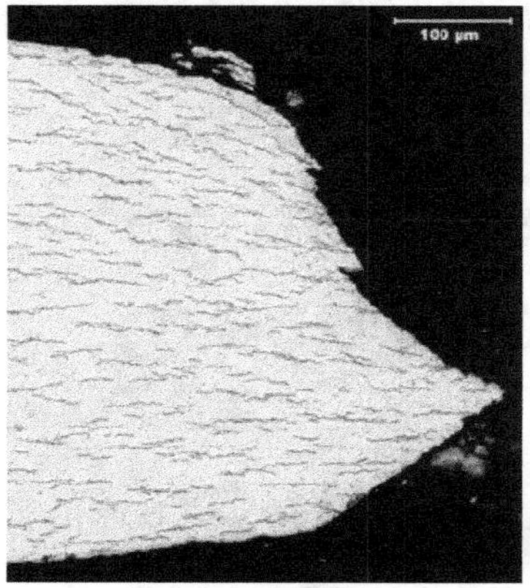

**Homogenized + Radial Treatment**
(circumferential + radial hydrides)

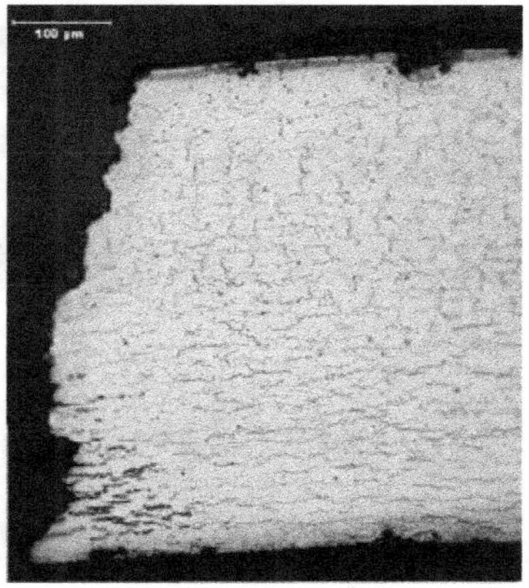

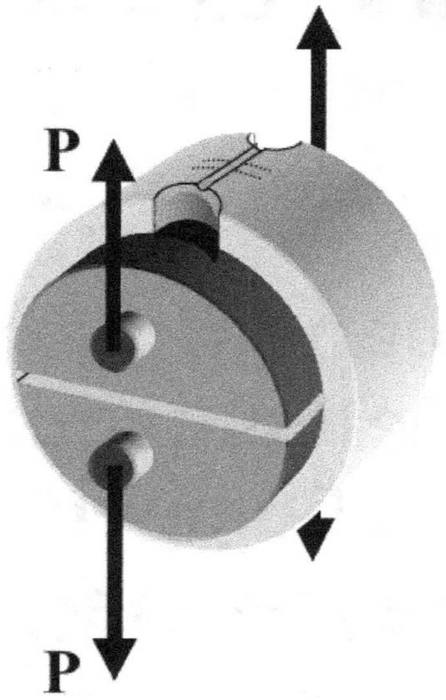

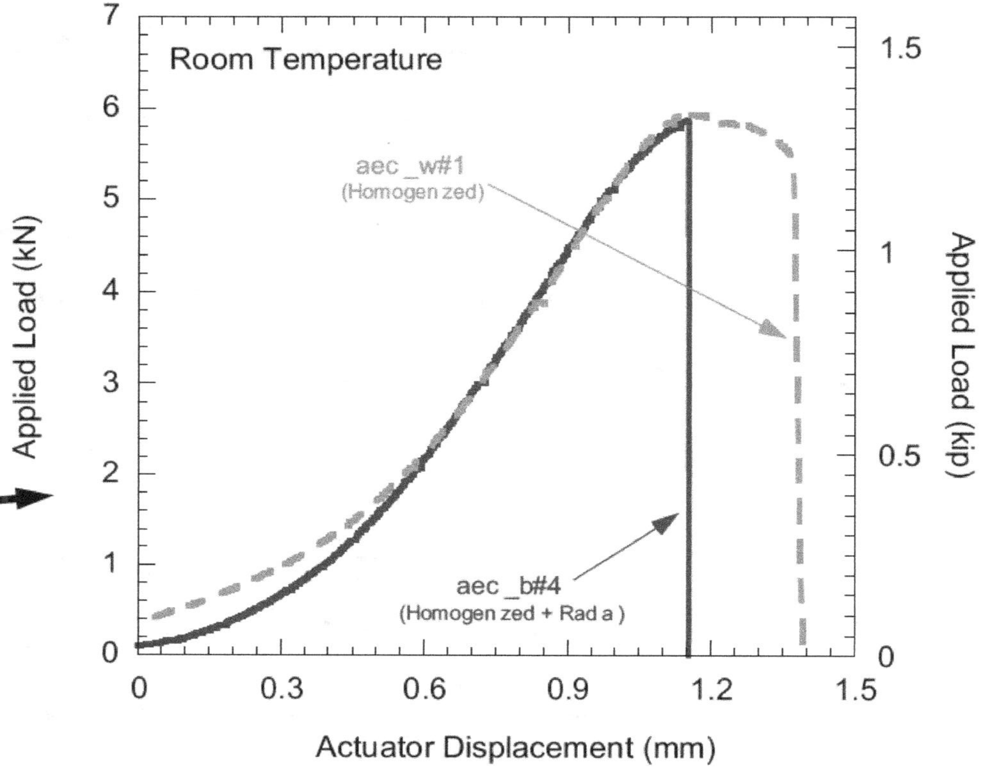

**Homogenized**
(all circumferential hydrides)

Room Temperature

aec_w#1
(Homogen zed)

aec_b#4
(Homogen zed + Rad a )

Applied Load (kip)

Applied Load (kN)

Actuator Displacement (mm)

P

P

27

# Summary

- Data and technical guidance regarding high-burnup-fuel cladding performance during pre-storage, storage, and post-storage operations are needed to support licensing of SNF transport, storage and transport-storage casks.

- A comprehensive testing program is on-going; future work is planned to address these needs with particular emphasis on the thermo-mechanical conditions which promote radial-hydride embrittlement.

- Creep and tensile testing will be augmented by ring-compression and impact screening testing in order to assess the reduction in cladding ductility and failure energy due to radial-hydride embrittlement.

# Mechanical Performance of High-Burnup Fuel Cladding during Cask Storage and Post-Storage Handling and Transportation

Robert S. Daum, Hanchung Tsai, Yung Liu, Saurin Majumdar, and Michael C. Billone
Energy Technology Division
Argonne National Laboratory
Argonne, Illinois, USA

## Abstract

The assessment of cladding performance during spent-nuclear-fuel (SNF) pre-storage operations, dry-cask storage, and post-storage handling and transport, including hypothetical accidents, is important in ensuring that sub-criticality is maintained, that radioactivity is contained, that cask external dose rates are limited, and that SNF assemblies can be safely retrieved, handled and transported at the end of dry-cask storage. Although cladding failure is not prohibited by federal regulations, the failure mode and extent of cladding failure have a significant effect on the possible reconfiguration of SNF within storage and transport casks. High-burnup SNF Zircaloy-4 (Zry-4) cladding from pressurized water reactors is more susceptible to failure and possible fuel dispersal than low-burnup Zry-4 due to higher hydrogen pickup, internal pressure, and corrosion level. A testing program has been developed to investigate these effects in high-burnup Zry-4 cladding, with particular emphasis on the conditions that promote radial-hydride precipitation and the effects of these radial hydrides on cladding integrity. Preliminary experimental results are presented from axial tensile and thermal creep tests, as well as conditions for radial-hydride precipitation.

## Introduction

Current U.S. regulations for storage (10 CFR 72) and transportation (10 CFR 71) of spent nuclear fuel (SNF) are designed primarily to maintain sub-criticality and to ensure that doses are less than regulatory limits, that the cask provides adequate fuel confinement and containment, and that the fuel is retrievable. As discharge fuel burnup increases, SNF Zircaloy-4 (Zry-4) cladding may become more susceptible to brittle failure under normal conditions and postulated accidents during SNF handling, storage, and transportation. Although cladding failure is not prohibited by federal regulations, such failure and gross fuel dispersal may compromise these regulatory requirements.

Specifically, hydride precipitation, corrosion (i.e., wall thinning), irradiation-induced defects, and possibly even post-reactor thermal creep will reduce cladding ductility and impact resistance. Speculated to be the most degrading, transfer-and-drying operations conducted from the SNF pool to the storage cask may result in thermo-mechanical conditions that promote radial-hydride-induced degradation of cladding ductility and impact resistance during severe loadings associated with hypothetical accidents. Such cladding would be more susceptible to brittle failure at the end of dry storage when temperatures are expected to be 150-250°C—below the ductile-to-brittle transition temperature (DBTT) for zirconium hydrides and within a range of temperature in which more hydrogen is precipitated as hydride as compared to temperatures at the beginning of storage.

In fact, Marshall and Louthan [1] showed that cold-worked Zircaloy-2 (Zry-2) was completely brittle at room temperature during uniaxial tensile testing when, regardless of total hydrogen content, as little as 50 weight parts per million (wppm) of hydrogen precipitated as radial hydride platelets, which were oriented between 50 and 90° to the principal tensile stress (where 90° is fully perpendicular to the principal tensile stress). Singh et al. [2] found that Zr-2.5%Nb alloy containing 100-wppm hydrogen as radial hydrides and subjected to uniaxial tension showed as little as 9% total elongation at temperatures <150°C but twice that at ≥150°C. This increase in elongation may be attributed to an increase in radial-hydride ductility, suggesting that the DBTT is between room temperature and 150°C. Likewise, Choubey and Puls [3] observed a ductile-to-brittle transition in Zr-2.5%Nb with radial hydrides as temperatures dropped below 100°C. Obviously, few data for Zry-4 are available to yield a clear understanding regarding the temperature effects of radial-hydride precipitation on ductility.

Given this radial-hydride-induced embrittlement, identifying the thermo-mechanical conditions that promote radial-hydride precipitation has been the subject of numerous studies. First, to understand hydride dissolution kinetics, McMinn et al. [4], Kearns [5], and Vizcaino et al. [6] experimentally determined the terminal solid solubility for hydride dissolution upon heating. As shown in Fig. 1, the solubility increases with temperature such that only 10 wppm is in solution at 200°C whereas approximately 300 wppm of hydrogen would be in solution at 450°C. Although their respective absolute values of solubility differed, McMinn et al. [4] and Vizcaino et al. [6] both observed an increase in solubility with irradiation, most likely due to hydrogen trapping by radiation-induced defects.

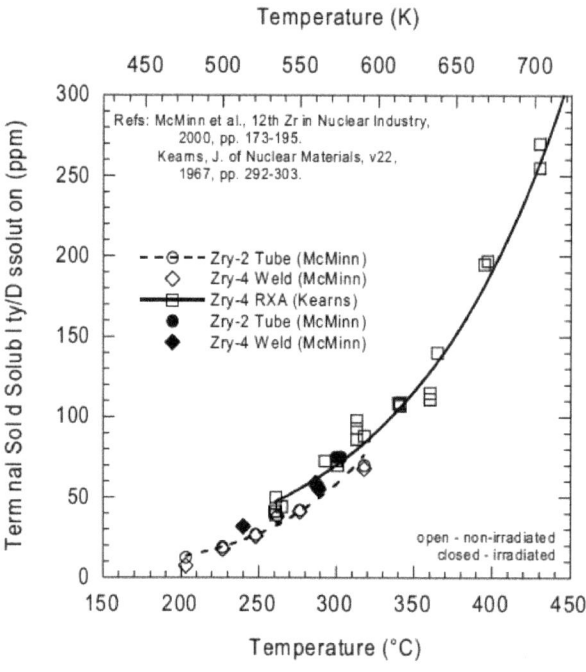

Figure 1 – Hydrogen terminal solid solubility for dissolution of hydrides in non-irradiated and irradiated Zircaloy alloys [4],[5].

Discharged high-burnup Zry-4 fuel cladding contains roughly 600-800 wppm hydrogen, and at room temperature, this hydrogen is predominantly precipitated as circumferentially oriented hydride platelets localized in the form of a rim toward the outer surface. During short-term drying operations, in which temperatures may be ≥400°C, it is likely that, based on the results of Fig. 1, >200 wppm of hydrogen will be in solution. This hydrogen in solution is available to precipitate as radial hydrides upon cooling;

however, the application of a *threshold* or *critical* cladding hoop stress at elevated temperatures is required to nucleate and grow these hydrides along the radial direction; if the hoop stress is lower than this *threshold* value, only circumferential-hydride precipitation occurs. In fact, Bai et al. [7] thermodynamically modeled this combination of stress and temperature that promotes radial-hydride nucleation and growth in recrystallized Zry-4. At 400°C, both the model and radial-hydride treatments of fully recrystallized Zry-4 sheet specimens show a *threshold* stress of approximately 95 MPa, whereas treatments of stress-relieved specimens suggest values as high as 180 MPa.

Ignoring the influences of composition, manufacturing histories, and test techniques for a moment, Singh et al. [2] exposed tensile specimens machined with tapered gauge sections from non-irradiated Zr-2.5%Nb pressure tubes containing ≈100 wppm hydrogen to radial-hydride treatments at 150-350°C. This study found that, when 50% of the specimen thickness exhibited radial-hydride precipitation, the corresponding *threshold* stress is approximately 130 MPa at 350°C. Using a similar technique, Leger and Donner [8] found a *threshold* stress of as low as 75 MPa at 300°C for non-irradiated Zry-2 pressure-tube material, but they speculated that, because of uncertainties in assessing the effects of residual stress due to specimen preparation, it may be even 15-20 MPa higher.

As for irradiated Zry-4, Einziger and Kohli [9] found ≈90% radial-hydride precipitation in low-burnup Zry-4 (<100 wppm hydrogen) after cooling to room temperature from 323°C under internal pressurization to 135-MPa hoop stress. Tsai and Billone [10] also found that, after accumulating 0.35% permanent creep strain, low-burnup Zry-4 (≈250 wppm hydrogen) cooled from 380°C under 190-MPa hoop stress resulted in mostly radial hydrides. Most recently, Chung [11] presented a review of stress-relieved Zry-4 cladding studies and concluded that this *threshold* stress is ≈80-100 MPa between 250 and 550°C. In addition to tensile hoop stress, many of these cited studies emphasize that residual stress may also play a significant role in hydride reorientation, and the effects of radiation-induced damage on residual stresses in irradiated Zry-4 may further exacerbate radial-hydride precipitation.

In comparison, the internal pressure of high-burnup-fuel Zry-4 rods is permitted to be up to but not exceed the reactor coolant pressure (roughly 15.5 MPa for pressurized water reactors), resulting in an applied hoop stress of 110 MPa at 330°C. Furthermore, limited numbers of peak power rods are permitted to be 1.3 times the coolant pressure and, therefore, may have an applied hoop stress of 145 MPa. If drying operations occur at temperatures upwards of 400°C, these hoop stresses may increase to as much as 160 MPa. Therefore, given these potentially *high* hoop stresses and the fact that few data are available, it is essential to determine the *threshold* stress for hydride reorientation in high-burnup Zry-4.

Based on this limited database, the current technical basis [12] limits peak cladding temperatures to ≤400°C during short-term operations (namely, drying operations) and normal storage conditions to mitigate the above conditions, which may potentially exacerbate cladding-failure propensity. Note that this technical basis also includes limiting hoop stresses in low-burnup SNF Zry-4 cladding to 90 MPa. Brown et al. [13] provided further details for this basis but, until more pertinent data to more thoroughly assess high-burnup cladding behavior during the complete cycle of pre-storage operations, storage and post-storage handling and transport becomes available, license applications for high-burnup storage, storage-and-transport, and transport casks are evaluated on a case-by-case basis.

Therefore, understanding the conditions that may promote radial-hydride precipitation, the associated mechanical performance, and resulting failure susceptibility of high-burnup Zry-4 cladding is essential for license application and evaluation of storage and transportation systems, as well as the handling of such systems and individual SNF assemblies. The data generated within the ANL test program are to be used both by applicants for high-burnup cask licenses and by cask-license evaluators. Preliminary results and a test plan, which includes ring-compression-ductility and ring-crush-impact screening tests, are also described.

**Experimental Procedures**

Experiments have concentrated on the microstructural characterization, mechanical-property testing, and radial-hydride treatment of stress-relieved Zry-4 (15x15 design) in a non-irradiated condition and irradiated to both low and high fuel burnups in Surry Unit No. 2 (referred as Surry) and H.B. Robinson Unit No. 2 (referred as HBR), respectively. Characterization of these materials has been previously reported elsewhere [10],[14],[15]. Table I summarizes these materials and properties.

Uniaxial tensile tests were conducted at room temperature and 400°C and at a nominal strain rate of 0.001/sec (0.1%/sec). Gauge sections were machined from defueled cladding using an electro-discharge machine and then either welded or transverse-pin-loaded to the tensile grips, as seen in Fig. 2. The welded-grip technique was found to have insufficient strength to couple HBR cladding to the load train, most likely due to excessive contamination of weldment by reactor-produced oxide and/or hydrogen. Isothermal creep tests were conducted using defueled, internally pressurized cladding tubes at temperatures of 360-400°C and initial hoop stress of 160-220 MPa. To promote radial-hydride precipitation, some creep specimens were cooled to room temperature at ≈2.5°C/min while maintaining internal pressure. References [10], [14], and [15] provide further details regarding these test techniques, specimen preparation, and test conditions.

Table I – Zry-4 cladding materials and properties.

| Cladding Material | Sn Content (wt%) | Average Rod Burnup (GWd/MTU) | Fast Neutron Fluence ($\times 10^{26}$ n/m$^2$; E > 1 MeV) | Peak Oxide Thickness (μm) | Peak H Content (wppm) |
|---|---|---|---|---|---|
| F ANP[a] | 1.30 | 0 | 0 | 0 | 1800[b] |
| Surry[c] | ≈1.5 | 36 | 0.7 | 30 | 350 |
| HBR | 1.41 | 67 | 1.4 | 110 | 850 |

[a]F ANP denotes the manufacturer, Framatome Advanced Nuclear Products.
[b]F ANP cladding was gas charged at 400°C to contents of 200 1800 wppm.
[c]Surry cladding was dry cask stored for ≈15 years after discharge from the SNF pool.

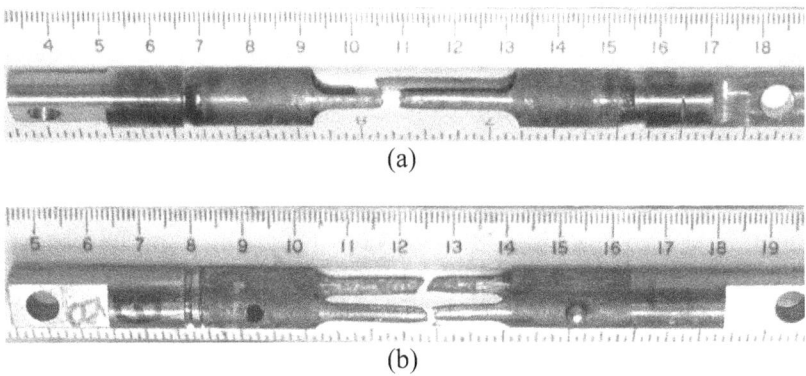

(a)

(b)

Figure 2 – Images of axial-tensile specimens after testing at 400°C showing tensile-grip techniques using (a) welded grips for Surry (thin corrosion layer and low hydrogen content) and (b) transverse-pin-loaded grips for HBR (thick corrosion layer and high hydrogen content); note, the left-side transverse pin has been removed in (b).

## Results and Discussion

Engineering stress versus plastic strain responses for F-ANP, Surry, and HBR cladding materials are shown in Fig. 3. The engineering mechanical properties are presented in Table II and plotted in Fig. 4. It is worth noting that Surry cladding was dry-cask stored for ≈15 years, during which time peak cladding temperature ranged between 415°C (in vacuum) and 350°C (in helium), and partial annealing of radiation damage may have occurred [14],[16].

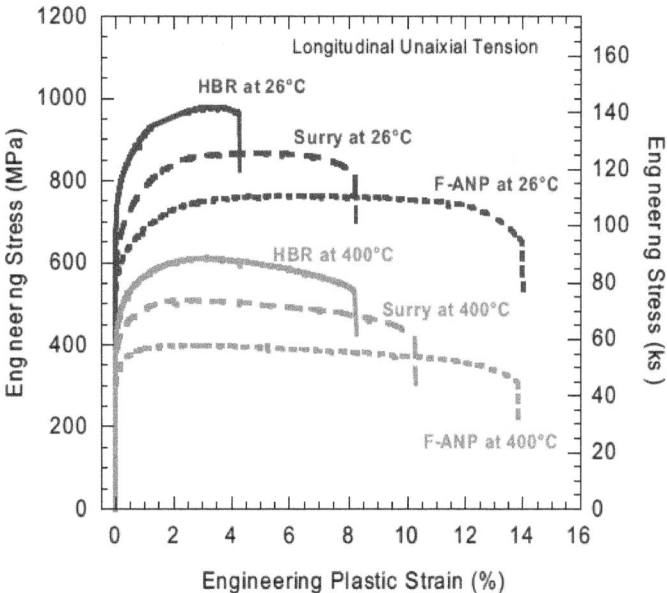

Figure 3 – Longitudinal (or axial) engineering stress vs. plastic strain response for non-irradiated (F-ANP), low-burnup-and-dry-cask-stored (Surry), and high-burnup (HBR) Zry-4 at room temperature and 400°C.

Table II – Average mechanical properties from longitudinal tensile testing of non-irradiated (F-ANP), low-burnup (Surry), and high-burnup (HBR) Zry-4 at a nominal strain rate of 0.1%/s.

| Cladding Material | Temp. (°C) | 0.2% Yield Strength (MPa) | Ultimate Strength (MPa) | Uniform Elongation (%) | Total Elongation (%) | Critical Strain Energy Density (MPa) |
|---|---|---|---|---|---|---|
| F ANP |  | 605 | 765 | 5.7 | 14.0 | 106 |
| Surry | 26 | 680 | 868 | 4.6 | 8.2 | 83 |
| HBR |  | 803 | 980 | 3.3 | 4.0 | 42 |
| F ANP |  | 355 | 400 | 2.8 | 13.8 | 55 |
| Surry | 400 | 433 | 512 | 2.1 | 10.2 | 54 |
| HBR |  | 490 | 612 | 3.1 | 8.3 | 56 |

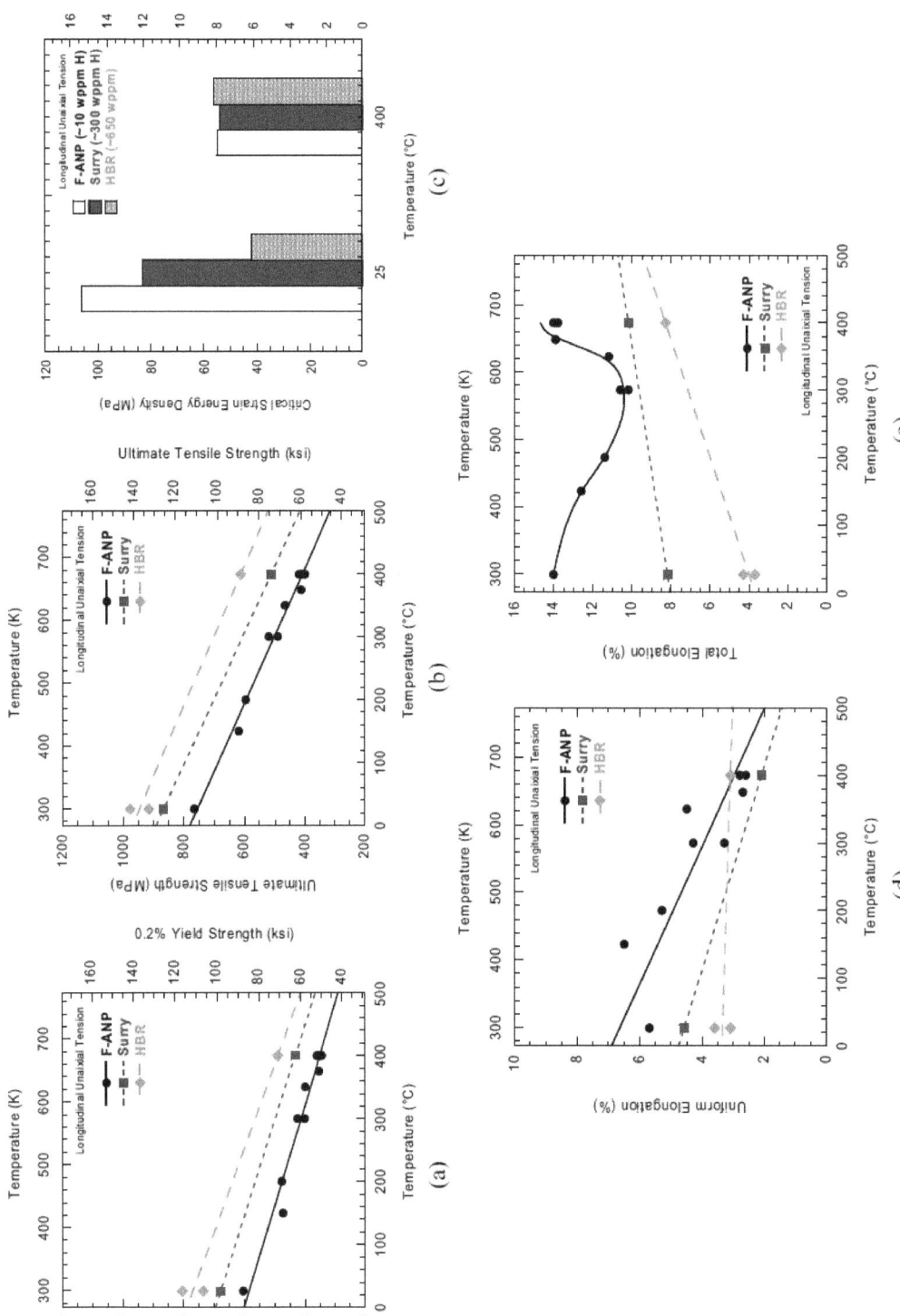

Figure 4 – Results of longitudinal (axial) tensile testing of irradiated (both Surry and HBR) and non-irradiated (FANP) Zry-4 cladding showing (a) yield strength, (b) ultimate tensile strength, (c) critical strain energy density, (d) uniform elongation, and (e) total elongation as a function of temperature.

34

As compared to non-irradiated F-ANP Zry-4, tensile properties of irradiated cladding show an increase in yield and ultimate tensile strengths and a decrease in uniform and total elongations at room temperature. The strength increase appears to be due mainly to radiation-induced hardening, while the ductility decrease appears to be due to both radiation- and hydride-induced embrittlement. In fact, fractography of an HBR specimen tested at room temperature shows transitions between brittle, mixed brittle+ductile, and predominantly ductile fracture, as shown in Fig. 5. When compared to the hydride microstructure and distribution (as seen in Fig. 4a), these transitions appear to coincide with decreases in hydride density and are consistent with other studies [17].

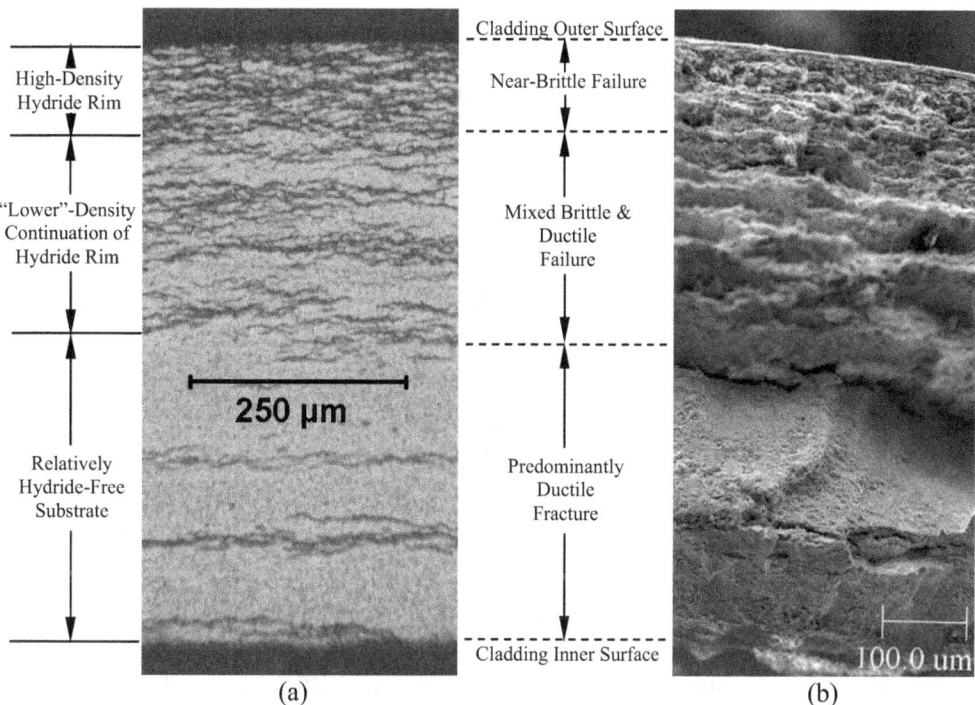

(a)                                                    (b)

Figure 5 – Post-test images of HBR specimen after axial-tensile testing at room temperature showing (a) hydride microstructure ≈0.5 mm from the fracture surface and (b) fracture-surface topography; note, corrosion layer was removed from specimen to facilitate specimen machining.

At 400°C, radiation-induced hardening is still evident but cladding embrittlement is less noticeable than at room temperature. In particular, the uniform elongation and critical strain energy density (i.e., integration area under the engineering stress vs. total strain) of all cladding tested in this study are essentially identical. Although fractography of 400°C specimens is pending, this enhancement in relative ductility at elevated temperatures is consistent with the increased ductility of the hydride microstructure.

Isothermal annealing tests of HBR cladding were conducted previously [10] and showed that strength properties (based on microhardness data) appear to recover by ≈75% and circumferential hydrides tend to homogenize across the cladding radius after 72 hours at 420°C. These results suggest that SNF drying operations may partially anneal radiation-induced hardening. The degree of ductility recovery and the effects on creep behavior with annealing remain to be investigated.

With regard to creep, testing of both Surry and HBR cladding have been reported elsewhere [10],[15],[16] and have shown that both cladding materials retain >3% creep capacity. In light of the above tensile results and neglecting any strength anisotropy, these creep tests were conducted at ≈0.4 times the yield strength. During cooling (≈2.5°C/min) from creep-test temperatures under ≈190-MPa hoop stress, radial-hydride precipitation was found to predominantly occur in both Surry [10] and HBR cladding, as shown in Fig. 6. Using the procedure developed by American Standard for Testing and Materials [18], radial-hydride fraction ($F_n$) was measured to be approximately 0.70 and 0.85 for Figs. 6a and 6b, respectively.

Upon examination of Fig. 6, the amount of hydrogen appears lower than expected as compared to that found in regions of the Surry and HBR rods adjacent to the creep specimens. Therefore, hydrogen contents were measured in radial-hydride specimens and compared to measurements from regions adjacent to those regions in Fig. 7. Surry hydrogen contents from regions adjacent to those in Figs. 6a and

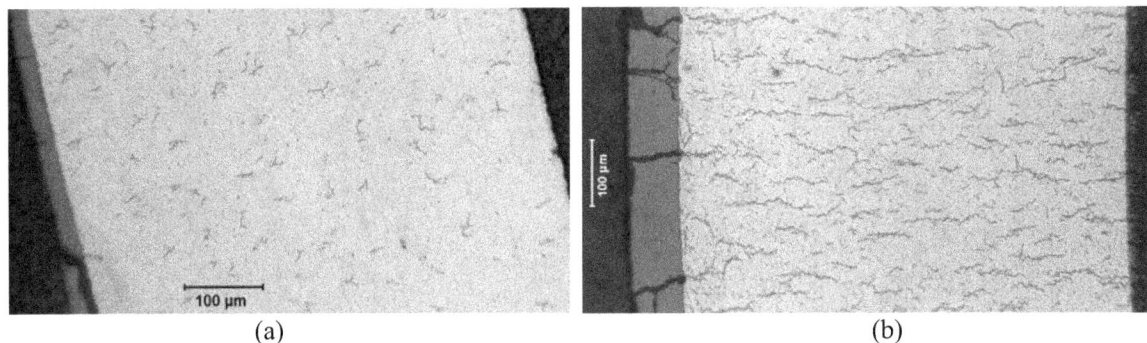

(a)                                   (b)

Figure 6 – Transverse micrographs showing radial-hydride precipitation after creep testing followed by cooling to room temperature for (a) Surry cladding (0.35% strain at 380°C/190 MPa) and (b) HBR cladding (3.5% strain at 400°C/190 MPa); note, corrosion layer at the left of images.

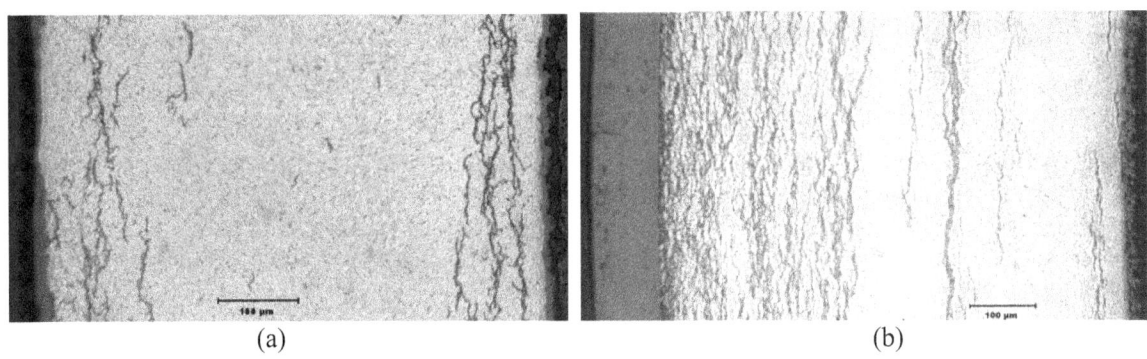

(a)                                   (b)

Figure 7 – Transverse micrographs showing mostly circumferential-hydride precipitation in (a) as-irradiated-and-dry-cask-stored Surry and (b) as-irradiated HBR cladding; note, corrosion layer to the left.

7a were found to be 130 and ≈250 wppm, respectively. Similarly, HBR hydrogen contents adjacent to regions of Figs. 6b and 7b were found to be 320 and ≈650 wppm, respectively. It was speculated that hydrogen migrated axially from the cladding specimen to welded components of the high-pressure gas system. In fact, the hydrogen content in one such HBR-specimen component (fabricated from Zircadyne 702 alloy) was found to be 210 wppm, as compared with 10 wppm for the as-received material, indicating that indeed hydrogen diffused into the component.

Redesign of the thermal creep test train and furnace is in progress to minimize the hydrogen loss. However, note that the hydrogen solubility of non-irradiated Zry-4 is ≈200 wppm at 400°C, as shown in Fig. 1. The absence of visible circumferential hydrides at the specimen midplane (with 320 wppm) suggests that high-burnup Zry-4 is capable of trapping about 100 wppm of hydrogen, which most likely precipitates during rapid cooling as sub-micron-size hydrides. It will be interesting to determine if this excess hydrogen precipitates as visible radial hydrides under the low cooling rates (≈4°C/hr) typical of drying-transfer-storage.

Upon closer inspection of Fig. 6b, we found that creep deformation caused the fracture of the corrosion layer, allowing subsequent oxidation of the cladding to occur, as seen in Fig. 8. This local oxidation appears to coincide with enhanced hydride precipitation, which may be the result of a stress concentration. Although this experiment was conducted under more aggressive conditions than those expected during storage, this may present conditions favorable to the onset of delayed hydride cracking upon severe mechanical loading like that during transportation or accident conditions. Furthermore, the high degree of radial-hydride *continuity* or *networking* through the cladding thickness in Fig. 6b is an interesting phenomenon, especially considering that this figure represents a total hydrogen content of 320 wppm in this specimen. For actual hydrogen contents of ≥600 wppm for high-burnup Zry-4, we expect an even higher density of hydrides but with a homogeneous mixture of radial and circumferential hydrides. Quantification of such hydride *continuity* may be a better metric to correlate to macroscopic ductility and will be the subject of continuing study.

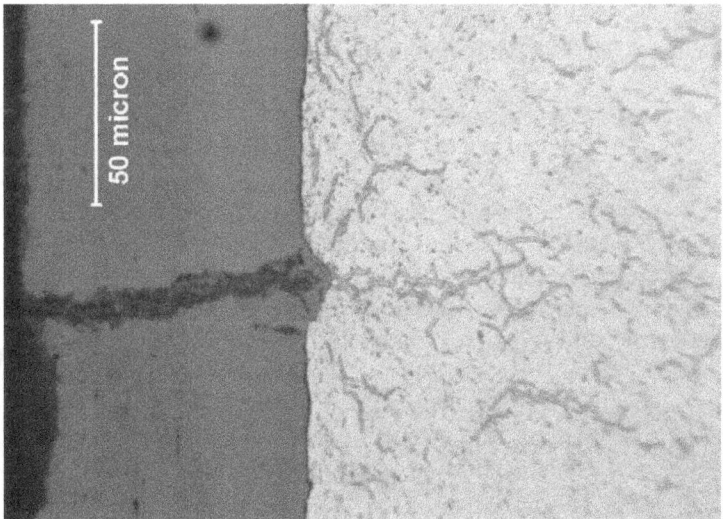

Figure 7 – Transverse micrograph of HBR creep specimen (3.5% permanent strain at 400°C/190 MPa) after cooling under stress showing corrosion-layer cracking and subsequent cladding oxidation and enhancement of hydride precipitation near crack; note, corrosion layer at the left of image.

37

**Future Work**

The preliminary axial-tensile, thermal-creep, and hydride-reorientation results have been used to develop a test plan to better understand the mechanical behavior of high-burnup Zry-4 cladding under drying-transfer, storage, and post-storage handling-transport conditions. In addition to tensile and creep tests of pool-stored high-burnup Zry-4, sealed specimens will be annealed for ≈3 days at 380-420°C and at hoop stresses of 0, 60, 90, 120, and 150 MPa and slow-cooled at ≈4°C/hr under decreasing pressure. Rings cut from these 100-mm-long samples will be subjected to ductility (diametral compression at 0.1%/s and 100%/s) and crush-impact failure-energy screening tests. These tests will be conducted at room temperature and 150°C. Metallographic examinations of post-annealing-treatment rings and post-ductility-test rings will be conducted to estimate the *threshold* stress to promote radial hydrides.

Additionally, rings will be cut directly from the discharge rods and annealed at identical temperatures and average hoop stresses using ring-stretch-tensile grips but then cooled at a much higher rate (≈2°C/min) to contrast hydride size and morphology to those produced by lower cooling rates of the sealed specimens. Because the ring-stretch-tensile grips will induce bending stress in the ring specimen, finite-element modeling will be used to estimate the *threshold* stress for radial-hydride precipitation.

The decreases in ductility and failure-impact energy will be correlated to the extent of radial hydride formation to map out cooling conditions – especially stress at 400°C – that are detrimental to high-burnup Zry-4 cladding integrity. Additional tests (e.g., fracture toughness) may be conducted on cladding subjected to these detrimental cooling conditions. Regardless of test technique, post-test specimens will be characterized, and the hydride microstructure will be quantified to correlate to ductility, impact energy, and/or toughness.

**Conclusions**

An experimental program is ongoing at Argonne National Laboratory to provide data to support license application and evaluation of SNF waste packages. In particular, the objectives of this program are to determine the mechanical behavior of SNF Zry-4 cladding materials and those conditions that promote radial-hydride precipitation. The following are the conclusions of this program:

1.  Results of tensile and creep testing suggest that high-burnup Zry-4 retains >3% uniform ductility between room temperature and 400°C, enveloping those temperatures of drying operations and long-term storage.
2.  Initial testing shows that radial-hydride precipitation ($F_n \approx 0.85$) occurs in high-burnup Zry-4 after fast cooling from 400°C to room temperature under 190-MPa hoop stress; these conditions and observations are consistent with other studies. Additional tests are planned to determine the *threshold* conditions (temperature [380-420°C], hoop stress [0-150 MPa], and cooling rate) that promote radial-hydride precipitation.
3.  Screening ring-compression tests at ≈0.1 and >100%/sec strain rates are planned to determine the relative change in ductility and failure energy due to radial-hydride precipitation on cladding. In addition, ring-crush impact tests are planned to determine impact-failure energy. These tests will be the prelude to a more comprehensive testing program.

**Acknowledgments**

The authors would like to thank Ralph Meyer, Harold Scott, Nancy Slater-Thompson, and Odelli Ozer for their support and technical guidance. This work was supported by the U.S. Nuclear Regulatory Commission, Office of Nuclear Regulatory Research, and the U.S. Department of Energy, Office of Civilian Radioactive Waste Management.

# References

[1]   R.P. Marshall and M.R. Louthan, "Tensile properties of Zircaloy with oriented hydrides," *Trans. of ASM*, Vol. 56, 1963, pp. 693-700.

[2]   R.N. Singh et al., "Stress-reorientation of hydrides and hydride embrittlement of Zr-2.5 wt% Nb pressure tube alloy," *J. of Nuclear Materials*, Vol. 325, 2004, pp. 26-33.

[3]   R. Choubey and M.P. Puls, "Crack initiation at long radial hydrides in Zr-2.5Nb pressure tube material at elevated temperatures," *Met. and Mat. Trans. A*, Vol. 25A, May 1994, pp. 993-1004.

[4]   A. McMinn, E.C. Darby, and J.S. Schofield, "The terminal solid solubility of hydrogen in zirconium alloys," *Zirconium in the Nuclear Industry: 12th Inter. Sym., ASTM STP 1354*, G.P. Sabol and G.D. Moan, Eds., American Society for Testing and Materials, West Conshohocken, PA, 2000, pp. 173-195.

[5]   J.J. Kearns, "Terminal solubility and partitioning of hydrogen in the alpha phase of zirconium, Zircaloy-2 and Zircaloy-4," *J. of Nuclear Materials*, Vol. 22, 1967, pp. 292-303.

[6]   P. Vizcaino, A.D. Banchik, and J.P. Abriata, "Solubility of hydrogen in Zircaloy-4: Irradiation induced recover and thermal recovery," *J. of Nuclear Materials*, Vol. 304, 2002, pp. 96-106.

[7]   J.B. Bai et al., "Hydride embrittlement in Zircaloy-4 plate: Part II. Interaction between the tensile stress and hydride morphology," *Met. and Mat. Trans. A*, Vol. 25A, 1994, pp. 1199-1208.

[8]   M. Leger and A. Donner, "The effect of stress on orientation of hydrides in zirconium alloy pressure tube materials," *Canadian Met. Quarterly*, Vol. 24 (3), 1985, pp. 235-243.

[9]   R.E. Einziger and R. Kohli, "Low-temperature rupture behavior of Zircaloy-clad pressurized water reactor spent fuel rods under dry storage conditions," *Nuclear Technology*, Vol. 67, 1984, pp. 107-123.

[10]  H. Tsai and M.C. Billone, "Cladding behavior during dry cask handling and storage," *2003 Nuclear Safety Research Conference*, NUREG/CP-0185, Washington, DC, October 20-22, 2003, pp. 71-83.

[11]  H.M. Chung, "Understanding hydride- and hydrogen-related processes in high-burnup cladding in spent-fuel-storage and accident situations," *2004 Inter. Meeting on LWR Fuel Performance*, Orlando, FL, September 19-22, 2004.

[12]  Interim Staff Guidance No. 11, Revision 3, issued by the Spent Fuel Project Office, U.S. Nuclear Regulatory Commission.

[13]  C. Brown et al., "Maximum cladding stresses for bounding PWR fuel rods during short term operations for dry cask storage," *2004 Inter. Meeting on LWR Fuel Performance*, Orlando, FL, September 19-22, 2004.

[14]  R.S. Daum, S. Majumdar, and M.C. Billone, "Mechanical properties of irradiated Zircaloy-4 for dry cask storage conditions and accidents," *2003 Nuclear Safety Research Conference*, NUREG/CP-0185, Washington, DC, October 20-22, 2003, pp. 85-96.

[15]  H. Tsai and M.C. Billone, "Thermal creep of irradiated Zircaloy cladding," *J. of Testing and Evaluation (ASTM)*, in-print (presented at the *14th Inter. Symposium on Zirconium in the Nuclear Industry*, Stockholm, Sweden, June 13-17, 2004).

[16]  R.E. Einziger et al., "Examination of spent PWR fuel rods after 15 years in dry storage," NUREG/CR-6831, ANL-03/17, September 2003.

[17]  R.S. Daum et al., "On the embrittlement of Zircaloy-4 under RIA-relevant conditions," *Zirconium in the Nuclear Industry: 13th Inter. Symposium, ASTM STP 1423*, G.D. Moan and P. Rudling, Eds., American Society for Testing and Materials, West Conshohocken, PA, 2002, pp. 702-719.

[18]  ASTM Standard B 811 – 97, "Standard specification for wrought zirconium alloy seamless tubes for nuclear reactor fuel cladding," *American Society for Testing and Materials*, West Conshohocken, PA, February 1998, pp. 920-935.

# Data Needs for the Transportation and Storage of High Burnup Fuel

RE Einziger, CL Brown, CG Interrante and GP Hornseth

US Nuclear Regulatory Commission
Spent Fuel Program Office

The storage and transportation of low to medium burnup spent nuclear fuel (<45 GWd/MTU) are mature, ongoing operations with a strong safety record. In order to improve nuclear reactor utilization, utilities are operating plants with fuel that is licensed for extended burnup. Fuel with burnups up to 62 GWd/MTU and possibly beyond will need to be stored and transported. In addition, events of recent years have raised the question of the safety of transport and storage casks that might come under terrorist attacks.

The NRC has been recently focusing on the storage and transportation of the higher burnup fuel because its cladding may have degraded mechanical properties relative to lower burnup fuel. The increased fluence and time-in-reactor have caused changes in the fuel pellet and cladding characteristics, In addition, newer cladding alloys are in use that have been developed to meet in-reactor performance standards.

During extended operation, the fuel is in the reactor for a longer time, hence more cladding oxidation occurs. Approximately 15-20% of the hydrogen generated during the oxidation process diffuses into the cladding, and for the most part accumulates at the outer cooler edge. Depending on the alloy, the hydrogen concentration in this outer rim of the cladding can reach above 600 wppm compared to 300 wppm or less in lower burnup fuel. As the temperature of the fuel is raised in the drying process, after cask loading, much of this hydrogen will go back into solution. Later cooling of the fuel will re-precipitate the hydrogen as hydrides as the solubility limit decreases with decreasing temperature. If the applied hoop stress due to the internal rod gas pressure is sufficient, radial hydrides will form. Radial hydrides may degrade the mechanical properties of the cladding and possibly lead to rod breach during transportation and storage conditions. At the current time, this critical stress is not well defined.

To meet the performance needs in-reactor, the vendors have developed a number of new alloys that have more corrosion resistance than Zircaloy at comparable burnups. The corrosion of these alloys produces less hydrogen thus making the potential for hydride generation and reorientation potentially less detrimental. A comparative data base on the creep behavior, fracture toughness, and other mechanical properties must be established by the licensee to determine if these alloys fall under the same guidelines for storage and transportation as the Zircaloy alloys.

At higher burnup, the fuel pellet forms a rim region, representing about 4-8% by volume of the fuel. This rim retains fission gas under high pressure, restructures to a submicron grain size, and has higher plutonium content than the body of the pellet. Very little is known about the behavior of this rim region under impacts that might be characteristic of a severe drop or terrorist attack. Since the fuel grains are already in the respirable size range, it is important to know the relative fracture and dispersal behavior of this fine-grained material compared to behavior of grains from lower burnup fuel that are 100 times larger.

To address the effects of these changes on the safe transportation and storage of spent fuel, expansion of the current data bases by the vendors and utilities is expected. The NRC has instituted a confirmatory data gathering activity to allow the staff to evaluate the adequacy of these data bases when license applications for the storage and transportation of high burnup fuels are submitted.

The NRC continues to seek data and analysis methods from the nuclear industry to support the safe storage and transportation of high burnup fuel. This paper will discuss data that are needed to evaluate the storage and transportation of high burnup spent nuclear fuel, and the NRC confirmatory programs to obtain data. Furthermore, the implications of the uncertainties on pending changes for Interim Staff Guidance ISG-11 Revision 3 will also be discussed.

# Data Needs for Storage and Transportation of High Burnup Fuel

U.S. Nuclear Regulatory Commission
Spent Fuel Project Office
Washington, DC

RE Einziger, CL Brown, GP Hornseth and
CG Interrante

October, 2004

# Summary

- Staff will issue changes to ISG-11 Rev 3 specifying acceptable conditions for storage and transportation

- Better data bases are needed to evaluate storage of high-burnup fuel, and newer alloy fuels

- Hydride reorientation is major uncertainty

- ANL is conducting confirmatory research (cosponsored by NRC, EPRI, DOE) on cladding annealing, critical hydride reorientation stress, and cladding creep of Zircaloy-4 high burnup cladding.

- Responsibility for providing an adequate data base to support licensing of storage and transportation of high-burnup fuel and newer claddings will rest on licensee

# Regulatory Requirements

- Dry Storage – 10CFR72
  - Ensure doses are controlled
  - Maintain subcriticality
  - Ensure confinement of spent fuel
  - No gross rupture of cladding or double confine fuel during storage
  - Maintain retrievability of spent fuel
- Transportation – 10CFR71
  - maintain a non-critical configuration

44

# Current ISG-11 Rev 3 limits for Storage

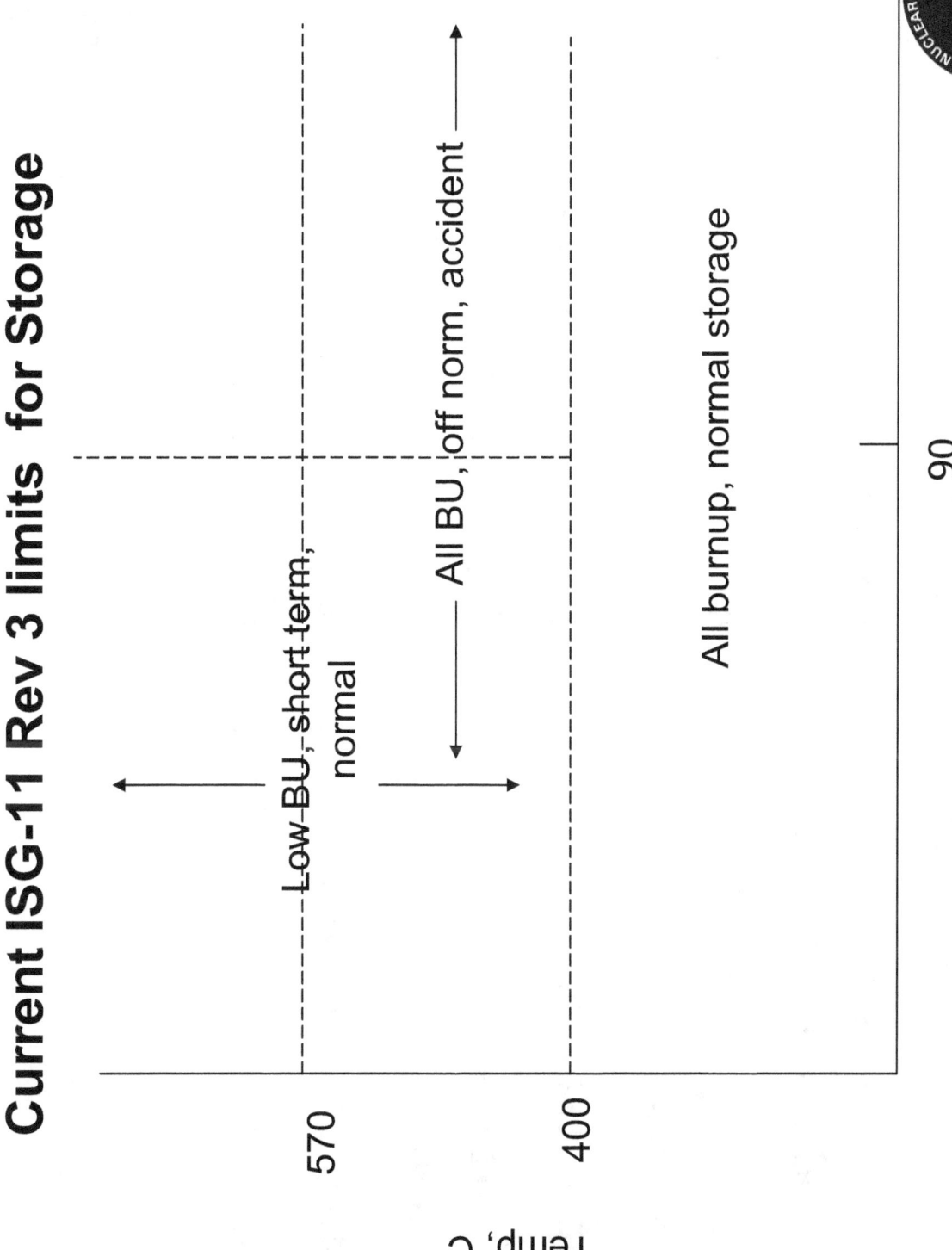

# Current Staff Guidance, ISG-11, Rev3

- Guidance for normal storage is independent of cladding material and burnup level

- Single temperature limit to minimize creep rupture and hydride re-orientation

- Hoop stress limit for LBF during short-term operations

- Minimize repeated thermal cycling - temperature differences less than 65°C

- Damaged fuel defined by ISG-1 guidance

- Transportation is handled on a case-by-case basis

# High Burnup Fuel Characteristics Affecting Rod

## Degradation

- Cladding

  - more oxidation

  - higher hydrogen content

  - new compositions

- Rod

  - higher fission-gas release ergo higher cladding stress

47

# Major Areas of Concern

## Basis for Revisions

- Hydride re-orientation

- Initial Fuel Rod Condition

- Data on mechanical properties and characteristics of irradiated HBF Zircaloy (for transport) and irradiated advance claddings (for storage and transport)

48

# Hydride Re-orientation

- Zircaloy HBF cladding has high hydrogen content

- Initially, circumferential hydrides concentrated on outer surface

- Temperature rise during vacuum drying causes hydrides to go into solution up to solvus.

- Upon cooling, hydrides re-precipitate as solubility limit exceeded

- Orientation of hydrides is a function of cladding stress

49

# Hydride Re-orientation

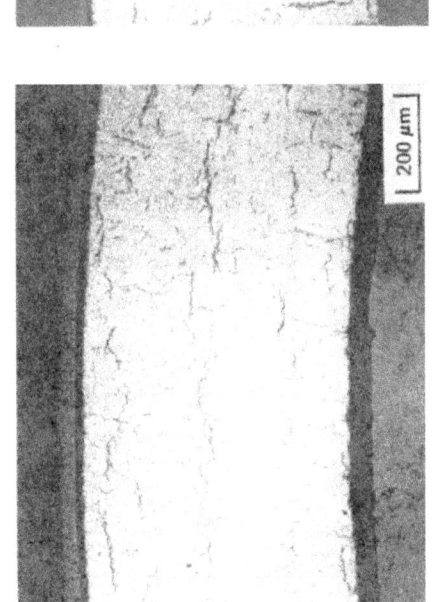

f(Temp, H conc, etc.)

Low Hoop Stress

High Hoop Stress

Circumferential Hydrides in Irradiated Zircaloy Cladding

Mixed Hydrides In Irradiated Zircaloy Cladding

Radial Hydrides In Irradiated Zircaloy Cladding

200 µm

200 µm

Photographs from Nuclear Technology, v. 67 (Oct. 1982) p. 107

# Applied hoop stress for hydride re-orientation as function of isothermal annealing temperature.

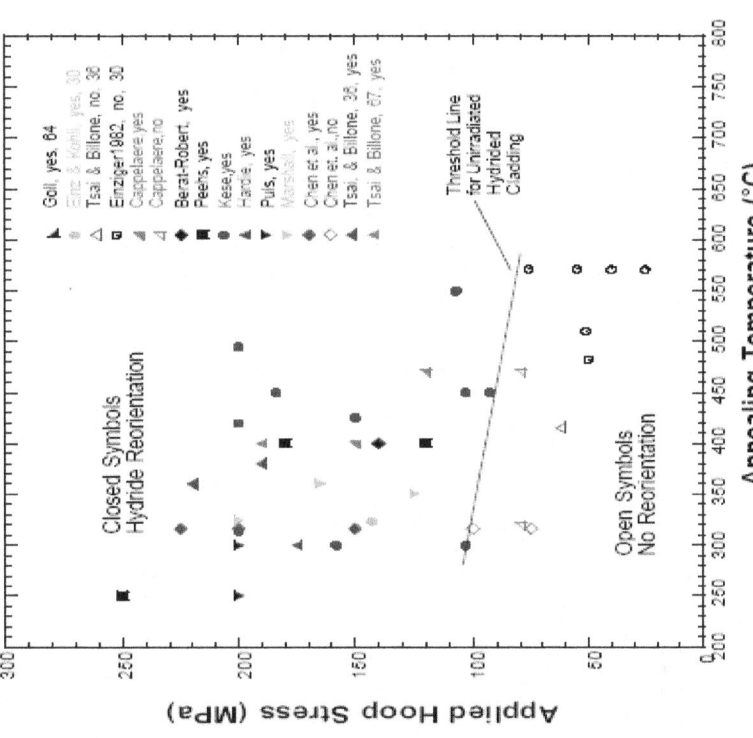

- Very sparse database
  - Mostly Unirradiated Material
  - Most obtained at high cooling rates
  - Most Obtained at high stress levels

  - Limiting stress for reorientation appears to increase with decreasing temperature.

- Zirc-2 may require less stress than Zirc-4

- No data on newer alloys

- Critical stress estimates vary from 80? to 120 MPa at 400 C

# Parameters Influencing Critical Hoop Stress

- Material Composition
- Cladding metallurgical Condition
- Cladding Physical Damage
- Burnup
- Maximum Temperature
- Cooling Rate
- Source of stress ?
- Initial hydride content

# Effect of Annealing Temperature on Hydride Re-orientation in PWR Cladding

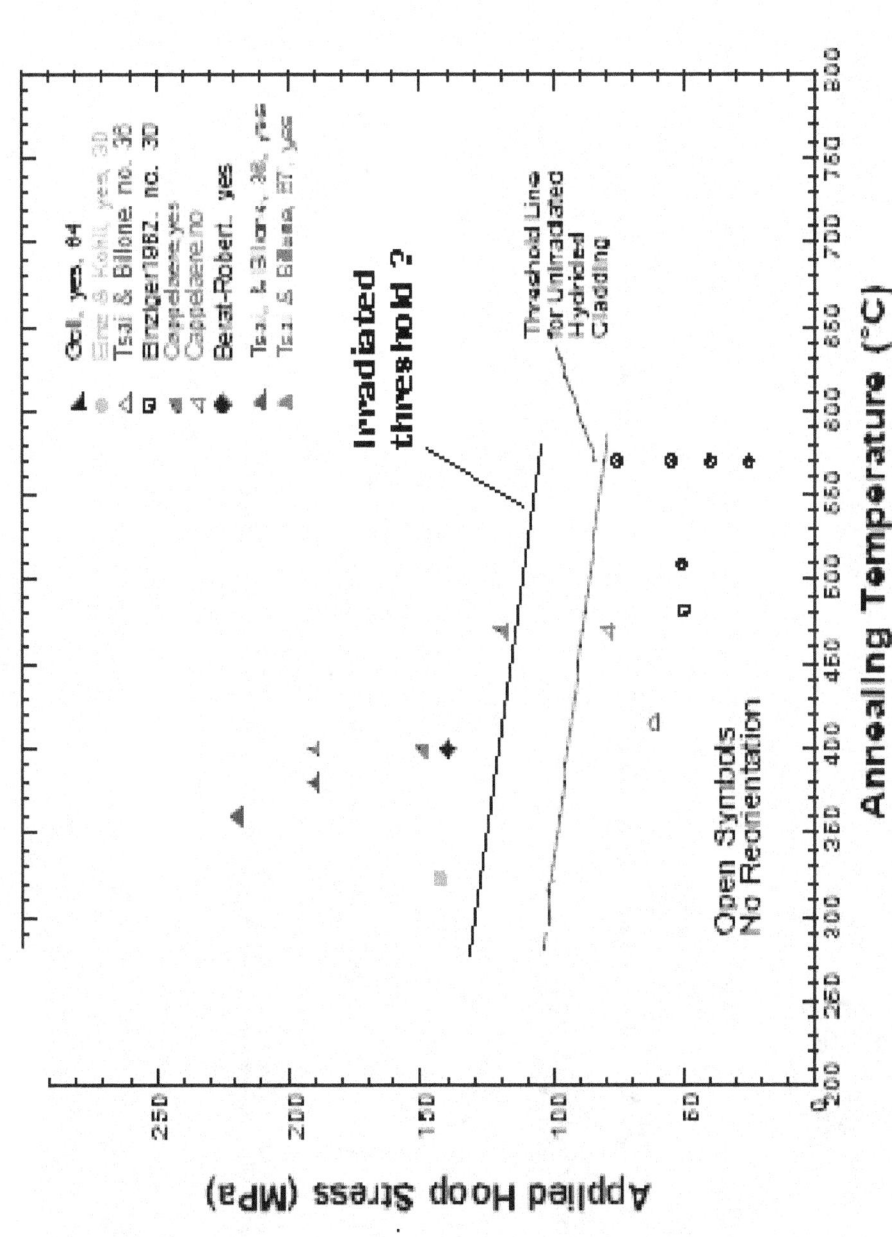

# Factors Driving the Cladding Hoop Stress Beyond 90 MPa in Drying, Storage, Transportation

- Severe cladding thinning due to CRUD or other mechanisms

- Operating at reactor limits of ~2200 psi

- Higher fission gas release at high burnup

- Gas generated in boron-containing rods

# How much hydride re-orientation is necessary to degrade mechanical properties?

- A lot of data on degradation of fracture toughness due to excess hydrides.
  - lots of scatter in data
  - depends on relation of hydride direction to stress vector.
  - depends on test method
  - circumferential hydrides have present lower risks for fracture
- Depends on property being considered
- Can be as low as 40 ppm

# Condition of the Cladding

- The expected condition of fuel rods as they come out of the reactor is the starting point for estimating storage and transportation degradation

- Condition of the fuel and cladding depends on:

  - manufacturing processes

  - in-reactor irradiation conditions

  - water chemistry in the reactor coolant

- Due to a variety of mechanisms operating in-reactor, fuel is falling outside the accepted data-base for post-irradiation fuel characteristics

# Condition of the cladding - How Many Intact but
## Damaged Rods?

- Is the database for fuel and cladding physical condition sufficient to describe the variety of fuel expected to be stored and transported?

- Failed Rods with defects larger than pinholes can be identified and separated for Special treatment if necessary

- What is the range of the Characteristics (wall thinning, gas pressure, etc) and number of fuel rods influenced by the same damage mechanism?

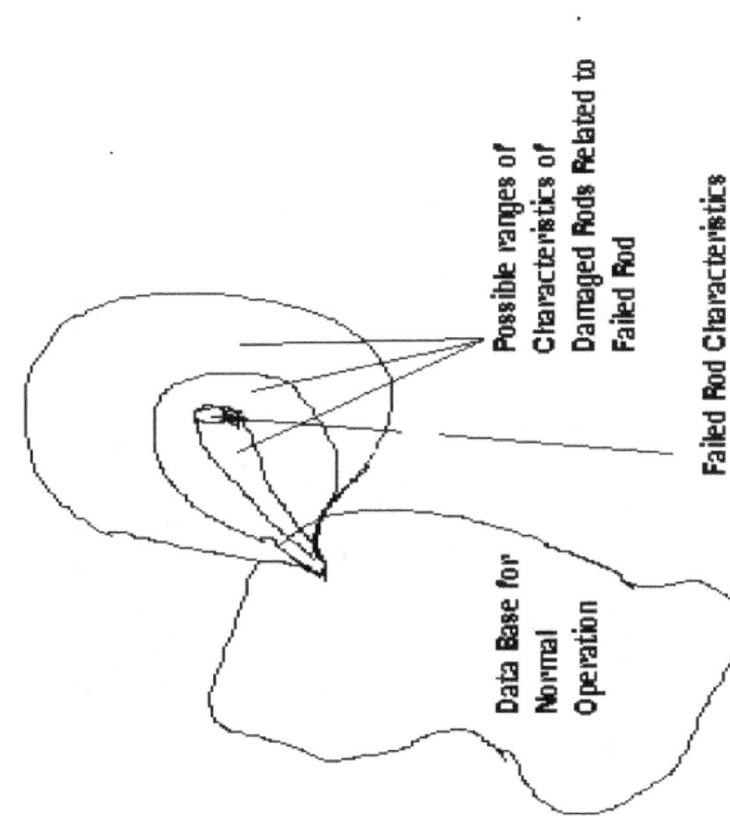

Possible ranges of Characteristics of Damaged Rods Related to Failed Rod

Failed Rod Characteristics

Data Base for Normal Operation

# Examples of New Fuel Cladding Conditions

### How do we classify damaged fuel?

- New CRUD patterns are being found that thins the cladding and adds hydrogen. Appears to be localized in reactor but sporadic on rods.

- Increased rate of unexplained failures as fuel is driven to higher burnup. How many additional unfailed rods are damaged? How do we identify damaged rods?

- PCI is making a comeback.

# How do we deal with damaged fuel that is not breached?

- Definition of damaged fuel

- Pre-storage inspection

- Calculation of results of breach during storage due to degraded initial condition

# Newer Claddings

- Staff has no mechanical data for advanced claddings
  - ZIRLO
  - M5
  - OPTIN

- Unknown if hydride re-orientation is a problem in newer claddings

60

# Proposed ISG-11 Limits for Storage

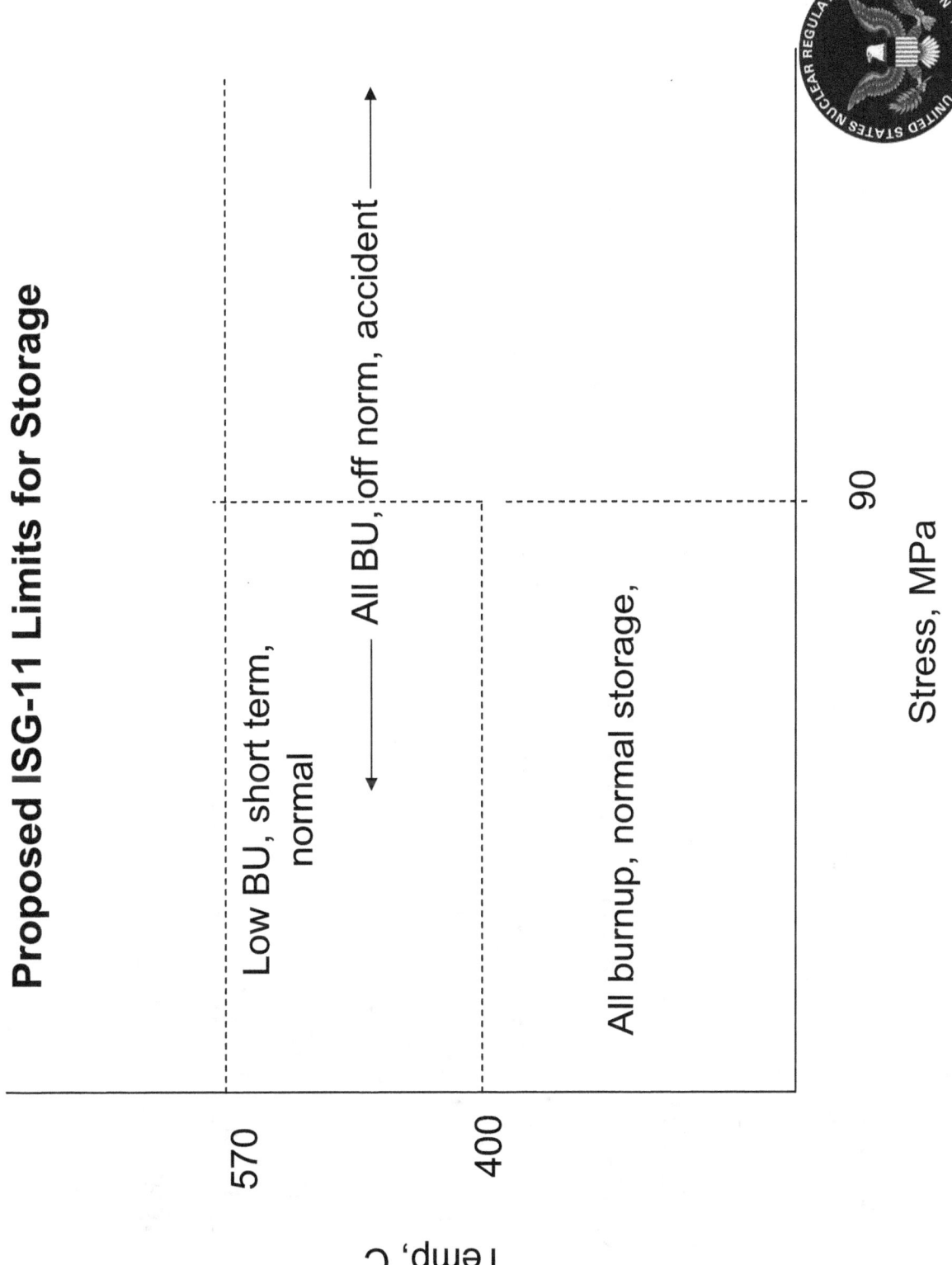

Low BU, short term, normal

All BU, off norm, accident

All burnup, normal storage,

570

400

Temp, C

90

Stress, MPa

# Research Needs for High-Burnup Fuel

Storage

- cladding creep-confirmatory data needed

- critical stress needed for hydride re-orientation

- effect of hydride re-orientation on strength and creep

- axial- and side-impact strength

- data on advanced cladding types

Transportation

- axial- and side-impact strength

- vibration- magnitudes and effects

62

# Research Needs for High Burnup Fuel

- ANL is currently conducting confirmatory research (co-sponsored by NRC, EPRI, and DOE) on cladding annealing, critical hydride reorientation stress, and cladding creep of Zircaloy-4 high burnup cladding.

  – Staff issued User Need Memo in 2004
  – Addresses the need for mechanical properties
  – Addresses the need to understand the phenomenon of hydride re-orientation
  – Addresses the need to understand cladding response in a transport accident.

# Confirmatory Research

The NRC sponsors research programs (many on a cooperative basis) to:

- to provide independent information to support the NRC's decision-making process

- to identify and characterize technical questions that may become important safety issues in the future.

- is designed to improve the agency's knowledge in areas where uncertainty in knowledge exists, where safety margins are not well characterized, and where regulatory decisions need, or will need to be confirmed.

- the development of sound technical bases allows focus on safety issues and more realistic decisions.

# Summary

- Staff will issue changes to ISG-11 Rev 3 specifying acceptable conditions for storage and transportation

- Better data bases are needed to evaluate storage of high-burnup fuel, and newer alloy fuels

- Hydride reorientation is major uncertainty

- ANL is conducting confirmatory research (co-sponsored by NRC, EPRI, DOE) on cladding annealing, critical hydride reorientation stress, and cladding creep of Zircaloy-4 high burnup cladding.

- Responsibility for providing an adequate data base to support licensing of storage and transportation of high-burnup fuel and newer claddings will rest on licensee

65

# Data Needs for Storage and Transportation of High Burnup Fuel

**RE Einziger, CL Brown, GP Hornseth and CG Interrante**
US Nuclear Regulatory Commission
Spent Fuel Project Office

## Abstract

*The NRC has recently focused on the storage and transportation of high burnup fuel because its cladding may have degraded mechanical properties relative to lower burnup fuel. The hoop stress at higher burnups may cause radial hydrides that degrade the mechanical properties of the cladding and possibly lead to fuel rod disruption during transportation and storage. At the current time, the critical stress for hydride reorientation is not well defined. The increased fluence and time-in-reactor change the cladding characteristics. New and unidentified modes of breach have been appeared in-reactor that might change the initial condition of the fuel being stored; hence affect its overall performance. To meet the performance needs in-reactor, the vendors have developed a number of new alloys. A comparative data base on the creep behavior, fracture toughness, and other mechanical properties must be established by the applicant to determine if these new alloys fall under the same guidelines for storage and transportation as the Zircaloy alloys.*

*Expansion of the current data bases by the applicants is expected to address the effects of these changes on the safe transportation and storage of spent fuel,. The NRC only sponsors confirmatory research to allow the staff to evaluate the adequacy of the applicant data. This paper will discuss data that are needed to evaluate the storage and transportation of high burnup spent nuclear fuel, and the NRC confirmatory programs to obtain data. Furthermore, the implications of the uncertainties on changes for Interim Staff Guidance ISG-11 Revision 3 will also be discussed.*

## I. Introduction

The storage and transportation of low to medium burnup spent nuclear fuel (<45 GWd/MTU) are mature, ongoing operations with a strong safety record. In order to improve nuclear reactor utilization, utilities are operating plants with fuel that is licensed for extended burnup. Fuel with burnups up to 62 GWd/MTU and possibly beyond will need to be stored and transported. The Nuclear Regulatory Commission (NRC) has been recently focusing on the storage and transportation of the higher burnup fuel because its cladding may have degraded mechanical properties relative to lower burnup fuel.

The increased fluence and time-in-reactor have resulted in larger hydride contents that, if reoriented to a radial direction, may degrade the cladding's fracture properties. In addition, newer cladding alloys are in use that have been developed to meet in-reactor performance standards but not necessarily storage and transport requirements. A comparative data base on the creep behavior, fracture toughness, and other mechanical properties must be established by the licensee to determine if these alloys fall under the same guidelines for storage and transportation as the Zircaloy alloys. Currently, some reactors are experiencing an increased failure rate of the fuel cladding, some of which are due to unidentified mechanisms. These mechanisms may affect both the condition of the fuel as it enters storage and the predicted in-storage behavior of the SNF.

To address the effects of these changes on the safe transportation and storage of spent fuel, expansion of the current data bases by the vendors and utilities is expected. The NRC continues to seek data and analysis methods from the nuclear industry to support the safe storage and transportation of high burnup fuel. This paper will discuss data that are needed to evaluate the storage and transportation of high burnup spent nuclear fuel, and the NRC confirmatory programs to obtain data. Furthermore, the implications of the uncertainties on pending changes for Interim Staff Guidance, ISG-11 Revision 3, will also be discussed.

## A. Regulations

Dry storage of spent nuclear fuel (SNF) is regulated by 10CFR 72. The regulations require that the dose levels to the public must be controlled, the system must remain subcritical, and the spent fuel has to be confined. Any change in the spent fuel condition that will compromise these functions is unacceptable. As a result, the regulation [10CFR 72.122(h)(1)] specifies that the fuel cladding must either be protected against gross degradation or be double confined during storage if already damaged. This is not per say to maintain the cladding integrity but rather to assure adequate confinement and sub-criticallity under normal storage conditions, and to minimize release of radioactive material to the storage environment within the sealed cask. An operational benefit is that the fuel will remain in a retrievable physical condition as intact assemblies.

Transportation of SNF is governed by 10CFR71. The only criterion on the fuel is that it remains in a sub-critical configuration. Currently, this is interpreted to mean that even in the worst configuration with the cask fully moderated, the fuel will remain sub-critical. The system must remain sub-critical even after an accident event such as the regulatory nine meter drop, impact on a blunt steel puncture pin, and fire. There is nothing specified in this regulation that requires the fuel to remain in a substantially intact configuration.

## B. Current Situation

In order to assure that the fuel stays in a configuration that will meet the goals indicated above, the NRC staff has issued guidance (ISG-11 Rev.3) with regard to suggested maximum storage temperatures, internal rod pressures and cask atmospheres. These guidelines are based on the expected initial condition of the fuel as it is taken out of the reactor water pool, and degradation mechanisms thought to be potentially active during dry storage, namely creep rupture of the cladding, hydride embrittlement due to hydride reorientation, and fuel pellet oxidation.

Fuel with cladding that has pinhole leaks or narrow cracks is currently considered to be intact since it maintains it structural integrity under normal storage. This is only true if the fuel is in a non-oxygenated atmosphere as the expected maximum temperature of the fuel will be substantially higher than the temperature where rapid oxidation of $UO_2$ occurs. During one of the revisions of ISG-11, the need for an inert storage atmosphere was inadvertently omitted.

The current temperature and stress guidelines are shown in Figure 1. These are only guidelines and applicants can propose other limits if the are backed by an adequate data base in the safety analysis report (SAR). The 400°C maximum temperature was based on limiting the creep stain to 1%. Creep strain correlations for low burnup Zircaloy were used for the determination. This temperature limit also insures a maximum concentration of ~220 wppm hydrogen in solution in the cladding. Currently this limit is acceptable for any burnup or fuel type. In addition, low burnup fuel (<45 GWd/MTU) could have a short term excursion to higher temperatures, such as in vacuum drying, if the cladding hoop stress is maintained below 90 MPa. The stress limit, based on the current best estimate of the critical stress, was imposed so that no excessive hydride reorientation would occur. During an accident the temperature of the cladding was restricted to 570°C, a temperature where Zircaloy-4 cladding of moderate burnup has been experimentally shown not to fail in six weeks although excessive creep did occur. Currently, the NRC staff has not imposed an upper cladding stress limit, for normal operation, on either high or low burnup fuel, or an upper temperature limit for short term excursions of low burnup fuel.

There are a number of reasons that there are currently no similar guidelines during the transportation of spent fuel:

- The duration is short and similar mechanisms may not be active.
- There are no mechanical and impact properties available in the open literature to enable the staff to predict the cladding behavior during transportation.

The performance of the fuel in relation to the requirement to maintain subcriticality during both normal and accident transport is currently evaluated on a case-by-case basis.

## C. High Burnup fuel Characteristics

The dividing point between high and low burnup fuel is 45 GWd/MTU. This was initially set because it was the upper limit of fuel performance data in conditions typical of dry storage. It also happened to be approximately the lower burnup limit where a number of changes in fuel characteristics occur that may significantly affect the performance of the fuel in dry storage.

The extent of cladding oxidation at higher burnups will depend on the composition of the zirconium alloy. Zircaloy corrosion increases rapidly after 45 GWd/MTU and becomes appreciable at higher burnups. High burnup PWR rods with Zircaloy-4 cladding can have oxide thicknesses in the range of 80 to 100 μm [SAB00]. Consequently, as the cladding thins, the stress can increase by up to 10%. The newer cladding compositions such as,

M5, and Zirlo, etc., were developed to resist corrosion during the extended irradiation [ISH00, MEY00]. The corrosion of these alloys is considerably less than Zircaloy at equivalent burnups. Even ZIRLO has attained oxide thicknesses above 80 μm in recent years [KNO03, KAI00] as the plants are experiencing higher duties than ever before. The trend is increasing due to the utility's desire to reduce fuel costs.

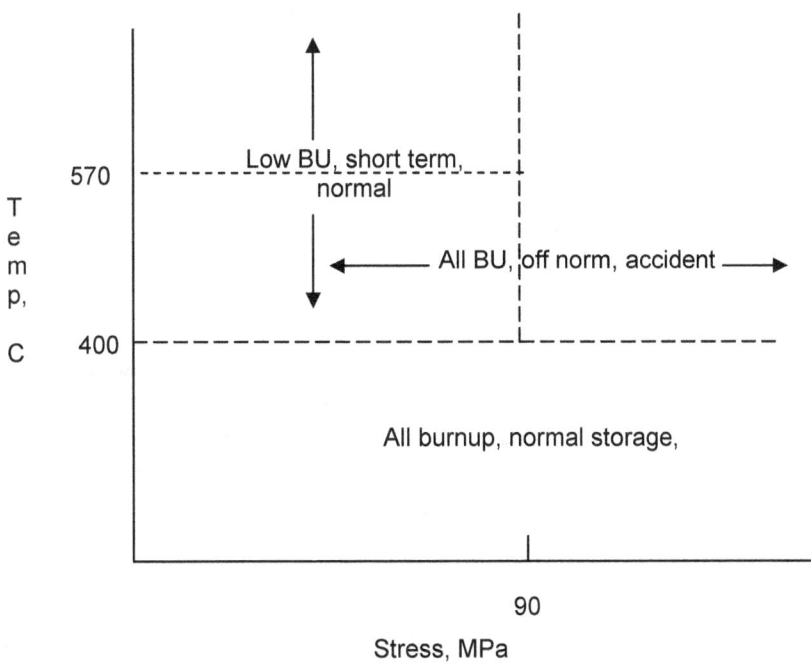

Figure 1 - Current ISG-11 Rev 3 Limits for Dry Storage. Limits are applicable to all cladding types.


As the oxide thickness increases the cladding loses ductility due to the absorption of 10-20% of the resultant hydrogen that forms hydrides in the outer layer of the cladding. The average hydrogen content in the cladding at ~30 GWd/MTU is ~100-200 wppm and it increases to the range of 600 wppm at 65 GWd/MTU. As the cladding cools, the excess hydrogen in solution precipitates as zirconium hydrides. Due to the texture in Zircaloy-4, these tend to be long, circumferential platelets. In Zircaloy-2 they take the form of shorter randomly orientated platelets due to the lack of texture in this alloy. As a result of the high hydride content, the high burnup cladding has a loss of ductility when there is a large amount of corrosion [ITA00, GAR96]

The newer zirconium cladding alloys were developed to reduce the corrosion potential and consequently the hydrogen content of the cladding. There are no publications that indicate that these alloys were developed to improve or even maintain the mechanical properties of the cladding. The data base of the mechanical properties for these alloys is too sparse to determine how they will behave in dry storage and transportation.

In PWR fuel and the newer pressurized BWR fuel, the fission gas release is 1-2% at burnups below 35 GWd/MTU. As the irradiation continues to higher burnup, the average release, from a batch of high burnup (50-62 GWd/MTU) discharged fuel rods, is between 3 and 7%. The fractional release continues to increase to about 28% at 100 GWd/MTU [MAN00]. The normal hoop stress on the cladding during storage and transportation is directly proportional to the internal pressure and hence increases with increased fission gas release. The significance of the increased stress is discussed later.

## D. Areas of Concern

There are three major areas of concern that are the basis for revision of ISG-11 Rev 3. These are hydride reorientation, initial starting condition of rods put into storage, and the mechanical properties of the newer cladding alloys.

As the burnup of the fuel increases, the hydrogen content of the cladding increases. After a portion of this hydrogen is taken back into solution during a temperature excursion such as vacuum drying or a fire accident, it will eventually reprecipitate as zirconium hydrides as the fuel cools. If the hydrides form in a radial direction due to the hoop stress in the cladding, the mechanical properties of the cladding and its ability to resist fracture when subjected to normal and accident loads during storage and transportation comes into question.

The limits on the conditions of storage (temperature, time, stress, and atmosphere) are recommended to reduce the number of unlikely fuel disturbances. The ability of the cladding to withstand impacts during storage and transportation will depend on its initial flaw structure as the fuel is removed from the reactor. There are several indications that in some fuel this condition may be significantly different than currently expected. This may be due to unknown mechanisms associated with in-reactor breaches, which occur due to a combination of plant uprates, challenging water chemistry, longer fuel cycles, etc. A significant change in the cladding flaw size distribution would result in a change in the recommended storage and transportation limits.

The current recommended storage guidance is based on the properties of low burnup Zircaloy cladding and indications of the expected high burnup performance of this alloy. ISG-11 Rev 3 allows fuel with advanced cladding to be stored. In order to evaluate the performance of these fuels under regulatory transport accidents and road vibration, data on pertinent properties is needed. Current data is insufficient to determine cladding performance, thus transport applications are considered on a case-by-case basis.

## II. Technical Uncertainties

### A. Hydride Reorientation

When the cladding corrodes in-reactor, 10-20% of the hydrogen generated by corrosion diffuses through the oxide layer into the cladding. Depending on the metallurgical processing of the Zircaloy, hydrogen in excess of the solubility limit at the precipitous will precipitate either as circumferential hydride platelets on the outer cladding surface (PWR) or as randomly oriented platelets (BWR). Additional hydrides form as the fuel cools in the reactor pool[1].

When the fuel cladding temperature is raised during short-term operations, either as part of the drying process when the fuel is put in dry storage, or in storage itself, in particular with high burn-up fuel, hydrogen will go back into solution at the level of the solvus for the temperature in question. At 400°C this is approximately 220 wppm. Upon cooling, after drying or in storage, the cladding is in a tensile hoop stress state. If this stress level is above a critical stress, as the precipitous is exceeded, the hydrides will precipitate in the radial direction. Figure 2 shows hydride reorientation as an effect of stress on the cladding. These radial hydrides, if in sufficient amounts, may degrade the mechanical properties of the cladding and possibly make the cladding susceptible to breach during normal and/or accident transportation events.

However, there is considerable uncertainty on the critical stress for hydride reorientation, the amount of radial hydrides required to degrade the mechanical properties, the effects of this degradation, and whether fuel rods can even attain hoop stresses above the critical level in storage or transportation.

### 1. Critical Stress

The data base for the reorientation of hydrides in Zircaloy as a function of the applied stress and temperature is given in Fig 3 from Chung [CHU04]. The data are rather sparse for rods where reorientation has not been observed, and there is considerable scatter in the data where reorientation has been observed. There are good reasons for these conclusions. The stress in most low burnup rods is in the 60 MPa or lower range, the hydrogen content is rather low (<100 wppm), and the cladding usually operates in a temperature range below 350°C. Consequently, hydrides in general are not a problem for low burnup fuel in storage or performance of fuel rods in reactor. As a result, they are rarely looked for. The scatter at the higher stresses in Figure 3 is great because there are so many parameters that influence the reorientation.

---

[1] Normally, these hydride precipitates are generally circumferentially oriented in PWR cladding, and randomly oriented in BWR cladding, unless the stress developed in-reactor was sufficient to allow radial formation.

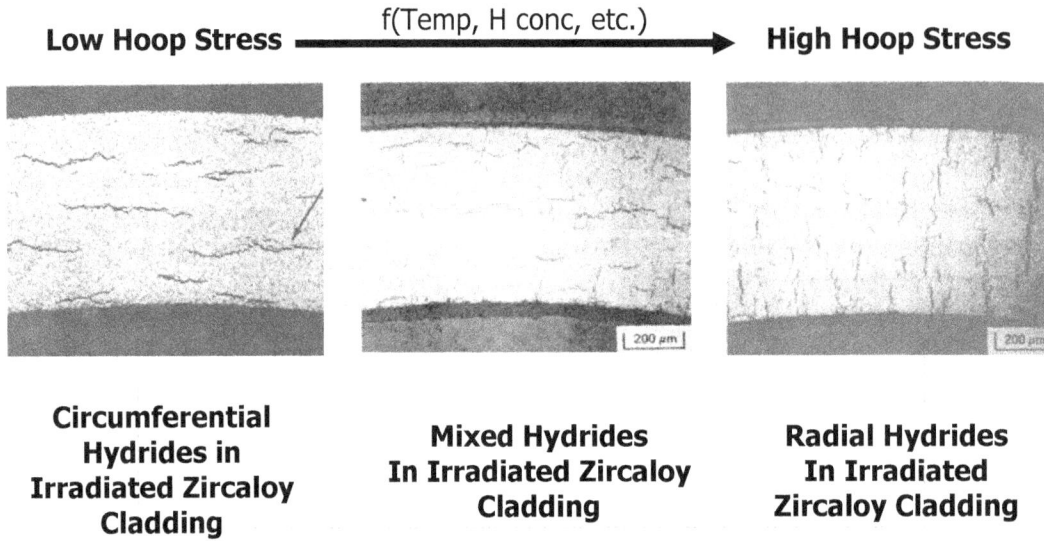

**Low Hoop Stress** f(Temp, H conc, etc.) **High Hoop Stress**

| Circumferential Hydrides in Irradiated Zircaloy Cladding | Mixed Hydrides In Irradiated Zircaloy Cladding | Radial Hydrides In Irradiated Zircaloy Cladding |

Photographs from Nuclear Technology, v. 67 (Oct. 1982) p. 107.

Figure 2 – Hydride Reorientation as a Function of stress Level. HB Robinson, ~30 GWd/MTU

Most of the data was obtained on unirradiated Zircaloy, most was for Zircaloy-4, and almost all the data was taken at rather fast cooling rates (>>4°C/h), essentially a quench (see [CHU04] for references to data). Likewise, most of the data is at stresses far above the stress expected even in high burnup fuel or fuel operated in-reactor under limiting conditions. While the threshold line shows a slight temperature dependence, the data is too scant to rigorously support this.

It is questionable how relevant this data is to spent fuel in storage. Spent fuel has been irradiated, cooled at a much slower rate, in many cases has a higher hydrogen content then the tested samples, and uses both Zircaloy-4, and -2 that have different metallurgical textures. There is no data for the newer advanced alloys. In spite of these deficiencies, it appears from Figure 3 that there is a stress level somewhere in the range of 90 MPa that is critical for hydride reorientation. This is in the expected range of stress for high burnup fuel. Therefore data are needed to establish a reasonable level of confidence in the critical stress for reorientation, especially for the higher burnup fuels.

There are many parameters that influence the critical hoop stress. These include: alloy composition and metallurgical condition, cladding physical damage, burnup, maximum temperature, cooling rate, source of stress, and initial hydride content. Ito [ITO04] studied the reorientation of hydrides in mid-burnup (46-54 GWd/MTU) Zircaloy-2 and Zircaloy-4 cladding. While in the temperature range of interest (420 to 360°C), the stresses where much higher than expected for SNF in dry storage. Likewise, the hydrogen content was lower than expected (~ 100 ppm for Zirc-2 and 100 to 300 ppm for Zirc-4) in today's SNF. Nevertheless, Ito observed that reorientation occurred in the Zircaloy-2 at lower stresses while it did not occur in Zircaloy-4. This is in agreement with Marshall [MAR67] who found that the degree of reorientation was highly dependent on the fabrication methodology of unirradiated Zircaloy. Zircaloy-4 is manufactured with a texture to discourage the formation radial hydrides while the Zircaloy-2 grains are randomly oriented.

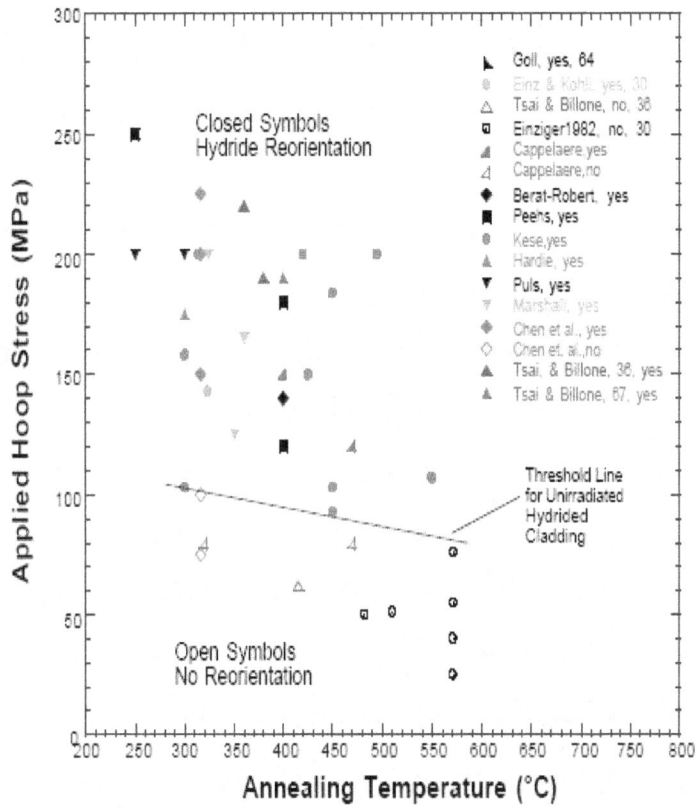

Figure 3 – Hydride reorientation as a function of temperature and stress. References to the original data can be found in Chung [CHU04]

Chan [CHA96] indicates that as the cooling rate increases, the fraction of radial hydrides decrease at any given stress. He also stated that a slower cooling rate could possibly lower the threshold stress by relaxing the internal residual stresses. In addition as the cooling rate drops, the radial hydrides have longer available time for coalescence to become larger and more continuous. Larger hydrides, formed at a lower cooling rate, have been observed in unirradiated Zircaloy-2 [ELL68]. As a result, at a lower cooling rate, the ductility will decrease sharply and less radial hydrides are necessary to cause this decrease [CHA96]. Although Chan makes his conclusions based on a model, it does indicate that a slower cooling rate, as expected in vacuum drying, dry storage, or the repository, may lower the critical stress considerably and must be considered in any testing to determine the threshold for formation of radial hydrides.

The reorientation data from irradiated fuel is shown in Fig 4. It indicates that the critical stress for irradiated material might be about 40 MPa higher than that for unirradiated Zircaloy. Based on calculations of terminal solubility, Ito [ITO04] projected that the critical stress for irradiated Zircaloy-4 might be higher than unirradiated Zircaloy-4. This would be beneficial but needs to be supported with data in the range of 80 to 150 MPa. Ito also predicted that the critical stress for irradiated Zircaloy-2 would be less than that for unirradiated cladding. Thus, both the effect of irradiation and cladding type must be considered in a testing program.

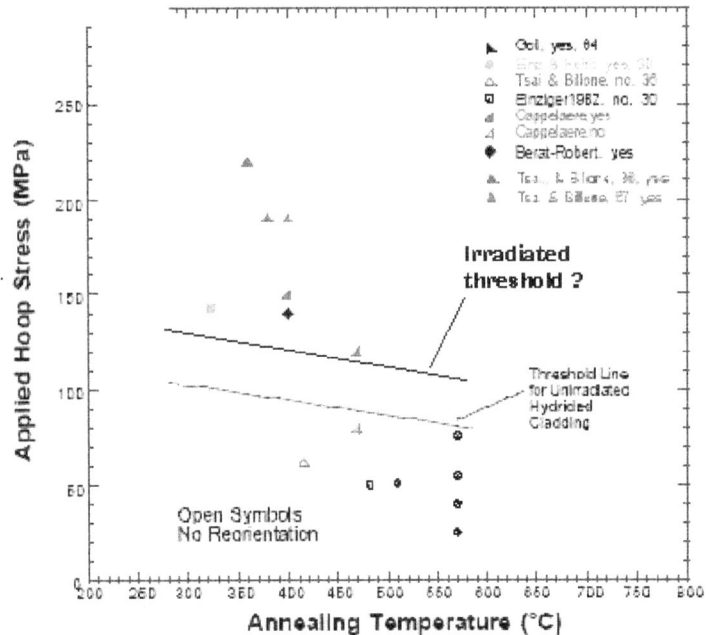

Figure 4 – Hydride Reorientation Data for Spent Fuel Cladding

## 2. Stresses in Spent Fuel

Modern fuel rods are prepressurised with He to increase the rate of heat transfer. During reactor operation, some of the fission gas, generated in the pellets, is released to the plenum  In addition the free volume in the rod decreases due to the creep down of the cladding caused by external pressure of the coolant. Earlier operating procedures ensured that the pressure inside the fuel rods was limited to the coolant pressure (~2200 psi) so as to preclude outward creep of the cladding. Recently, the pressure in the rods has been allowed to exceed the system pressure, up to the cladding lift off pressure (~3000 psi). Rods that operate close to the system or liftoff pressures in reactor have a cladding stress at 400°C of ~130 and 180 MPa, respectively. The stress at 570°C will be higher.

When the rod is placed in dry storage, the cask pressure is only 2 to 5 atmospheres. This in turn puts the cladding under a tensile hoop stress governed by the pressure in the rods.

The stress, for thin walled cylinders, is given by the formula: $\sigma = P \cdot D/2t$, where P is the internal rod pressure, D is the diameter of the rod, and t is the thickness of the cladding. The changes in cladding diameter will have little direct effect on the stress but will affect the internal rod pressure due to an increase of the rod internal volume. The stress can be increased by any in-reactor mechanism, such as larger fission gas release (fgr) or other sources of gas generation that increases the rod pressure or decreases the cladding thickness. A study by Brown et al [BRO04] indicated that due to either larger fgr as the burnup increases or generation of gas from integral boron absorbers, there are a number of rod designs that could have in excess of 90 MPa hoop stress at a temperature of 400° C. The rods were assumed to have normal gas release and cladding thinning due to corrosion for the burnup of interest.

Rods have incipient cracks on the inside cladding surface caused by interaction of the fuel and the cladding during reactor operation. These act as stress risers and effectively reduces the local thickness of the cladding. Although a number of attempts have been made to theoretically determine this distribution [SAN92], it has never been measured, and will be ignored for the time being. Due to corrosion, the cladding forms a hydride-rich layer on the external surface. This layer can be expected to have little strength and should be considered as cladding wastage as required in ISG-11 Rev 3. Any in-reactor mechanisms that thin the cladding wall also need to be considered. For example, excessive CRUD buildup (see next section) can thin the walls of selected rods up to 30% with a subsequent increase in stress during storage and transportation.

While the majority of fuel rods will have stress in storage and transportation that are in the 60-70 MPa range, more rods will approach or exceed a cladding stress of 90 MPa as burnup is increased. The cladding stress can be considerably higher than 90 MPa if: 1) the rod pressures approach system or lift-off pressures, or 2) there are unexpected events in the reactor. The applicant will have to evaluate the range of rod stresses expected in-storage to determine if hydride reorientation is applicable to a particular fuel loading in a cask.

## 3. Degradation of Mechanical Properties

There are no publicly available mechanical properties, specifically fracture toughness properties on irradiated Zircaloy with radial hydrides, or any of the newer alloys. As a result, there is considerable uncertainty on the amount and morphology of the hydrides that are necessary to degrade the fracture toughness. Attempts have been made to model the precipitation phenomena and show that too few radial hydrides are formed to be of concern. These models have depended upon the difference between the solvus and precipitous of hydrogen for unirradiated Zircaloy [KAM96]. If the precipitous measured by Vizcaino [VIZ02] for irradiated Zircaloy-4 were used instead, even less radial hydrides would be expected. Until the amount or fraction of radial hydrides that are necessary to degrade the mechanical properties have been established, for the temperatures of interest, these calculations will be irrelevant.

Marshall and Louthan [MAR53] found that the ductility of unirradiated Zircaloy-2 became nil if it contained as little as 40 wppm of radial hydride. Yagnik [YAG04] found that at room temperature unirradiated Zircaloy-4 lost about half its ductility with 30 ppm radial hydrides and 70% with 70 ppm radial hydrides. No loss of ductility was measured at 300°C, which points out the importance of data for temperatures of interest to specific applications.

Rashid [RAS01] reviewed and compiled fracture toughness data on zirconium alloys. Data obtained on hydrided and irradiated samples were included, but were very limited. Little of the data was prototypical of SNF. There were a number of reasons for this: 1) the alloy composition, 2) hydrogen charging method, or 3) fluence or test temperature was outside the applicable range. The limited data that were applicable indicated a reduction in fracture toughness by about a factor of two, at 250 ppm hydrogen. In this review, there was no indication of the orientation of the hydrides with respect to the stress.

## B. Initial Fuel Condition

The performance of the fuel assembly and fuel rods during storage and transportation, and the end condition of the fuel rods is estimated using a number of steps:

- Identify the degradation mechanisms active during storage and transportation.
- Determine the initial condition of the rods and assembly as they are placed in storage.
- Apply the degradation mechanisms to the initial fuel condition to determine the performance and end condition.

Other than the effects of annealing and hydride reorientation that are currently under investigation, no active degradation mechanisms have been identified. The range of the initial condition of the fuel is based on post-irradiation at-pool and hot-cell examinations. As a result, the current ISG-11 Rev 3 limits were recommended. If fuel breaches in-reactor with larger than a pinhole or tight crack, either the rod has to be identified and removed from the assembly, or the assembly must be canned in a secondary container. The expected condition of the fuel rods as they come out of the reactor is the starting point for estimating degradation during storage and transport. If the condition of the fuel rods falls outside the expected range of initial conditions then application of the degradation mechanisms may result in unacceptable performance during storage and transportation.

Recently, the rate of cladding breaches in the reactors has increased [YAN04] and unexpected behavior, such as the build up of excessive CRUD have occurred. These have been attributed to a combination of rod manufacturing process, in-reactor irradiation conditions, and water chemistry of the reactor coolant [MUT04, SCH04]. In some cases, the cause of rod breach has not been identified. As a result fuel characteristics are falling outside the accepted data-base for post-irradiation fuel characteristics. The breached rod doesn't pose a problem as it will be handled in the appropriate manner depending on the characteristics of the breach, but as illustrated in Figure 5, it is unknown how many additional rods may also have degraded properties due to the same active mechanism. Additional rods may have degraded characteristics (wall thinning, high internal pressure etc.), but not to the point of breach, due to the same active mechanism. The range of degraded rods needs to be determined, in order to decide the appropriate conditions for storage and transportation.

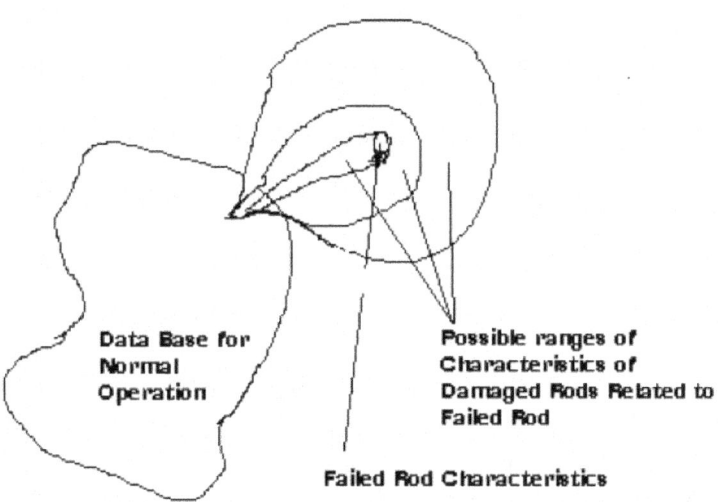

Figure 5 – The characteristics of a breached rod fall outside the range of expected fuel rod characteristics.

Some examples of fuel rod characteristics that fall outside the normally accepted range follow:

- CRUD buildup – Excessive CRUD buildup has led to breach in some rods due to localized overheating and corrosion of the cladding. Some of the adjacent rods have been found to have thinned cladding walls and other rods with the buildup have no thinning. Currently, the procedure is to examine every rod with the excessive CRUD buildup. The cause has not been definitively identified [KEY04, TRO04].
- Breaches with unknown causes have occurred primarily in optimized fuel designs with thinner cladding [YAN04].
- Higher FGR and hydrogen pickup than predicted by the codes [YAN04].
- Pellet Cladding Interaction (PCI) – This mechanism, thought to have been eliminated is making a comeback as fuel is being driven harder (longer reactor cycles). It has been observed in at least six plants even though the codes indicated the stress was within acceptable range [YAN04, SCH04].

- Initial radial hydrides – Foster [FOS02] and Seibold [SEI04] have found that radial hydrides can form in BWR cladding, while in-reactor, due to a stress caused by an expanding oxide layer on the outer surface, or by pellets with missing surfaces causing a stress from the cladding inside. This has resulted in long splits at low burnup.

The staff is undecided, at this point, how to deal with damaged but unbreached fuel:
- It could be incorporated into the definition of damaged fuel and be subject to assembly canning.
- Redefine undamaged fuel to include cladding with breaches larger than a pinhole.
- Require a pre-storage inspection of fuel from cores where unexplained breaches or abnormal fuel behavior has been observed, or
- Require calculations of the consequences of additional breaches in storage and transportation, if a maximum degradation is assumed to have occurred in additional rods.

No specific timetable has been set to reconcile this situation.

## C. New Cladding Types

The newer vintages of cladding are starting to approach the high burnup range. These claddings were developed to reduce the cladding corrosion, and hence the hydrogen content, below that of Zircaloy at equal burnups. They do an excellent job in this respect. In order to reduce the corrosion, slight changes in the alloy composition were made. As a result the cladding mechanical properties may be different then those of the Zircaloy alloys. For example, data are available indicating that these alloys creep more than Zircaloy [JUL04].

No data exists on the mechanical properties of the newer cladding alloys, at any burnup, especially in a hydrided state expected at high burnup. As a result, it is very difficult to predict the behavior of these alloys under hypothetical accident conditions during transportation or even under normal conditions of transport (e.g. fatigue from cyclic vibration stress). A limited set of mechanical properties of hydrided irradiated Zircaloy, particularly hydride reorientation threshold, and fracture thoughness is being obtained [VIR04]. If the properties of high burnup irradiated Zircaloy indicate the safe configuration of the fuel during transport, then only a data base that shows that Zircaloy bounds the newer alloys would be necessary. If the properties are inferior to Zircaloy then the applicant would have to provide a data base that would be sufficient to support calculations of acceptable fuel performance.

## III. Acquisition of Data

### A. Research Needs

The staff has identified data needed to evaluate the safety and retrievability of SNF during storage and transportation. The current temperature limit for storage of 400°C is based on limiting the creep deformation to 1%. Low burnup creep correlations for Zircaloy are used for this determination. High burnup creep correlations for both Zircaloy and other cladding alloys are needed to confirm this temperature or to establish new limits for other alloys.

A number of situations can be postulated where the cladding stress in a high burnup rod will exceed 100 MPa. As this fuel cools, excess hydrogen will precipitate if the stress is above the critical value. Based on current data, the critical stress is thought to be in the range of 90 MPa. This value can depend on a number of parameters including irradiation level and cooling rate. Values and uncertainty in the critical stress, obtained at sufficiently slow cooling rates, are needed for all PWR and BWR high burnup cladding alloys.

If the critical stress is in the range where hydride reorientation is plausible during storage, then the number of radial hydrides needed to degrade either the fracture toughness or axial elongation must be determined. This data is needed for all high burnup cladding. The fracture toughness and axial elongation are also necessary for evaluation of spent fuel behavior under normal and accident conditions.

To meet these needs the Spent Fuel Program Office (SFPO) staff issued a user need memo to Research in 2004 [VIR04] that addresses the need for mechanical properties of cladding and an understanding of the hydride reorientation phenomena in irradiated fuel. This user need memo also requests data to understand the cladding response in a transport accident. Argonne National Laboratory (ANL) is currently conducting confirmatory research co-sponsored by Electric Power Research Institute (EPRI), NRC, and Department of Energy (DOE) to address questions on cladding annealing, critical stress for hydride reorientation and cladding creep of Zircaloy-4 high burnup cladding. Even with this program, the responsibility for obtaining sufficient data to support an applicant's position lies with the applicant.

### B. Confirmatory Research

The Structural and Materials Section in SFPO reviews the materials aspects of licensing applications for storage and transportation casks. One aspect of that review is the expected behavior of the fuel and cladding under the conditions proposed by the licensee. The NRC sponsors confirmatory research on fuels and cladding materials with two principal objectives:
- To provide independent information to support NRC's decision making process.

- To identify and characterize technical questions that may become important safety issues in the future.

This confirmatory research is designed to improve the agency's knowledge and capabilities in areas where uncertainty in the knowledge exists, where safety margins are not well characterized, and where regulatory decisions need or will need to be confirmed.

The primary data base must be supplied by the applicant. It needs to be sufficiently complete to support their positions and/or uncertainties claimed in the application, and using benchmarked codes. This includes positions taken by risk assessment, bounding calculations, or empirical extrapolation. An insufficient data base could result in either a return of the application, a delay in the review, or a rejection of the application.

## IV. Changes to the ISG-II Rev 3

As a result of the data needs and uncertainties discussed above, a number of changes are being considered for ISG-11 Rev 3. The major ones shown in Figure 6 are a 90 MPa stress limit for normal storage at all burnup levels, and a maximum temperature of 570°C for all short term operations with low burnup fuel. In addition two conditions inadvertently removed from Rev 3 requiring a reflood analysis and an inert atmosphere (for the temperature limits to be applicable) will be reinstated. This guidance will continue to apply at burnup levels licensed by Nuclear Reactor Regulatory (NRR) branch of NRC. The staff will continue to be receptive to any other temperature and stress limits proposed by applicants for proposals that are adequately supported by relevant data.

As in the current case, transportation of high burnup Zircaloy clad fuel and any burnup level fuel clad with other zirconium alloys will continue to be handled on a case by case basis.

## V. Conclusions

The SFPO materials staff have reviewed ISG-11 Rev 3 and identified a number of data needs. These needs have been evaluated based on the uncertainty they introduce when providing guidance for acceptable conditions for the storage and transportation of high burnup fuel. In particular:

- Hydride reorientation continues to be the major uncertainty, with the variables of cooling rate, cladding composition, and stress being of principal concern. At what stress does it occur? How much reorientation is required to degrade the cladding mechanical properties? What will be the effect of the degraded properties on the ability of the cladding to meet the requirements of 10CFR Part 71 and part 72?
- A data base on the properties of the newer cladding alloys at high burnup is needed. Are the properties bounded by Zircaloy properties?
- The rising breach rate of fuel cladding in-reactor coupled with breaches by unidentified mechanisms raises the question of the number of damaged rods in the

same cores. How much additional damage is there in unbreached rods? How extensive is it in the SNF population. Methods for dealing with these rods have not been determined.

- While the NRC does sponsor some confirmatory research, the responsibility of providing a suitable data in support their arguments lies with the applicant.

Until these data needs are met, a number of changes to ISG-11 Rev 3 for storage are being contemplated. Transportation of high burnup Zircaloy clad fuel and fuel, of any burnup, clad in the newer alloys will continue to be treated on a case by case basis.

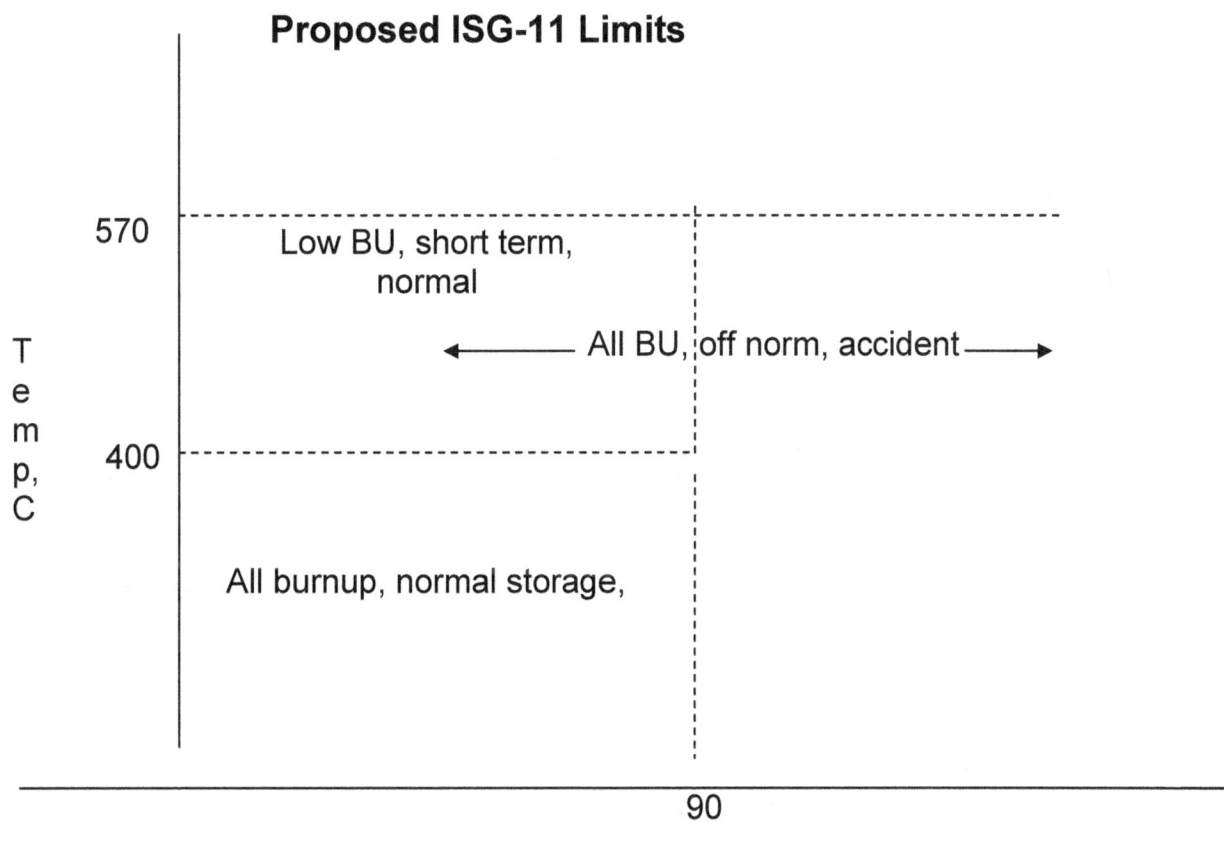

Figure 6 – Proposed changes to ISG-11 Rev 3

## VI. References

[BRO04] C. Brown, J. Guttmann, C. Beyer, and D. Lanning, Maximum Cladding Stress for Bounding PWR Fuels Rods During Short Term Operations for Dry Cask Storage in

*Proceedings of 2004 International Meeting on LWR Fuel Performance,* Orlando, FLA Sept 2004.p 459

[CHA96] KS Chan, A micromechanical Model for Predicting Hydride Embrittlement in Nuclear Fuel Cladding Material, J. Nuc Mater, **227** (1996) pp 220-236

[CHU04] HM Chung, Understanding Hydride- and Hydrogen-Related Processes in High-Burnup Cladding in Spent Fuel Storage and Accident Situations, in Storage in *Proceedings of 2004 International Meeting on LWR Fuel Performance,* Orlando, FLA Sept 2004.p 470

[ELL68] CE Ellis, Hydride Precipitates in Zirconium Alloys, J Nucl. Mater. **28**, (1968) pp 129-151

[FOS02] JP Foster et, al., Creep, Hydride Orientation and Oxide Spalling Issues Associated with Fuel Cladding Impact on High Burnup Fuel Dry Storage, in Proc of ANS General Meeting, Ft Lauderdale, FLA, Nov 2002

[GAR96] Garde, A.M., G.P. Smith, and R.C. Pirek, 'Effects of Hydride Precipitate Location and Neutron Fluence on the Ductility of Irradiated Zircaloy-4," *Proc. of 11ᵗʰ Int. Symp. On Zirconium in the Nuclear Industry,* ASTM STP 1295, 1996, pp 407-430

[ISH00] S Ishimoto et.al., 'Developent of New Zirconium Alloys for Ultra-High Burnup Fuel', in *Proceedings of the ANS International Topical Meeting on Light Water Fuel Performance,* Park City, Utah, April 2000

[ITA00] N. Itagaki et al., "Experience and Development of BWR Fuel Supplied to NFT" in *Proceedings of the ANS International Topical Meeting on Light Water Fuel Performance,* Park City, Utah, April 2000

[ITO04] K Ito, K Kamimura, and Y Tsukuda, Evaluation of Irradiation Effect on Spent Fuel Cladding Creep Properties, in Storage in *Proceedings of 2004 International Meeting on LWR Fuel Performance,* Orlando, FLA Sept 2004.p 440

[JUL04] B Julien et.al., Performance of Advanced Fuel Product Under PCI Conditions, in *Proceedings of 2004 International Meeting on LWR Fuel Performance,* Orlando, FLA Sept 2004. p 323

[KAI00] Kaiser, R.S., W.J. Leech, and A.L. Casadei, "The Fuel Duty Index (FDI)- A new Measure of fuel Rod Cladding Performance," in *Proceedings of the ANS International Topical Meeting on Light Water Fuel Performance,* Park City, Utah, April 2000, pp 393-400.

[KAM96] BF Kammenzind et al., Hydrogen Pickup and Redistribution in Alpha-Annealed Zircaloy-4, *Zirconium in the Nuclear Industry, 11ᵗʰ International Symposium,* ASTM STP 1295, 1996, pp 338-370

[KEY04] TA Keys et al., Fuel Corrosion Failures in the Browns Ferry Nuclear Plant, in *Proceedings of 2004 International Meeting on LWR Fuel Performance,* Orlando, FLA Sept 2004. p 229

[KNO03] Knott, R. et al. "Advanced PWR Fuel Designs for High Duty Operation' in *Proceedings of ENS TOPFUEL 2003* March 2003, Track 3.

[MAN00] R. Manzel, and C.T. Walker. "High Burnup Fuel Microstructure and Its Effect on Fuel rod Performance. in *Proceedings of the ANS International Topical Meeting on Light Water Fuel Performance,* Park City, Utah, April 2000

[MAR63] RP Marshall and MR Louthan, Tensile Properties of Zircaloy with Oriented Hydrides, Trans. Fo ASM, **56**, (1963), p 693

[MAR67] RP Marshall, Influence of Fabrication History on Stress-Oriented Hydrides in Zircaloy Tubing, J. Nucl. Mater., **24**, (1967), pp 34-48

[MEY00] R.O. Meyer, NRC Activities Related to High Burnup, New Fuel Types, and Mixed Oxide Fuels, in *Proceedings of the ANS International Topical Meeting on Light Water Fuel Performance,* Park City, Utah, April 2000

[MUT04] M Mutyala, Westinghouse Fuel direction, in *Proceedings of 2004 International Meeting on LWR Fuel Performance,* Orlando, FLA Sept 2004. p18

[RAS01] YR Rashid, RO Montgomery, and WF Lyon, Fracture Toughness Data for Zirconium Alloys, Application to Spent Fuel Cladding in Dry Storage, EPRI, Palo Alto, CA: 2001. 1001281.

[SAB00] Sabol, G. P. et.al., "In-reactor Fuel Cladding Corrosion Performance in Higher Burnups and Higher Coolant Temperatures, in *Proceedings of the ANS International Topical Meeting on Light Water Fuel Performance,* Portland, Oregon 1997, pp 397-404

[SAN92] TL Sanders et al, A method for Determining the Spent-Fuel Contribution to Transport Cask Containment Requirements, SAND90-2406, 1992.

[SCH04] RJ Schneider et al, Recent GNF Fuel Experience, in *Proceedings of 2004 International Meeting on LWR Fuel Performance,* Orlando, FLA Sept 2004. p 25

[SEI04] A Seibold and RS Reynolds, Performance of Framatome ANP BWR Fuel Rods, in *Proceedings of 2004 International Meeting on LWR Fuel Performance,* Orlando, FLA Sept 2004. p 249

[TRO04] R Tropasso, J Willse, and B Cheng, Crud-Induced Cladding Corrosion Failures in TMI-1 Cycle 10, in *Proceedings of 2004 International Meeting on LWR Fuel Performance,* Orlando, FLA Sept 2004. p 339

[VIR04]   MJ Virgilo to AC Tadani "User Need Memorandum – Assessment of High Burnup Fuel Cladding Integrity Performance Under Accident Conditions"   March 4, 2004, NRC Adams accession # ML040650621

[VIZ02] P Vizcaino, AD Banchik, and JP Abriata, Solubility of Hydrogen in Zircaloy-4: Irradiation Induced Increase and Thermal Recovery, J. Nucl. Mater., **304**, (2002), pp 96-106

[YAN04]   R Yang, O Ozer, K Edsinger, B Cheng, and J Deshon , An Integrated Approach to Maximizing Fuel Reliability, in *Proceedings of 2004 International Meeting on LWR Fuel Performance,* Orlando, FLA Sept 2004. p 11

# Perspective on requirements for spent fuel storage and transportation

Albert Machiels
Electric Power Research Institute, Inc.

A significant regulatory milestone was achieved when the Spent Fuel Project Office published Interim Staff Guidance (ISG) 11, Revision 2 entitled: *Cladding Considerations for the Transportation and Storage of Spent Fuel.* The acceptance criteria specified in Rev. 2 were supplemented by an additional set of acceptance criteria for low burnup fuel in Rev. 3 published in November 2003.

In both revisions, the acceptance criteria address dry storage only, but not transportation. This can be considered unusual for the following two reasons:

1. Most systems to be used for high-burnup spent fuel are intended to be dual-purpose. Clearly, the acceptance criteria for loading spent fuel in such dual-purpose systems should envelop both storage and transportation applications.
2. The most limiting considerations that led to the acceptance criteria for dry storage were actually based on transportation, i.e., on limiting the potential formation of radial hydrides in the spent-fuel cladding during dry storage in order to minimize potential degradation of the cladding mechanical properties used in analyzing transportation accidents.

Presently, the U.S. regulations (Part 71) do not have specific criteria with regard to performance of cladding under hypothetical accident conditions.[1] However, the configuration of the spent fuel in the transportation package after an accident is an input to the shielding and criticality evaluation, as well as possibly to the confinement evaluation. Clearly, if it can de demonstrated that no significant damage occurs either to the spent-fuel itself (no re-configuration), or to the package (no potential for moderator ingress), the criticality analysis, generally considered as a key driver from a regulatory perspective, would be greatly simplified.

Confirmatory and new experimental work is being conducted at ANL with the participation and funding of several organizations, including the US NRC, US DOE, EPRI, and the fuel vendors. This work is expected to demonstrate the conservative, but appropriately realistic, technical basis, which resulted in the acceptance criteria for storage contained in Rev. 2/3 of ISG-11, and its applicability to transportation applications. Concurrently, modeling of spent-fuel performance under impact loading conditions is a necessary activity for both guiding the experimental work and getting the most value from its results.

These activities are, or will be, supplemented by other generic efforts:

1. Risk assessment of criticality event during transportation
2. Implementation of full-burnup credit

This three-prong approach (risk assessment, fuel cladding performance, and burnup credit) is expected to support the contention that transportation risk minimization is a direct function of the reduction in the

---

[1] However, Part 71 contains cladding performance criteria for normal conditions of transport.

number of shipments. If this is indeed the case, the use of high-capacity packages should generally be the preferred implementation path for dual-purpose systems.

# Perspective on Requirements for Spent Fuel Storage and Transportation

Albert Machiels

Sr. Technical Manager

EPRI

Nuclear Safety Research Conference 2004

October 26, 2004

Washington, DC

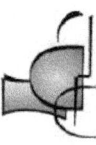

# Presentation Outline

- Introduction
- Risks in Transportation of Spent Fuel
- Probability of Criticality Event
- ISG-11, Revision 2
- Radial Hydrides
- Creep
- ISG-11, Revision 3
- Toward ISG-11, Revision 4?
- Conclusion

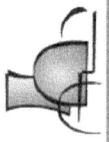

# Introduction

- Dry Storage
  - As of January 1, 2003: 362 loaded dry storage systems
  - Expected by January 1, 2005: ~200 additional loaded systems
  - Existing regulatory acceptance criteria (ISG-11, Rev. 3) applicable to spent fuel currently licensed by the NRC for commercial plant operations (no additional burnup restrictions)

- Transportation
  - Not currently occurring on a routine basis
  - Fuel burnup remains an important consideration
    - Less than 40-45 GWd/MTU → Approved applications
    - Greater than 45 GWd/MTU → ISG-19

- Issues
  - Realization of the value invested in dual-purpose (storage & *transportation*) systems
  - Qualification of storage-only systems for one-time *transportation*

- Main regulatory issue for transportation: criticality

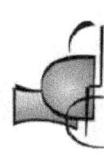

# Tractor-trailer Broadsided by a Train
## Locomotive Traveling at 80 miles per hour

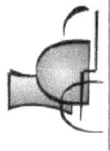

# Risks in Transportation of Spent Fuel

| | Normal Conditions | Accident Conditions |
|---|---|---|
| Criticality | §71.55(d)<br>"Preclude" | §71.55(e)<br>"Preclude" |
| Radiological | §71.47(d)<br>§71.51(a)(1)<br>"Small" | §71.55(d)<br>§71.51(a)(2)<br>"Very Small"[1] |
| Non-radiological | "___" | "Potential for fatal and nonfatal injury" |

[1] Single shipment incident-free dose risks greatly exceed (>$10^3$-$10^4$) single shipment accident dose risks [Ref. NUREG/CR-6672, page E-6]

EPRI

# Risks in Transportation of Spent Fuel
(continued)

- Risk minimization drives to a reduction in the number of shipments while maintaining (1) preclusion of criticality, and (2) low levels of exposure

- Maximum size of a transportable package

  - Crane lifting capacities at nuclear power plant sites

  - Practical weight limitations for unrestricted railroad or truck movement

- Given a practical upper bound on large cask total system weight (i.e., ~250,000 pounds), the issue is then to optimize the capacity, or number of fuel assemblies, of the package

- Regulatory acceptance of technologies relying on prevention of water inleakage (moderator exclusion) and (full) burnup credit is key in enabling full package capacity (especially true for spent fuel from PWR), and therefore, for achieving lower levels of risks associated with the transportation of spent fuel

# Probability of a Criticality Event During Transportation of Spent Fuel

- At the WRSM 1999, a "back-of-the-envelope" estimate of the frequency of a criticality event during transportation was presented [D. Lancaster et al.]

  - Likelihood of a rail cask accident with a greater than 2% strain coupled with a concurrent submersion

    - Under the rail shipping scenario of the Modal Study [NUREG/CR-4829], " ... this type of accident is estimated to occur once every ten million years." That is $10^{-7}$/year.

    - A more recent study, Reexamination of Spent Fuel Shipment Risk Estimates [NUREG/CR-6672], led to the conclusion that 'relative to the Modal Study result, expected accident population dose risks for both rail and truck are further decreased by about two orders of magnitude"

  - Likelihood of a critical configuration in the presence of water

    - Only possible by assuming misloadings of fresh fuel (typically, greater than two, see EPRI 1003418)

    - Estimated to be highly unlikely: $\sim 10^{-6}$ to $10^{-10}$/cask (actually, no fresh fuel assembly in the pool during cask loading campaign)

  - Likelihood of a criticality event: $<10^{-13}$-$10^{-17}$/year!

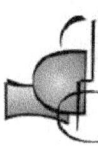

EPRI

93

# Probability of a Criticality Event During Transportation of Spent Fuel (continued)

- Use of risk information would likely result in "more realistically conservative" (this means less conservative!) approaches to the treatment of criticality *when dealing with spent fuel*

  – NRC guidance is certainly moving in that direction

    • "Fresh fuel assumption" → "Actinide-Only Burnup Credit" → "Full-burnup credit" (burden of proof?)

    • ISG-19 allowing testing for demonstrating moderator exclusion for accident conditions

    • Other approaches for moderator exclusion?

  – At a minimum, enabling technologies should be in place to support spent fuel shipments in order to achieve greater levels of overall public safety by minimizing the number of shipments!

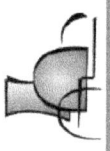

# Probability of a Criticality Event During Transportation of Spent Fuel (continued)

- What is the impact of fuel reconfiguration?

- Probability of a severe enough accident concurrent with inleakage of water does not change: <$10^{-7}$/year

- Given an accident, what is the probability of a critical event if fuel reconfiguration is involved?

  – Original fuel assembly geometry: close to best geometry for criticality

  – Obtaining more reactive configurations are difficult to achieve in the real, physical world

  – Administrative margin provides "defense-in-depth"

- Bottom line: Fuel reconfiguration is highly likely to further decrease the probability of a criticality event

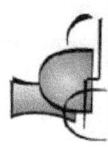

# ISG-11, Revision 2 (July 2002)

- Regulatory concerns about the formation of radial hydrides led to the technical bases for effectively resolving the issue of dry storage of spent, high burnup fuel

  - Concerns were driven by the potential effects of radially oriented hydrides on the cladding mechanical properties that are relevant for assessing the impact of transportation accidents on spent fuel integrity

    - In particular, the 400°C peak cladding temperature limit and the restrictions on thermal cycling were directly driven by hydride re-orientation phenomena

    - In addition, the 400°C peak cladding temperature limit would minimize changes in cladding yield strength; as a result, cladding hoop stresses would remain well below yield strength, far away from the onset of plastic instability

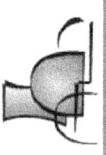

# Circumferential and Radial Hydrides

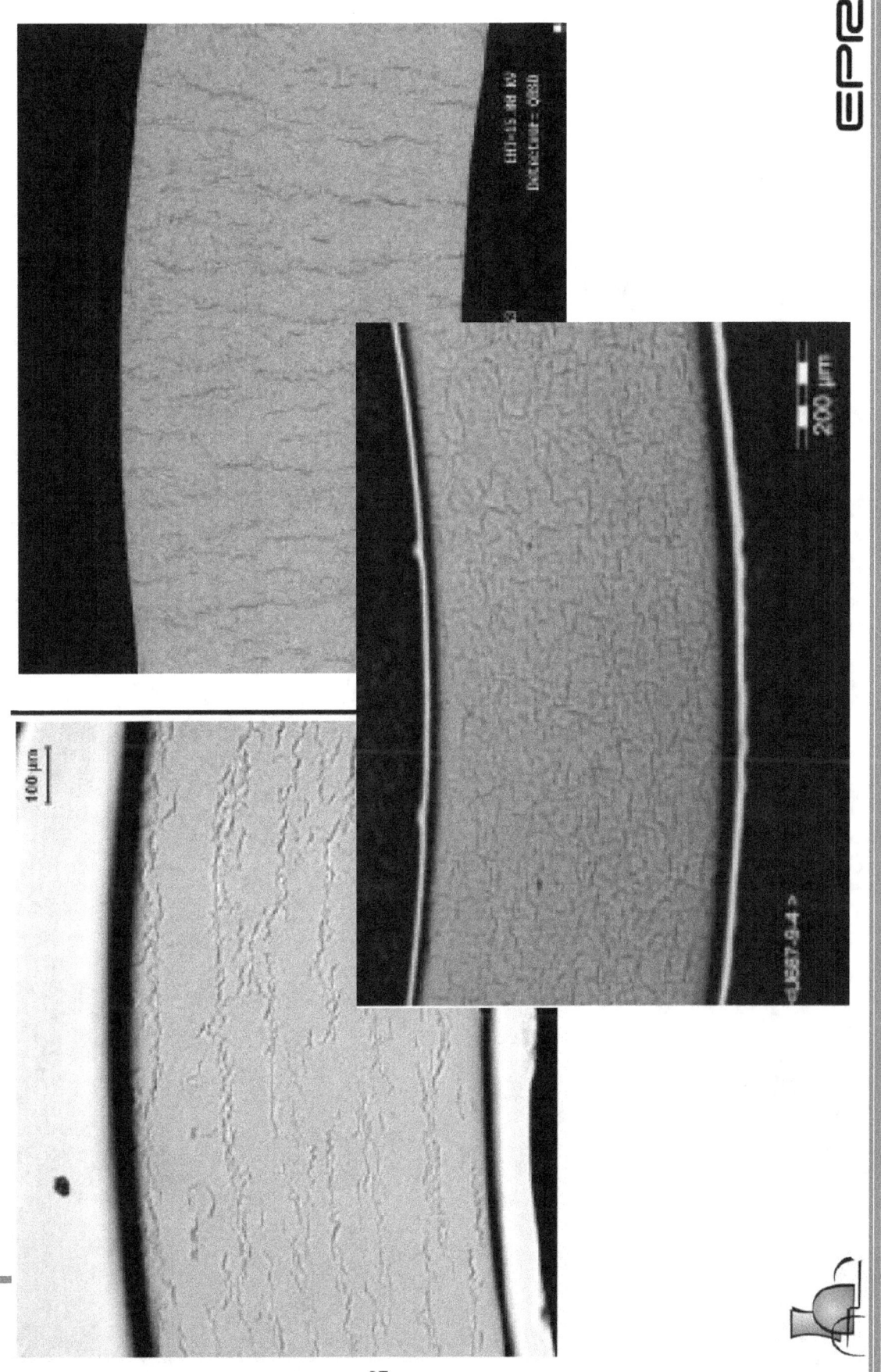

# Impact of Radial Hydrides: Structural Failure Mode III (*SAND90-2406-III Side-Drop*)

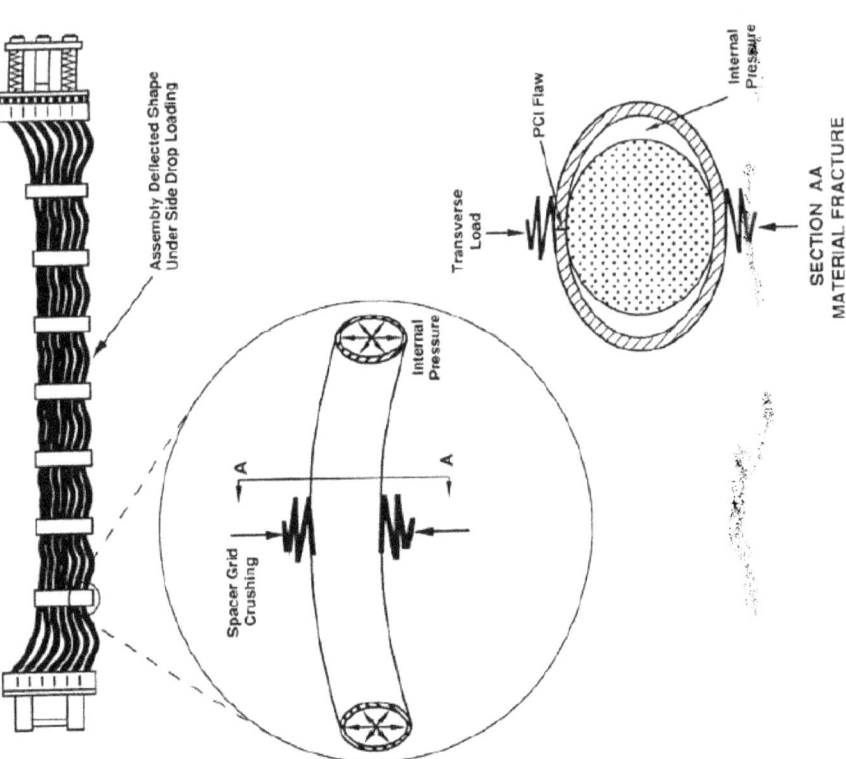

Assembly Deflected Shape Under Side Drop Loading

Internal Pressure

Spacer Grid Crushing

A

A

PCI Flaw

Internal Pressure

Transverse Load

SECTION AA
MATERIAL FRACTURE

Figure III-28.  Cladding Material Fracture Failure Mode

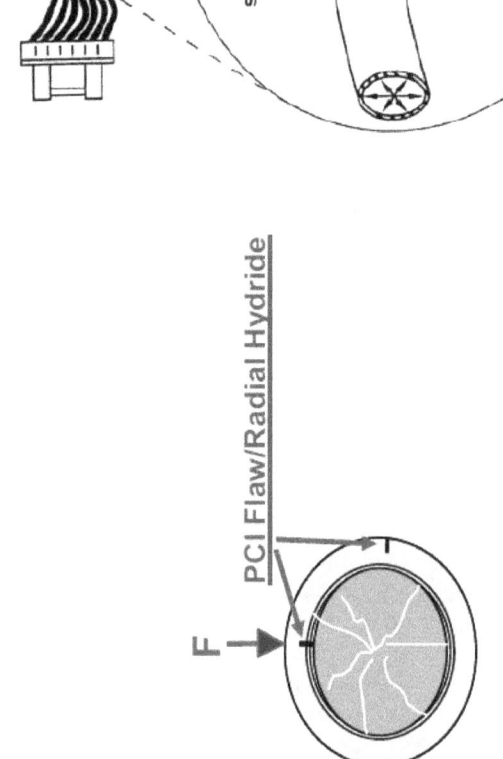

PCI Flaw/Radial Hydride

F

*Artist view of deformed assembly (accident condition simulation)*

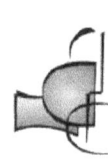

EPRI

# Hydride Precipitation Model (Based on C.E. Ells and M. Puls' Previous Work) – Calibration

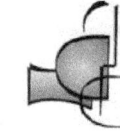

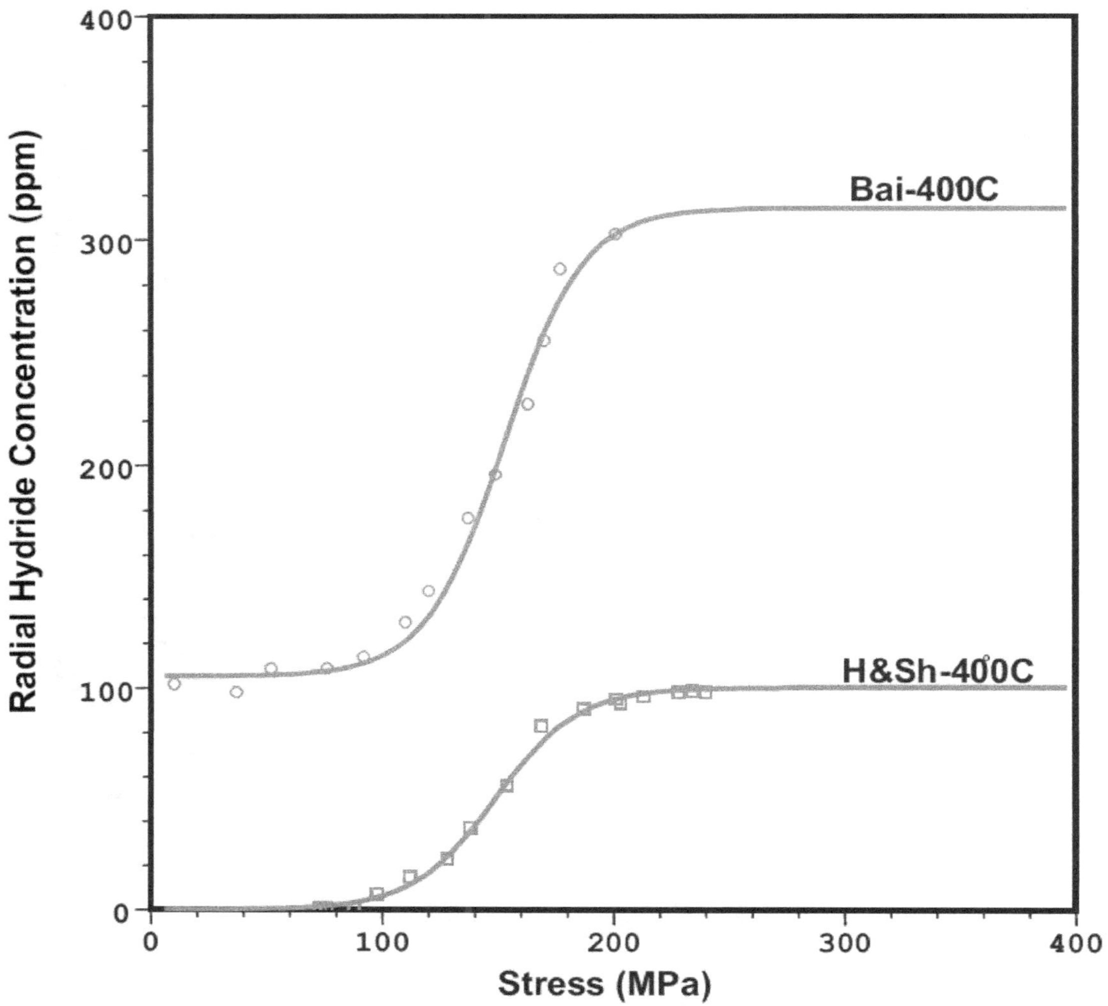

EPRI

99

# Hydride Precipitation Model –
# 40-year Dry Storage Simulation

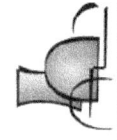

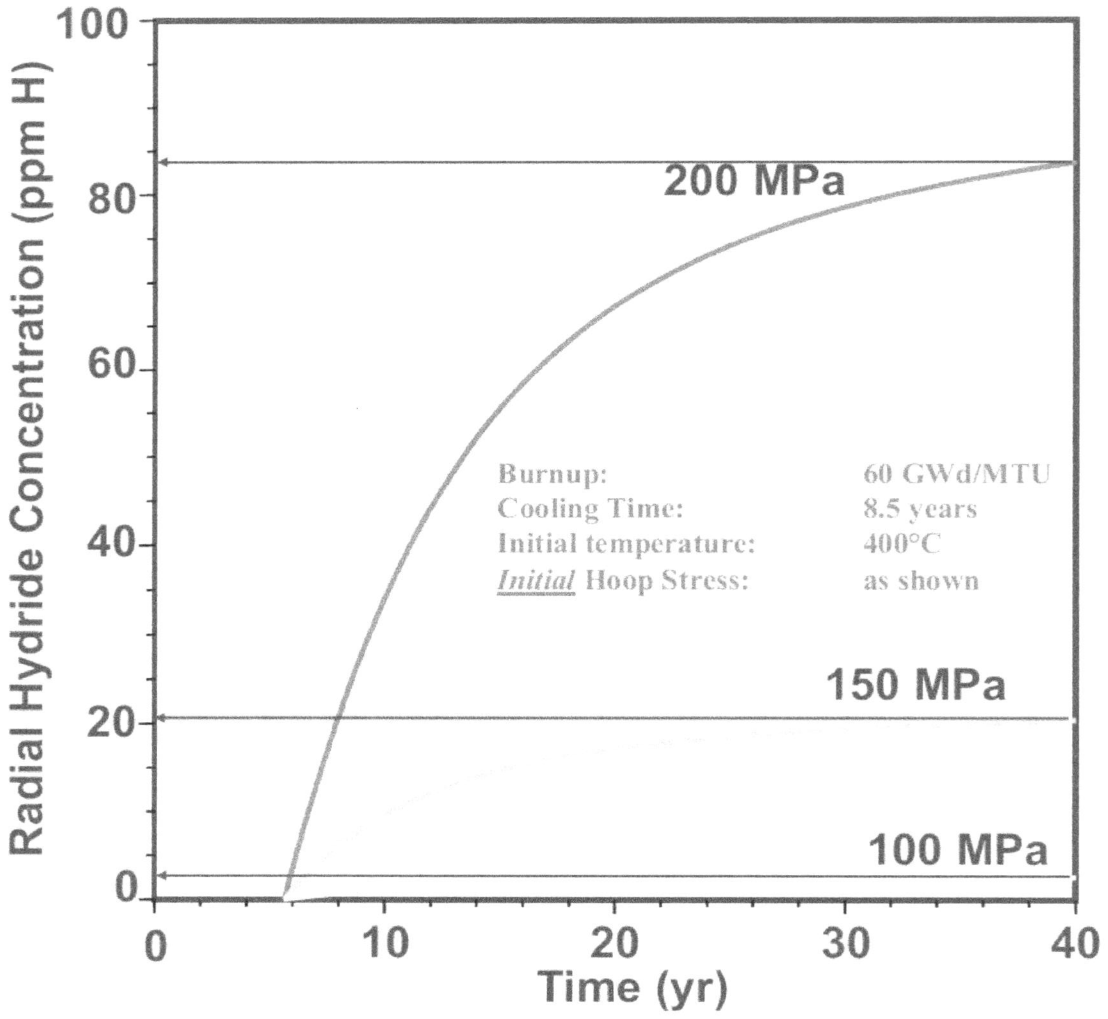

**Radial Hydride Concentration (ppm H)** vs **Time (yr)**

200 MPa

150 MPa

100 MPa

Burnup: 60 GWd/MTU
Cooling Time: 8.5 years
Initial temperature: 400°C
*Initial* Hoop Stress: as shown

# Creep

- Occurrence of creep rupture during dry storage is a remote possibility *as long as cladding hoop stresses remain below cladding yield strength* throughout the storage duration (avoid plastic instability associated with "tertiary" creep)

  – Cladding yield strength decrease is a function of temperature and time at temperature

  – Cladding strain is not a relevant figure-of-merit

- Should excessive creep occur, ruptures would not be classified as "gross degradation"

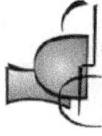

# Application Case #31 [EPRI Report 1003531]: 19.1 MPa at 400°C (24-hr drying up to 440°C)

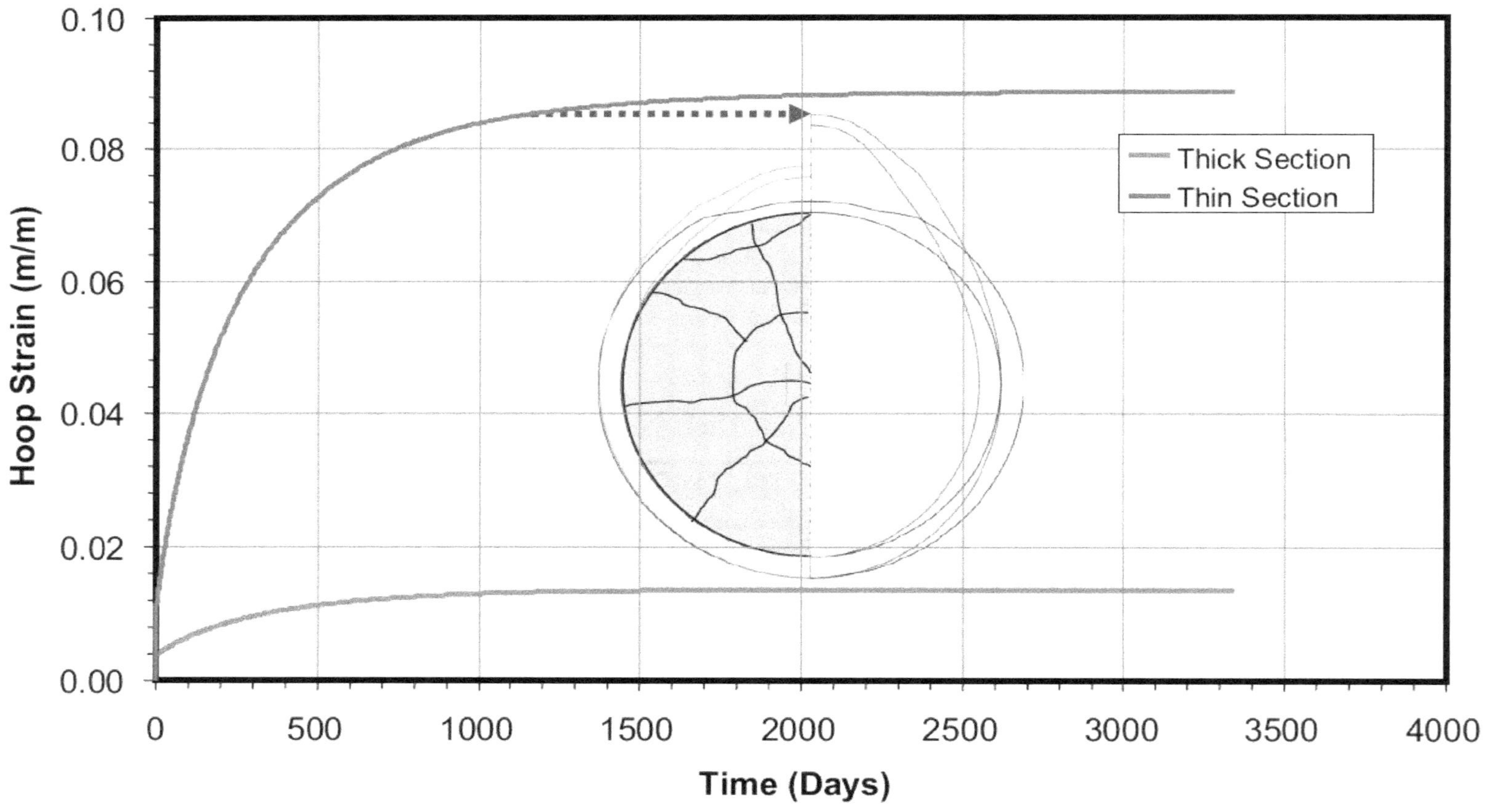

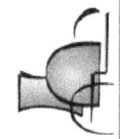

120 μm Oxide; Hydride Lens 50% of Thickness, 60° wide

EPRI

102

# Perspective on "ISG-11, Rev. 2" Requirements

- Rev. 2 is fairly simple and devoid of requiring an abundance of calculations with regard to temperature history, cladding hoop stress history, creep strain, etc.

- Instead, the 400°C limit and some restrictions on thermal cycling specify the upper limits for

  – Limiting the formation of radial hydrides to a low level, and

  – Preventing creep rupture

- Revision 2 = Simplicity

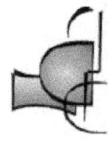

# ISG-11, Revision 3 (November 2003)

- Rev. 3 provides a broadening of the acceptance criteria (still limited to dry storage)

- Main addition:
  - For "low-burnup" fuel, a higher short-term temperature may be used
    - Additional acceptance criteria for low-burnup fuel: Temperature limit is the temperature at which the calculated, best-estimate cladding hoop stress is ≤90 MPa

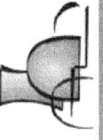

# Toward ISG-11, Revision 4?

- Generic acceptance criteria for the *transportation* of high-burnup fuel

  – Key concern: Will there be any fundamental change(s) in the guidance provided in Revision 3?

- Technical issue to address: Potential fuel reconfiguration under accident conditions for input into the criticality evaluation [per 10 CFR 71.55(e)]

- Is a Rev. 4 covering transportation necessary?

  – ISG-19 (published in May 2003)

105

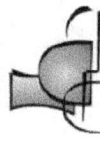

## (1) APPROVALS BASED ON RECONFIGURED FUEL

| Approach | Characteristics | Objective |
|---|---|---|
| Criticality Assessment of Bounding or Credible Reconfigured Fuel Geometries Based on Criticality Assuming Water Inleakage | 1. Postulate bounding fuel configurations for criticality.<br><br>2. Evaluate criticality and credibility of bounding configurations based on basic structural and material behavior.<br><br>3. Reduced reliance on material properties of high burnup fuel cladding and failure criteria.<br><br>4. Criticality analyses of reconfigured fuel from criticality bounding configurations. | With water inleakage, demonstrate subcriticality of defined set of credible or bounding fuel configurations based on criticality. |
| Criticality Assessment of Reconfigured Fuel Geometries Based on Actual Structural and Material Behavior Assuming Water Inleakage | 1. Need material properties of high burnup fuel cladding and failure criteria.<br><br>2. Requires nonlinear finite element analysis of fuel assemblies and fuel rods under drop impact conditions.<br><br>3. Failure modes and fuel rod failure distributions to be addressed (probabilistic approach to the distribution of material properties among fuel rods).<br><br>4. Develop credible fuel reconfiguration geometries.<br><br>5. Criticality analyses of reconfigured fuel from structural analysis results. | With water inleakage, demonstrate subcriticality of credible fuel configurations based on actual structural and material behavior .<br><br>This requires extensive data for irradiated hydrided cladding material properties for high burnup fuels. These data are currently not available. Therefore it is judged that this approach is currently not practical. |

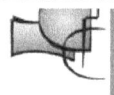

# Toward ISG-11, Revision 4? (continued)

- Ongoing R&D
  - Confirmatory R&D
    - Extent of hydride re-orientation *under prototypical dry storage time-temperature histories*
    - Creep deformation of high-fluence claddings *with elevated concentrations of hydrogen*
  - Developmental R&D
    - Analytical estimates of the extent of fuel reconfiguration
    - Materials properties (including advanced claddings) that support the analytical estimates

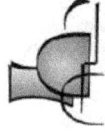

# Toward ISG-11, Revision 4? (continued)

- Will the outcome matter?
  - No significant extent of reconfiguration
    - For example: <1% of the rods affected (or some other relevant criterion)
    - ISG-19 can be greatly simplified
  - Significant extent of reconfiguration:
    - Discussion
      - More emphasis on moderator exclusion approaches (testing, as presently in ISG-19, or some other approach), or
      - Informed approach on the impact of fuel reconfiguration on criticality

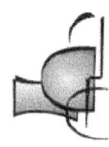

# Perspective - Conclusion

- High degree of confidence in the technical basis underlying ISG-11, Rev. 2/3 for storage

  - Regulatory concerns about radial hydride formation were justified, and ISG-11, Rev. 2/3 took these concerns into account (i.e., _transportation_ considerations became limiting for dry _storage_)

  - Formation of radial hydrides is expected to be very small under actual dry storage conditions

    - The 400°C temperature limit is a _peak_ cladding temperature

    - Analytical calculations of the peak cladding temperature are conservative (limiting loading conditions; conservative heat transfer correlations)

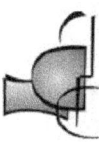

# Perspective – Conclusion (continued)

- The estimation of fuel reconfiguration is probably more appropriately related to ISG-19

  – The scope of ISG-11 should be limited to storage

- Risks are driven by the number of shipments

  – Non-radiological risks dominate given the structural strength of today's spent-fuel transportation packages

  – Maximizing capacity of the transportation package is key

- Probability of criticality event under accident conditions is exceedingly small

  – Fuel reconfiguration is an input into the criticality evaluation

- Bottom line:

  – Estimating the extent of fuel reconfiguration should provide an interesting piece of information

  – Best-estimate approach is appropriate, given the weak linkage between results to be obtained and risk significance

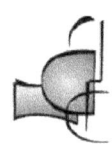

# LOCA Integral Test Results for High-Burnup BWR Fuel

## Yong Yan, Tatiana A. Burtseva, and Michael C. Billone

Argonne National Laboratory (ANL), Argonne, IL 60439

LOCA integral tests with high-burnup BWR fuel have been conducted at ANL to provide NRC and industry with data for assessing the applicability of the LOCA embrittlement criteria in 10 CFR50.46 to high-burnup fuel rods. Similar tests are planned for high-burnup PWR fuel. These criteria limit peak cladding temperature to 2200°F (1204°C) and maximum oxidation (equivalent cladding reacted, ECR) to 17% during high temperature steam oxidation to ensure residual ductility during ECCS quench (T $\geq$ 135°C) and during possible post-LOCA seismic events (T $\approx$ 100°C). Appendix K specifies the use of the Baker-Just (BJ) correlation for calculating reaction rate and ECR. NRC Regulatory Guide 1.157 (1989) allows for the use of best-estimate correlations (e.g. Cathcart-Pawel [CP]) for calculating reaction rate and ECR. At 1204°C, 17% BJ-ECR = 13% CP-ECR. In anticipation of the degrading effects of high-burnup operation on the cladding, NRC Information Notice 98-29 (1998) specifies that ECR should be based on total oxidation (corrosion plus transient steam oxidation). For PWR Zircaloy-4 (Zry-4) cladding, high-burnup operation results in peak corrosion layers of $\approx$100 $\mu$m, corresponding to 8-10% ECR, and peak hydrogen concentrations of 600-800 wppm. The ANL LOCA integral tests, which are conducted with fueled specimens, are designed to improve our understanding of the behavior of high-burnup fuel exposed to a LOCA transient, as well as to provide data for the assessment of the LOCA embrittlement criteria.

LOCA integral test results are reported for fueled high-burnup BWR specimens. These results are compared to baseline data for zirconia-pellet-filled, nonirradiated Zry-2 cladding specimens exposed to the same tests. Four LOCA integral tests have been conducted with specimens from Limerick BWR fuel rods at 56 GWd/MTU. In the as-discharged condition, the Limerick cladding (Zr-lined Zry-2 GE-11 9×9 design) has a corrosion layer of $\approx$10 $\mu$m and a hydrogen content of $\approx$70 wppm. The specimens were internally pressurized with helium to a gauge pressure of $\approx$8.3 MPa at 300°C. During heating in steam at 5°C/s, the internal pressure rose to $\leq$9 MPa prior to burst at $\approx$750°C. The full LOCA sequence (Fig. 1) calls for heating in steam at 5°C/s to 1204°C, holding for $\leq$5 minutes at 1204°C ($\leq$20% CP-ECR), slow-cooling at 3°C/s to 800°C and bottom-flooding to quench the cladding from 800 to 100°C.

The ICL#1 test specimen was ramped to burst in argon and slow cooled. The ICL#2 specimen was exposed to the LOCA test sequence with the exception of quench. The nondestructive results from these tests indicated more similarities than differences between high-burnup specimens and non-irradiated specimens. The ICL#3 specimen achieved partial quench (800°C to 470°C) before failure of the quartz chamber that surrounded the specimen. The full LOCA sequence with quench was demonstrated in the ICL#4 test (see Fig. 1). Nondestructive examinations included photography and profilometry for all 4 specimens and gamma scanning for the ICL#3 and #4 specimens. Destructive examinations were performed on ICL#2 and #3 specimens to determine oxide layer thickness, fuel morphology, and axial profiles of hydrogen and oxygen concentration. Oxide-layer thickness and oxygen-content results indicate two-sided oxidation in the ballooned-and-burst region of both high-burnup and nonirradiated specimens. The axial profile of hydrogen pickup for ICL#2 and #3 specimens is shown in Fig. 2 and compared to the data for nonirradiated Zry-2 cladding. For nonirradiated specimens, the hydrogen pickup was low in the burst region and very high at 70-90 mm above and below the burst mid-plane. For high-burnup-fueled cladding, the hydrogen peak was towards the burst mid-plane. Because of the large secondary hydriding from the cladding inner surface, significant degradation of post-quench ductility (PQD) is expected for the ICL#4 ballooned region, even at 100-135°C. A 5th hot-cell integral test, with a 2-minute hold time at 1204°C, is in progress to determine if the 17% BJ-ECR criterion is sufficient to protect the ballooned region from embrittlement due to steam oxidation and hydrogen pickup at 1204°C.

In the uniform burnup region (within grid spans 2-5), the high-burnup PWR cladding for the next set of LOCA tests differs from BWR cladding in terms of corrosion layer thickness (≈40 to 100 μm) and hydrogen content (≈400 to 800 wppm). For test planning purposes, the separate effects of hydrogen on diametral-compression PQD have been investigated with prehydrided, nonirradiated 15x15 Zry-4 cladding rings after oxidation at 1204°C and quench. For as-received (≈10 wppm H) cladding that was oxidized at 1204°C, the ductile-to-brittle-transition CP-ECR was 8% at room-temperature, 12% at 100°C, and 14% at 135°C. In contrast, cladding with 400-to-800 wppm hydrogen exhibited significant embrittlement, even after moderate oxidation at 1204°C. Samples prehydrided to 400-800 wppm and oxidized at 1204°C to 8% CP-ECR exhibited no ductility. With anticipated secondary hydrogen update from the cladding inner surface, the embrittlement ECR is expected to be <<17% for high-burnup PWR specimens subjected to LOCA integral tests at 1204°C, even if the ECR is determined by the sum of the corrosion layer and the BJ-calculated transient ECR. The baseline data from prehydrided cladding are being used to plan the PWR hold-times at 1204°C such that the embrittlement ECR can be determined effectively in the ballooned and non-ballooned regions of the PWR LOCA integral test specimens.

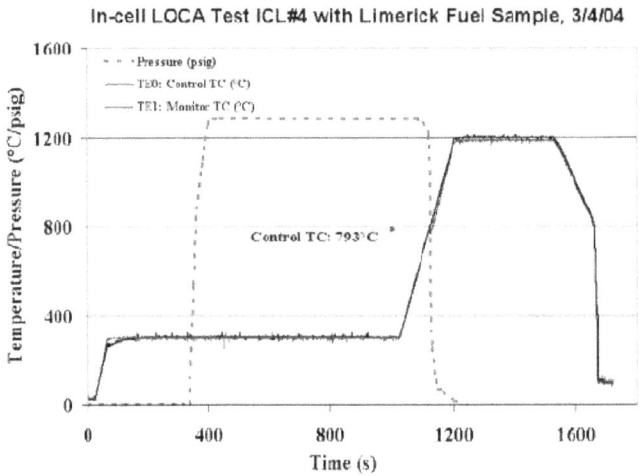

Fig 1. Temperature and pressure histories for LOCA integral test (ICL#4) with high-burnup BWR fuel.

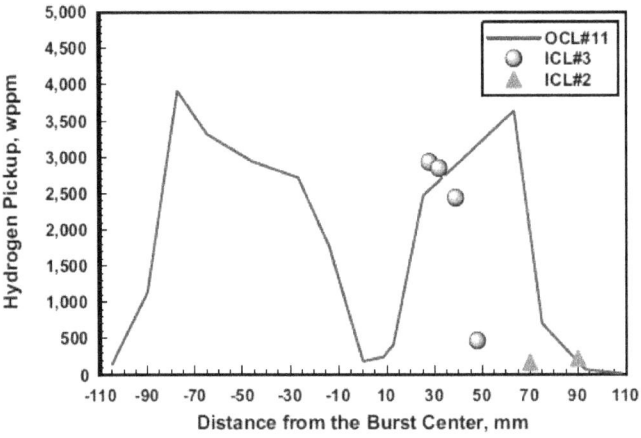

Fig 2. Axial profile of hydrogen pickup in high-burnup BWR cladding used in LOCA integral tests ICL#2 and ICL#3 (5 minutes at 1204°C); OCL#11 results are for nonirradiated Zry-2 cladding.

# LOCA Integral Test Results for High-Burnup BWR Fuel

Y. Yan, M. Billone, T. Burtseva, and H. Chung

Energy Technology Division

NSRC–2004, Washington, DC

October 25–27, 2004

## Argonne National Laboratory

A U.S. Department of Energy
Office of Science Laboratory
Operated by The University of Chicago

# *Outline*

- **Post-test Characterization of High-Burnup BWR LOCA Integral Test Samples**

    - Ballooning and burst (profilometry, photography)
    - Secondary hydriding (LECO H determination)
    - Steam oxidation (LECO O determination, metallography)
    - Gamma scanning (fuel relocation)
    - SEM examination (fractography)

- **Plans for H.B. Robinson LOCA Integral Tests**

    - Baseline post-quench-ductility (PQD) results for 15x15 Zry-4
        - ➢ *Offset strain vs. ECR for RT, 100°C and 135°C*
        - ➢ *Offset strain at 135°C vs. H content at 5% and 7.5% ECR*
    - Proposed test conditions and sequence

Nuclear
Regulatory
Commission

Pioneering
Science and
Technology

# Background

- ## Licensing Issues

  - 10 CFR 50.46 embrittlement criteria for maintaining residual ductility in Zircaloy (Zry) and ZIRLO cladding

    - ➤ *Temperature limit: peak cladding temperature (PCT) ≤1204°C*
    - ➤ *Oxidation limit: effective cladding reacted (ECR) ≤17%*

  - Additional Issues

    - ➤ *ECR calculation: Baker-Just (Appendix K) vs. Best-Estimate (Reg. Guide 1.157, 1989)*
    - ➤ *Inclusion of corrosion layer in ECR calculation to account for high-burnup effects (IN 98-29, 1998)*

  - Confirmation needed for high-burnup fuel

- ## High-Burnup Fuel Rod Segments

  - Limerick 9×9 Zry-2 BWR rods at 56 GWd/MTU

    *Corrosion layer ≈ 10 μm; H-content ≈ 70 wppm*

  - H.B. Robinson 15×15 Zry-4 PWR rods at 67 GWd/MTU

    *Corrosion layer ≤ 110 μm; H-content ≤ 800 wppm*

Nuclear
Regulatory
Commission

Pioneering
Science and
Technology

115

# LOCA Integral Test Sequence & Time

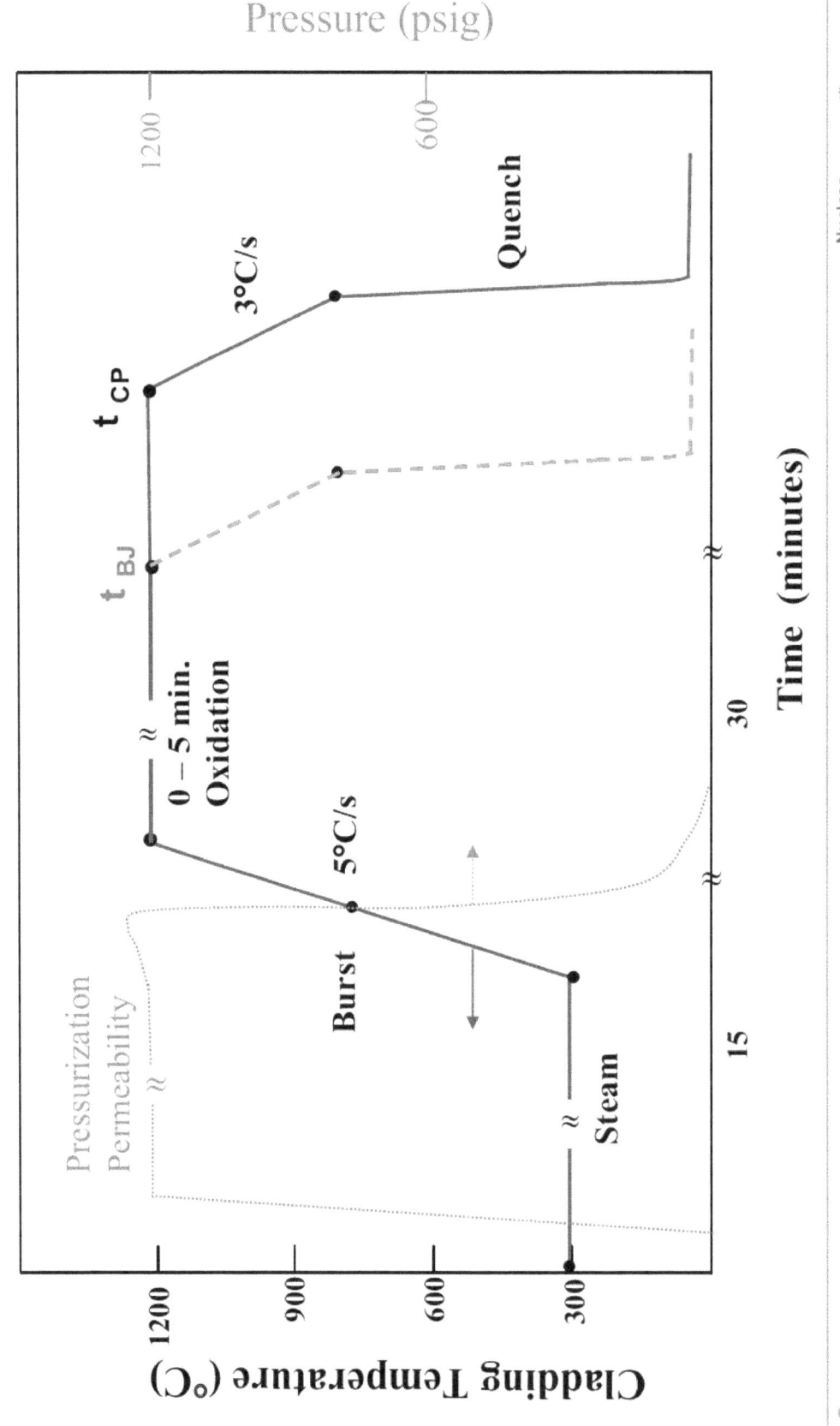

## *LOCA Integral Test Results for Non-Irradiated Zry-2 (1204°C for 5 min)*

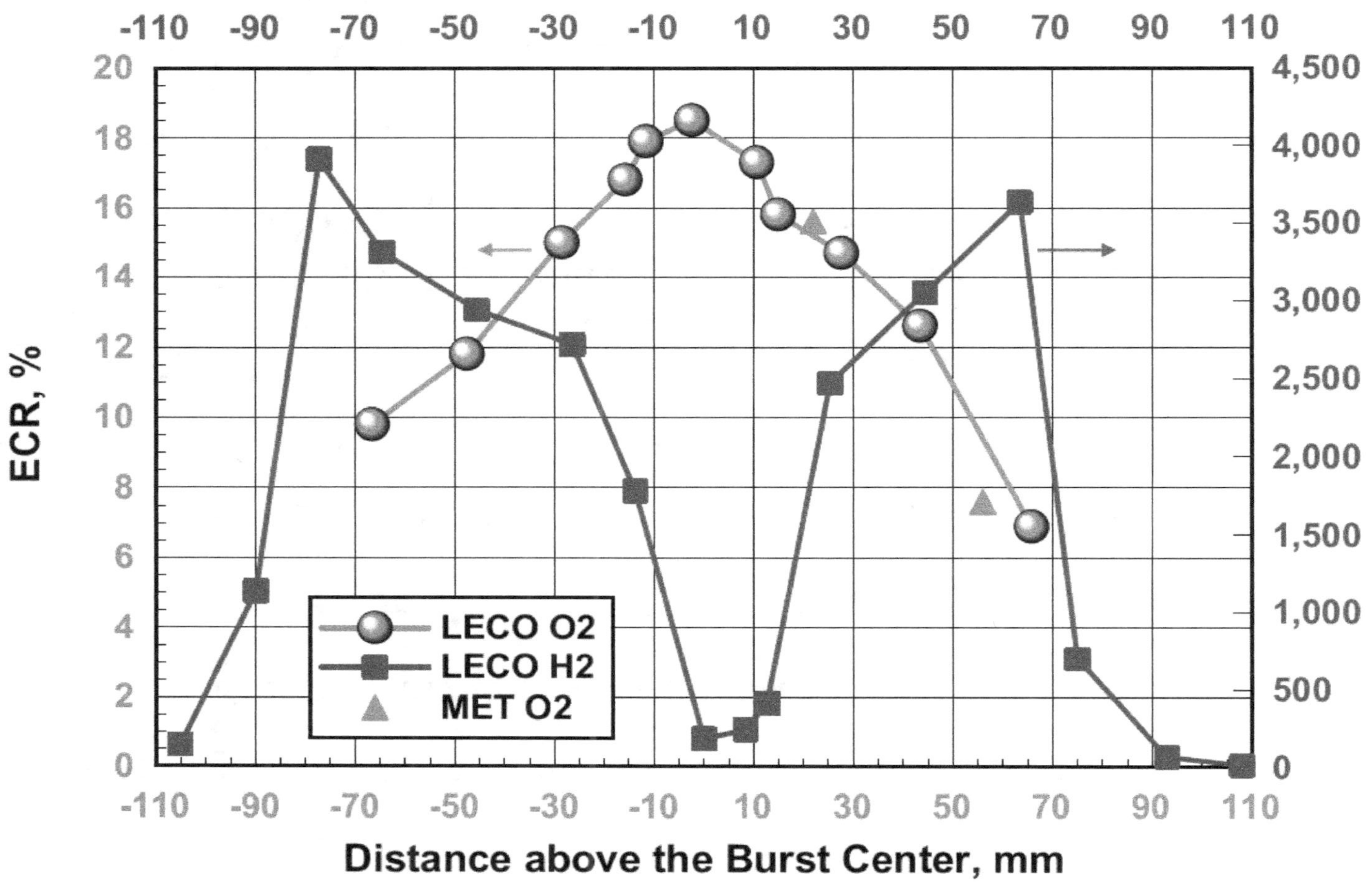

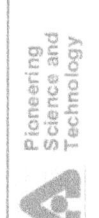

ECR, %
Hydrogen Pickup, wppm
Distance above the Burst Center, mm
LECO O2
LECO H2
MET O2
-110
-90
-70
-50
-30
-10
10
30
50
70
90
110
0
2
4
6
8
10
12
14
16
18
20
0
500
1,000
1,500
2,000
2,500
3,000
3,500
4,000
4,500


Pioneering Science and Technology

Nuclear Regulatory Commission

# LOCA Tests with High-Burnup BWR Fuel

- **In-cell LOCA Test ICL#1 (August 2002)**
  - Ramp-to-burst Test Conducted in Argon

- **In-cell LOCA Test ICL#2 (September 2002)**
  - LOCA Sequence with 5-minute Oxidation at 1204°C and Slow-furnace Cooling

- **In-cell LOCA Test ICL#3 (December 2003)**
  - 5-minute Oxidation at 1204°C Followed by Quench at 800°C (quartz tube failed at 480°C)

- **In-cell LOCA Test ICL#4 (March 2004)**
  - Full LOCA Sequence (5-minute Oxidation at 1204°C) with Quench at 800°C

# Temperature and Pressure of In-cell LOCA Tests

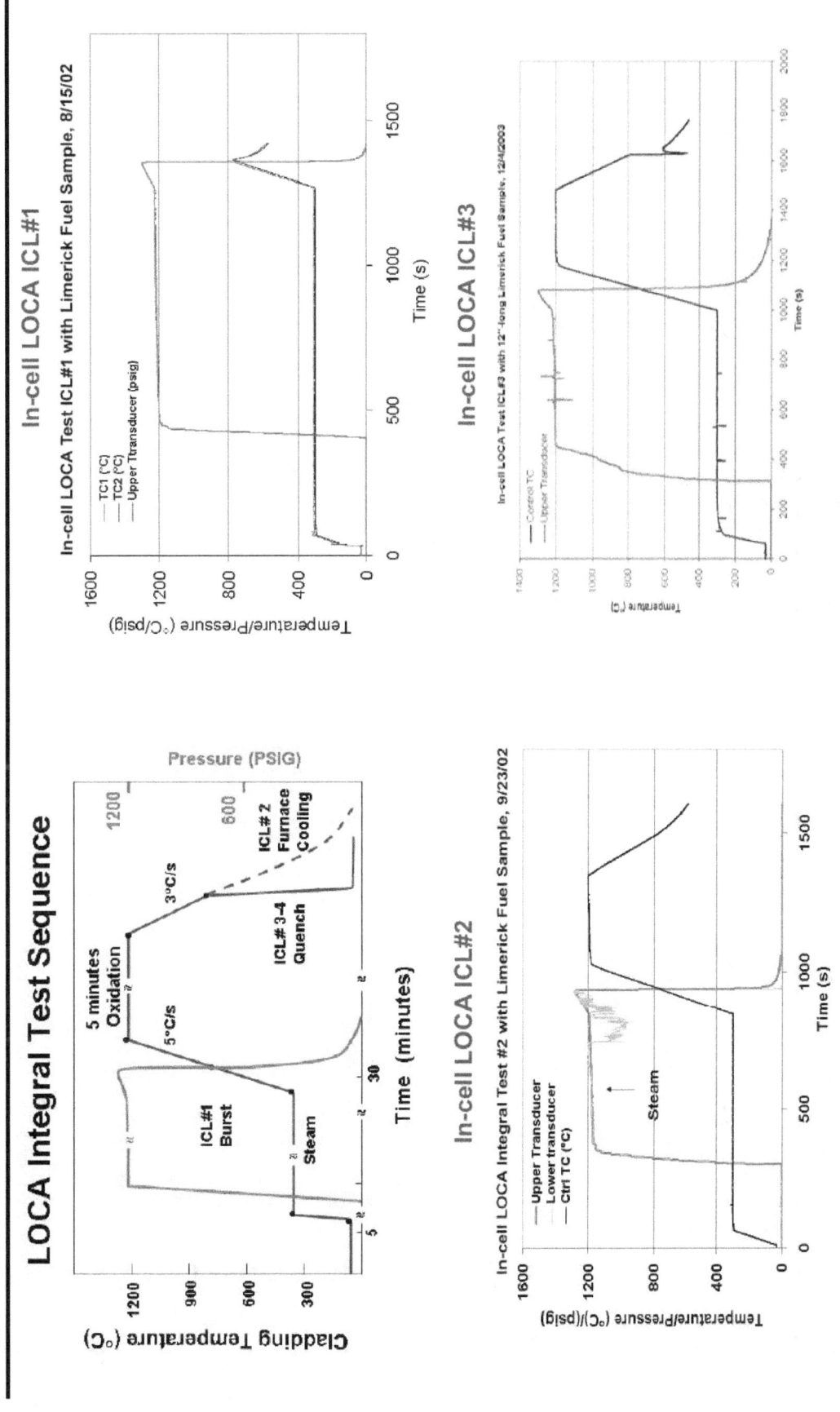

## LOCA Integral Test Sequence

### In-cell LOCA ICL#1

In-cell LOCA Test ICL#1 with Limerick Fuel Sample, 8/15/02

### In-cell LOCA ICL#3

In-cell LOCA Test ICL#3 with 12"-long Limerick Fuel Sample, 12/4/2003

### In-cell LOCA ICL#2

In-cell LOCA Integral Test #2 with Limerick Fuel Sample, 9/23/02

Pioneering
Science and
Technology

Nuclear
Regulatory
Commission

# Temperature and Pressure of In-cell LOCA Test ICL#4

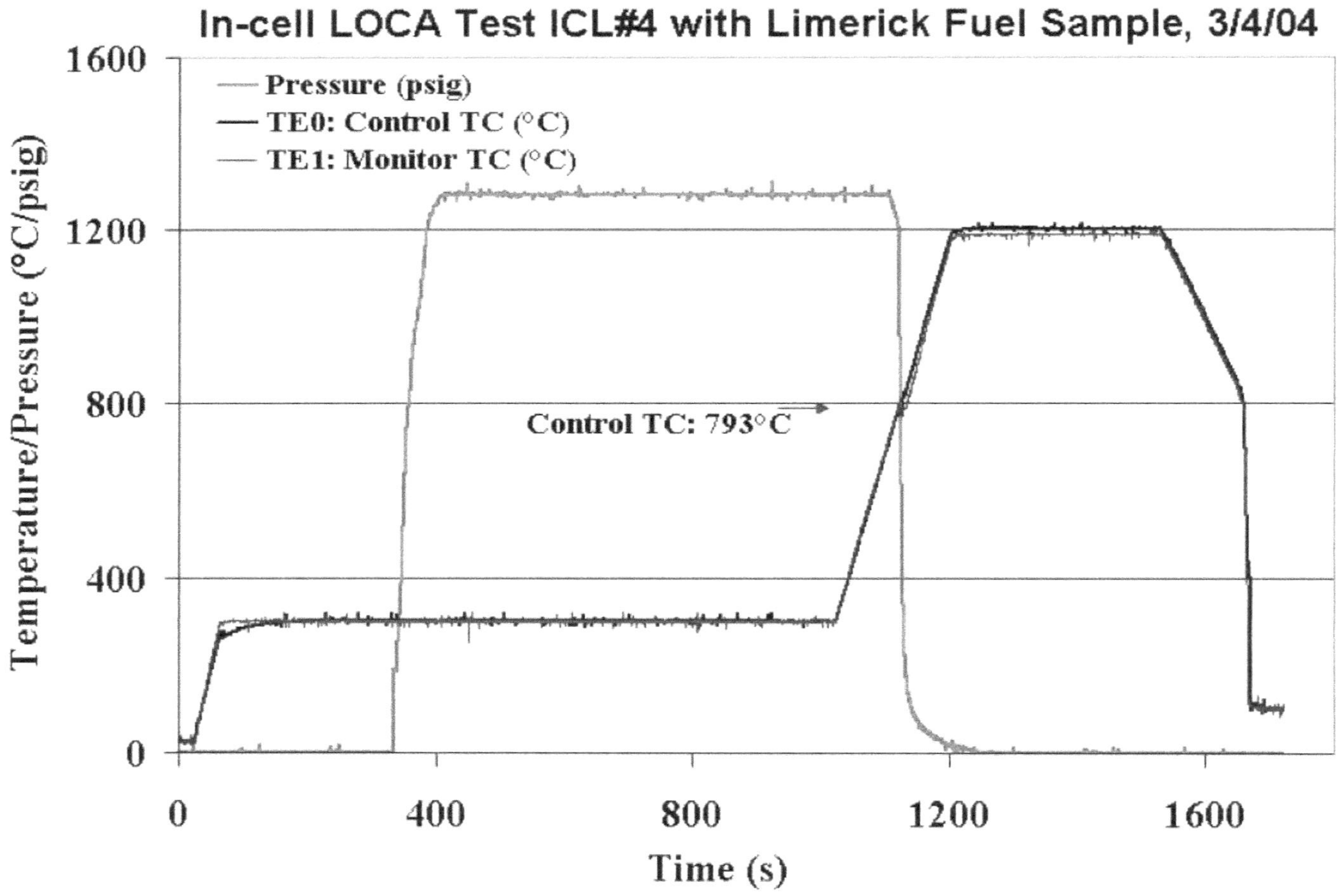

## In-cell LOCA Test ICL#4 with Limerick Fuel Sample, 3/4/04

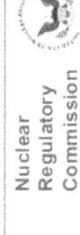

# Burst Openings for Ramp-to-Burst Tests

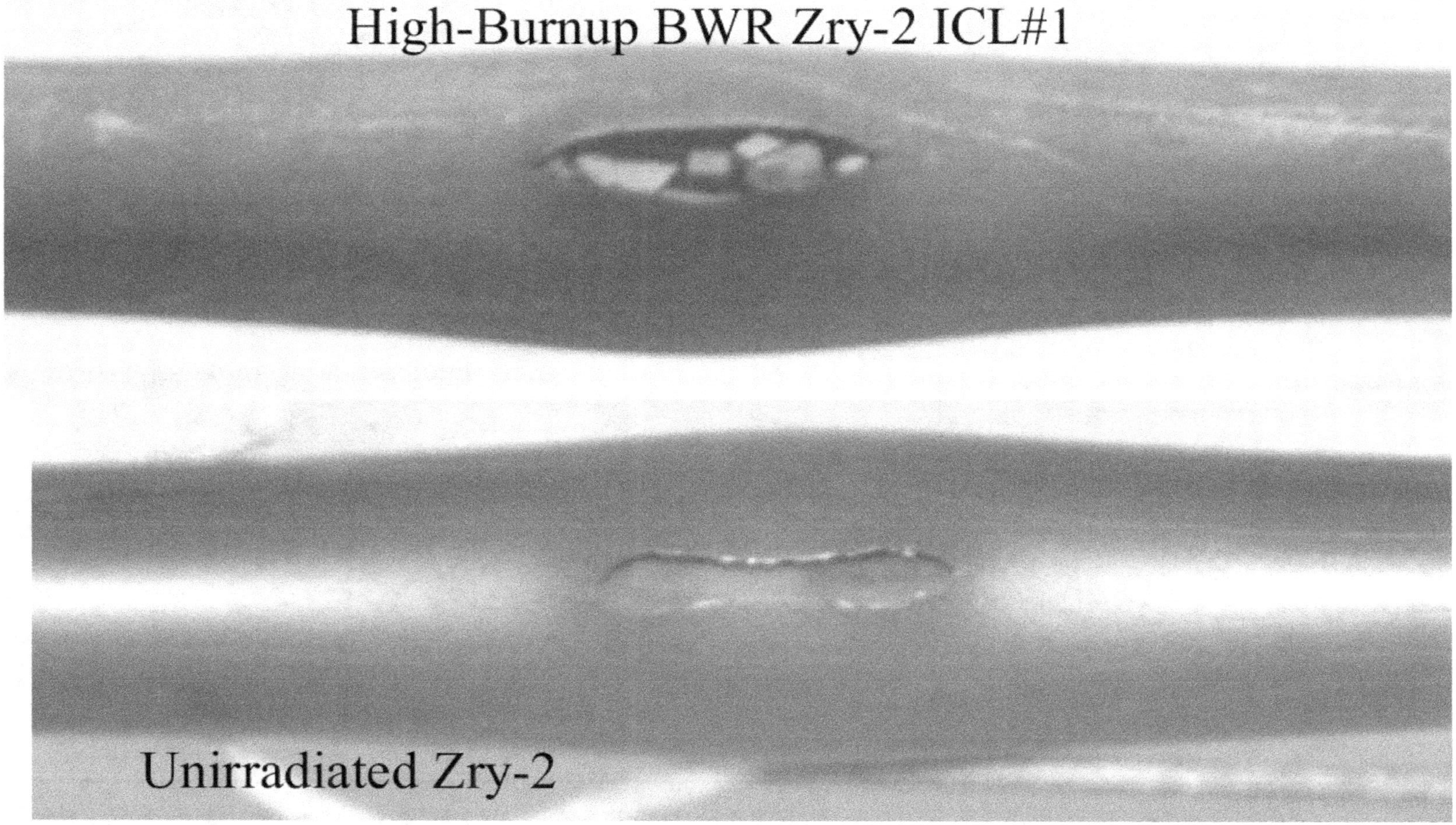

### High-Burnup BWR Zry-2 ICL#1

### Unirradiated Zry-2

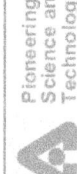

# Balloon and Burst Regions for High-burnup Tests

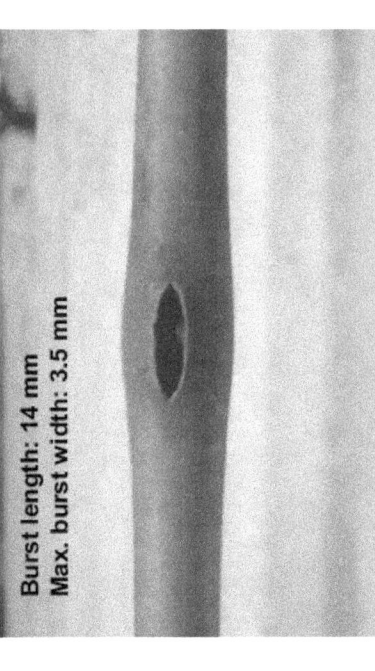

Burst length: 13 mm
Max. burst width: 3.0 mm

ICL#1: Ramp-to-Burst test conducted in argon

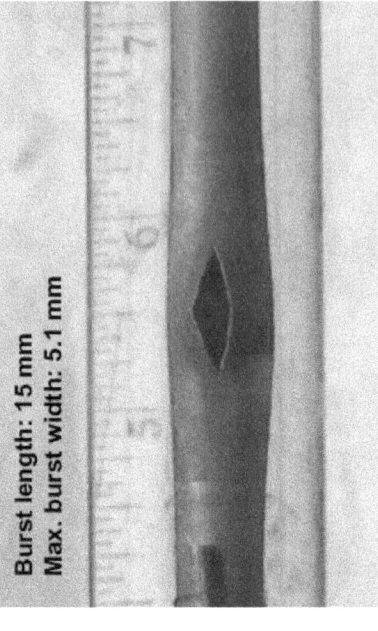

Burst length: 14 mm
Max. burst width: 3.5 mm

ICL#2: LOCA sequence with 5-minute oxidation at 1204°C and slow-furnace cooling

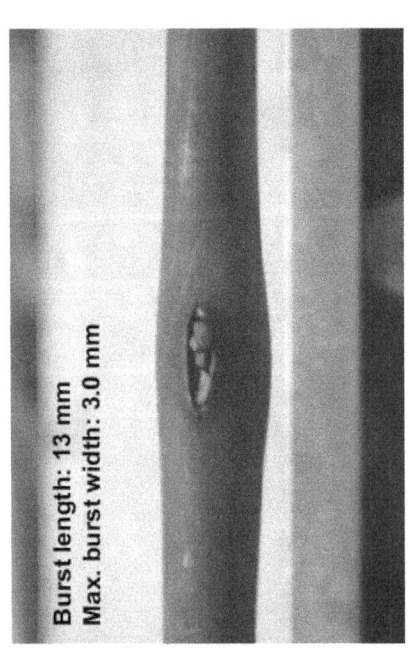

Burst length: 11 mm
Max. burst width: 4.6 mm

ICL#3: 5-min. oxidation at 1204°C fFollowed by quench at 800°C (quartz tube failed at 480°C)

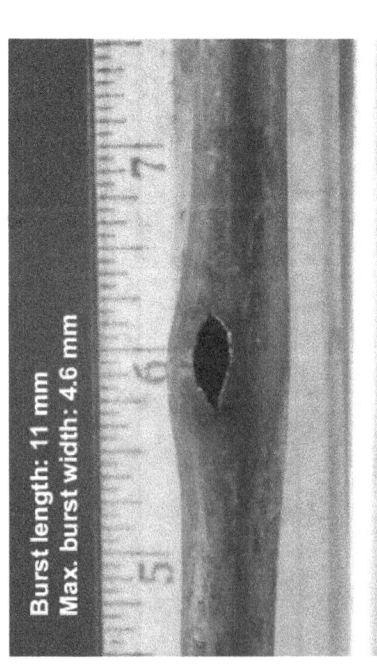

Burst length: 15 mm
Max. burst width: 5.1 mm

ICL#4: Full LOCA sequence (5-minute oxidation at 1204°C) with quench at 800°C

# High Mag. Images of LOCA Burst Cross-Section

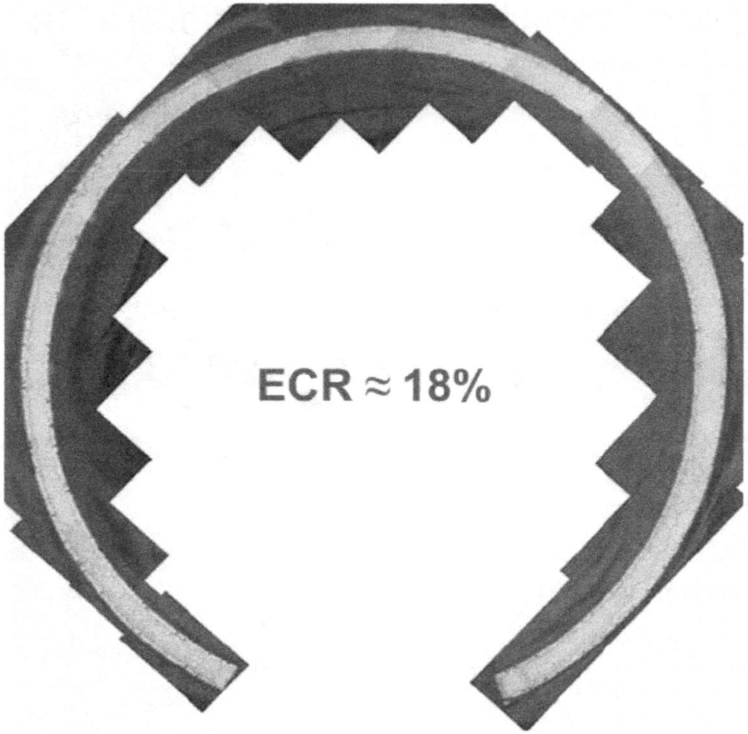

ECR ≈ 18%

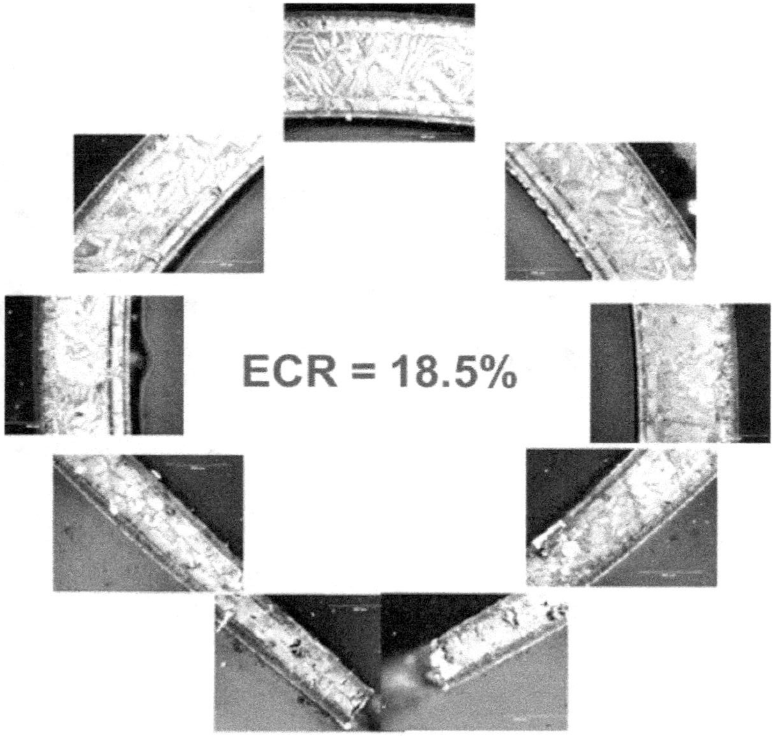

Sample ICL#3 (High burnup)
Burst mid-plane (1204°C, 5 min)
Strain: 36% - 52%

ECR = 18.5%

Sample OCL#11 (Unirrad. Zry-2)
Burst mid-plane (1204°C, 5 min)
Strain: 37% - 56%

Pioneering
Science and
Technology

Nuclear
Regulatory
Commission

# Post-test Characterization for ICL#3 Specimen

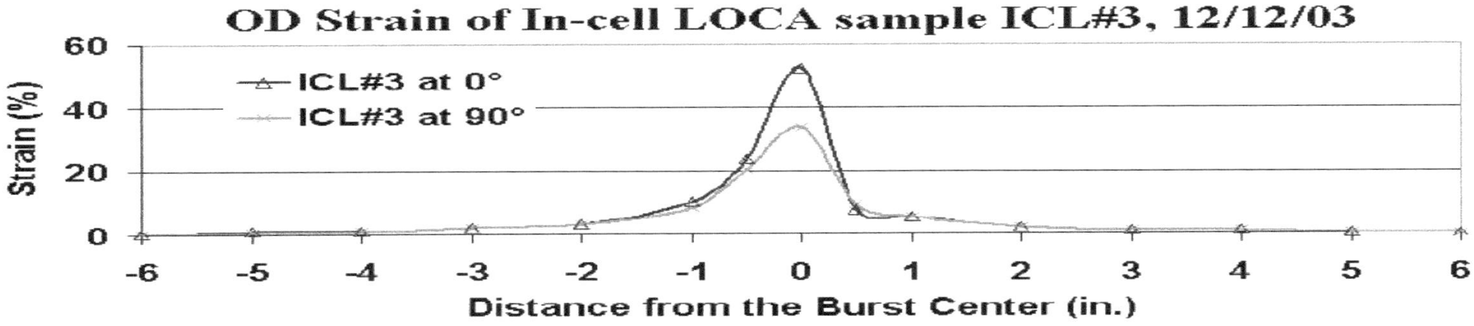

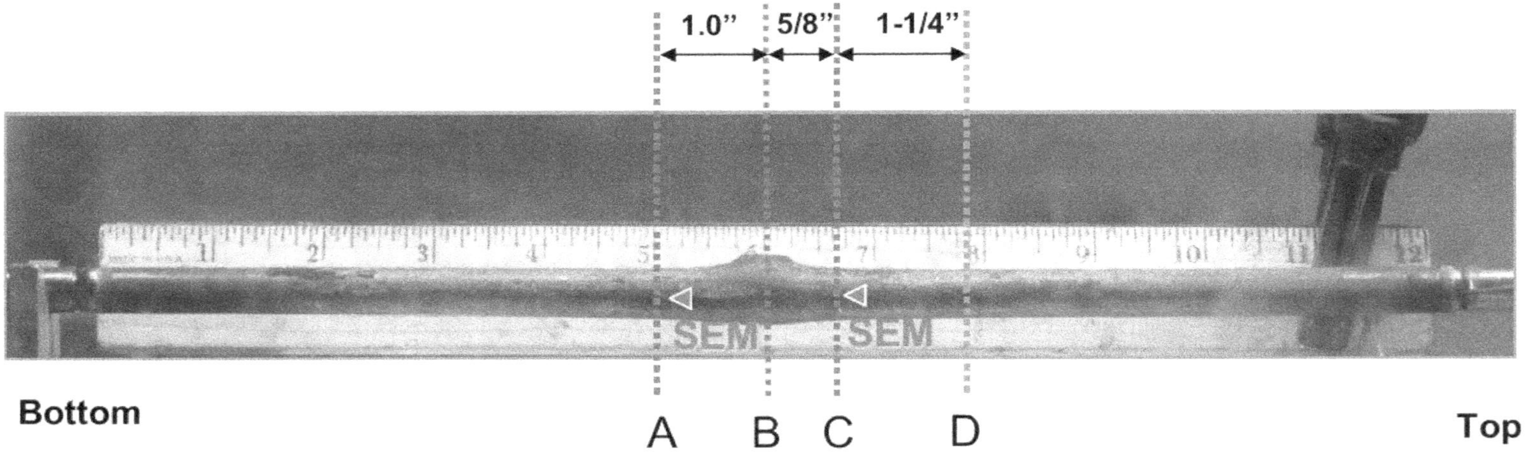

## OD Strain of In-cell LOCA sample ICL#3, 12/12/03

Strain (%): 0, 20, 40, 60

Legend:
— ICL#3 at 0°
— ICL#3 at 90°

Distance from the Burst Center (in.): -6, -5, -4, -3, -2, -1, 0, 1, 2, 3, 4, 5, 6

Bottom · A B C D · Top

Sample ICL#3 was broken at locations A, B and C during the sample handling before the sectioning was performed at location D.

Nuclear
Regulatory
Commission

Pioneering
Science and
Technology

# Cladding Cross-sections in ICL#3 Burst Region

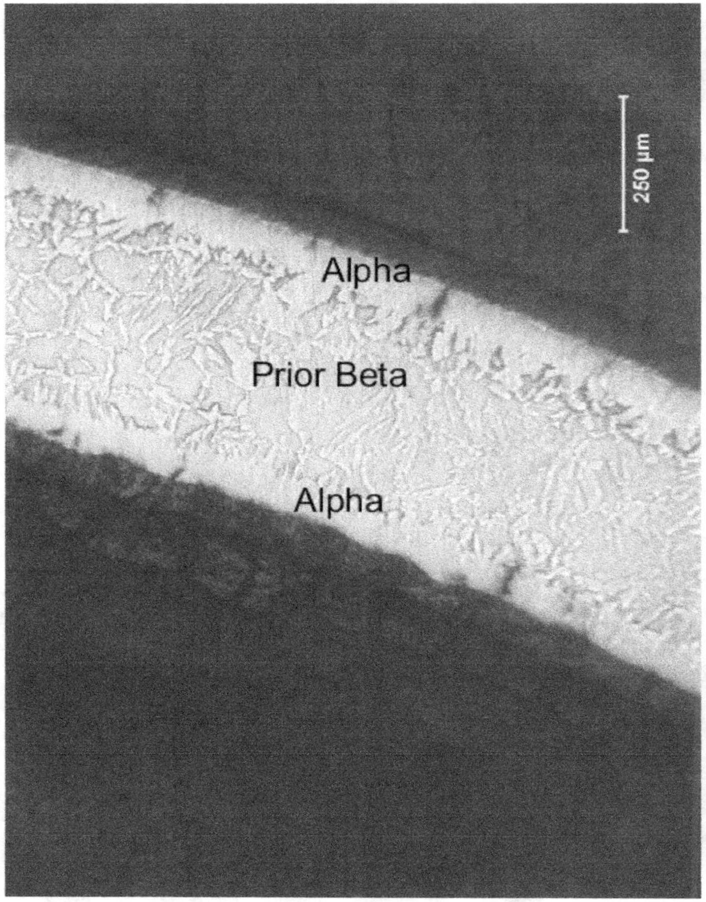

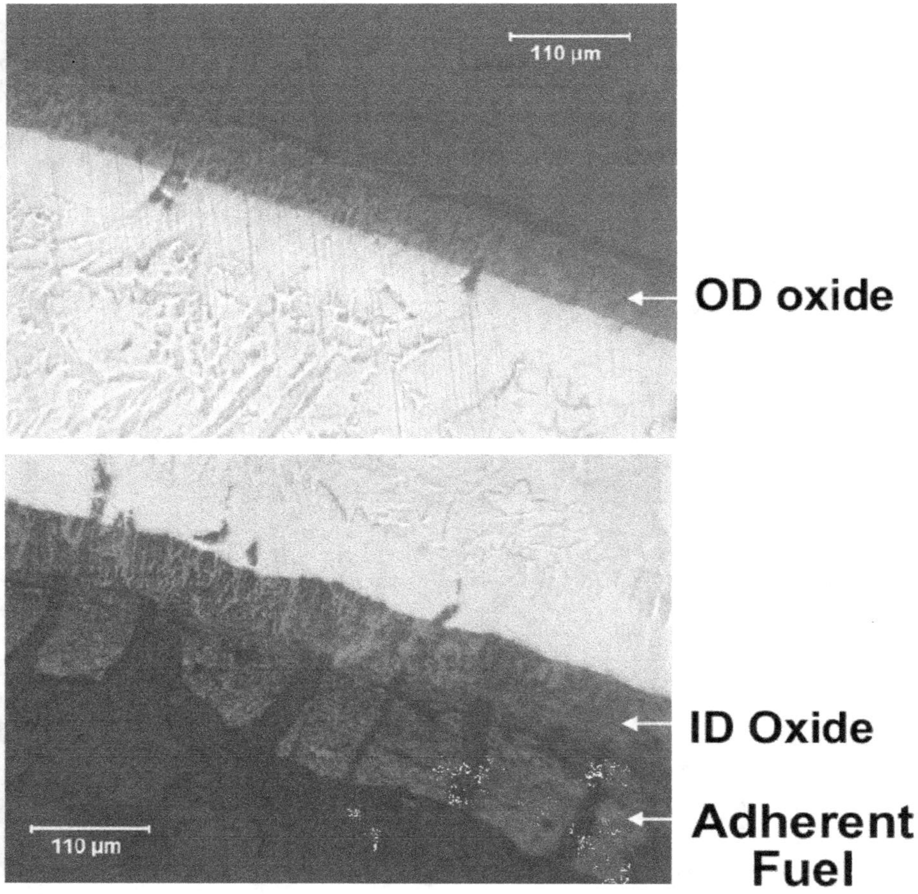

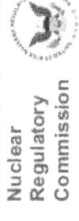

OD oxide

ID Oxide

Adherent Fuel

Alpha

Prior Beta

Alpha

110 μm

110 μm

250 μm

Nuclear
Regulatory
Commission

Pioneering
Science and
Technology

# ICL#3 Brittle Failure at ≈20 mm above Mid-plane (C)
## Similar Results were Obtained at ≈20 mm below (A)

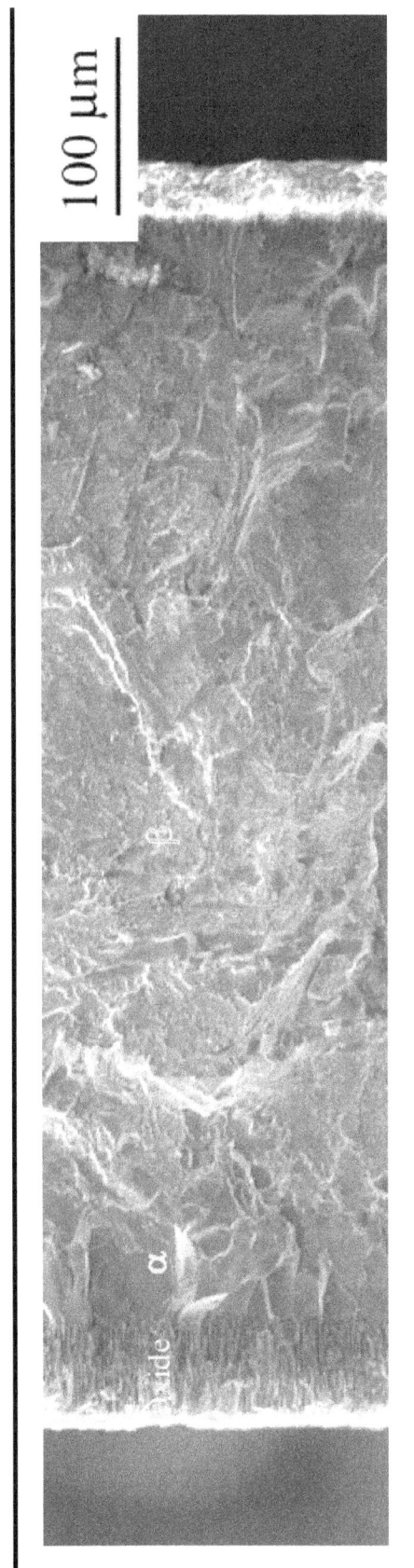

100 μm

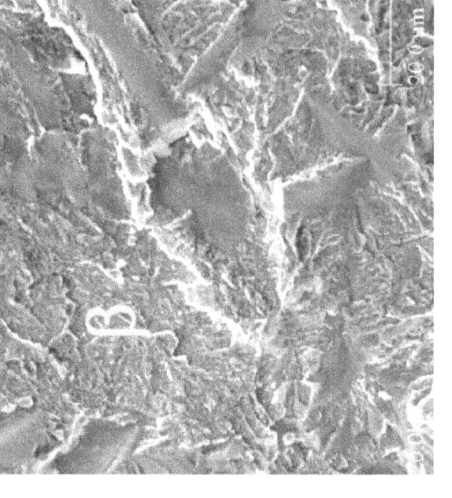

**Prior Beta**

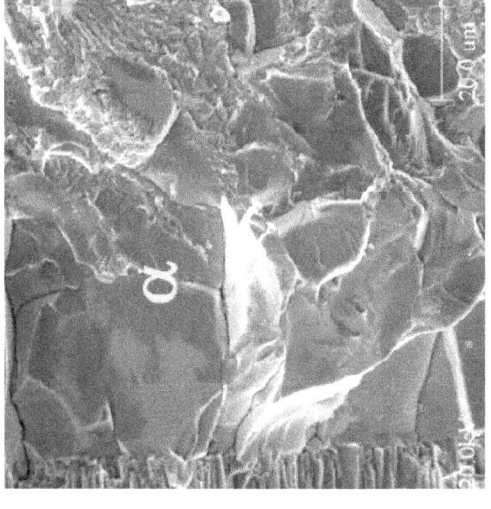

**Alpha**

**Oxide**

# Gamma Scan Profiles for ICL#3 and ICL#4 Specimens

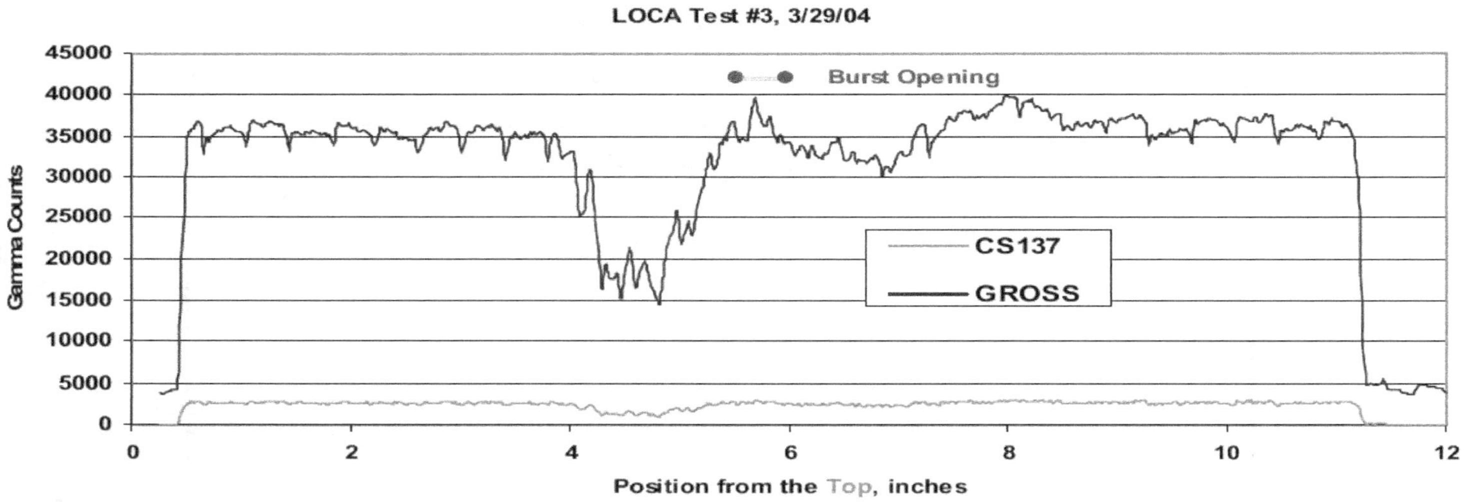

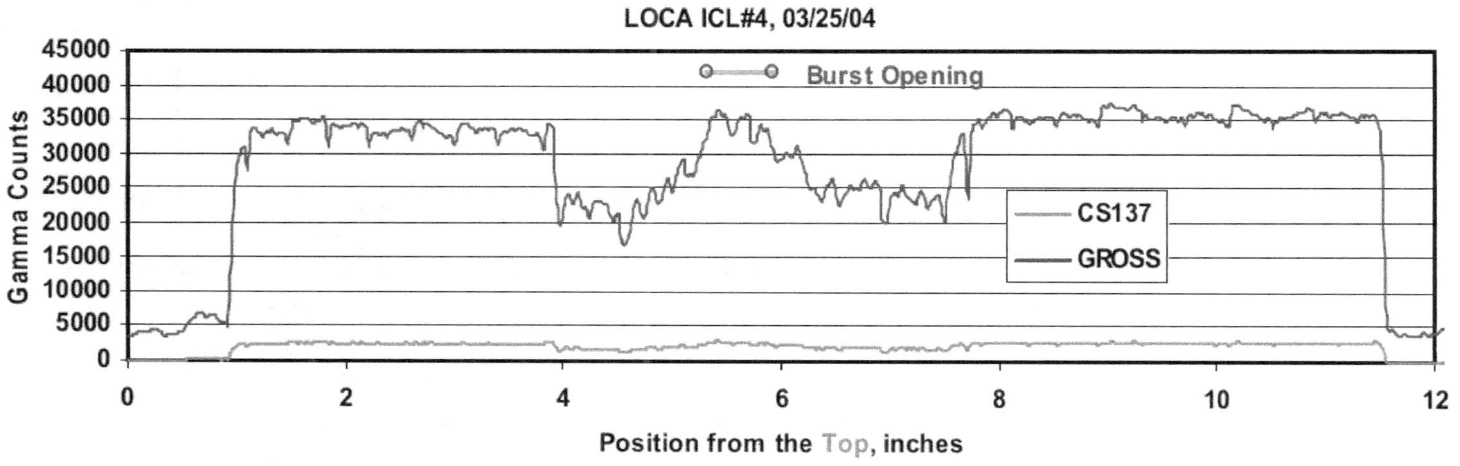

# *H and O Analyses of ICL#2 and ICL#3*

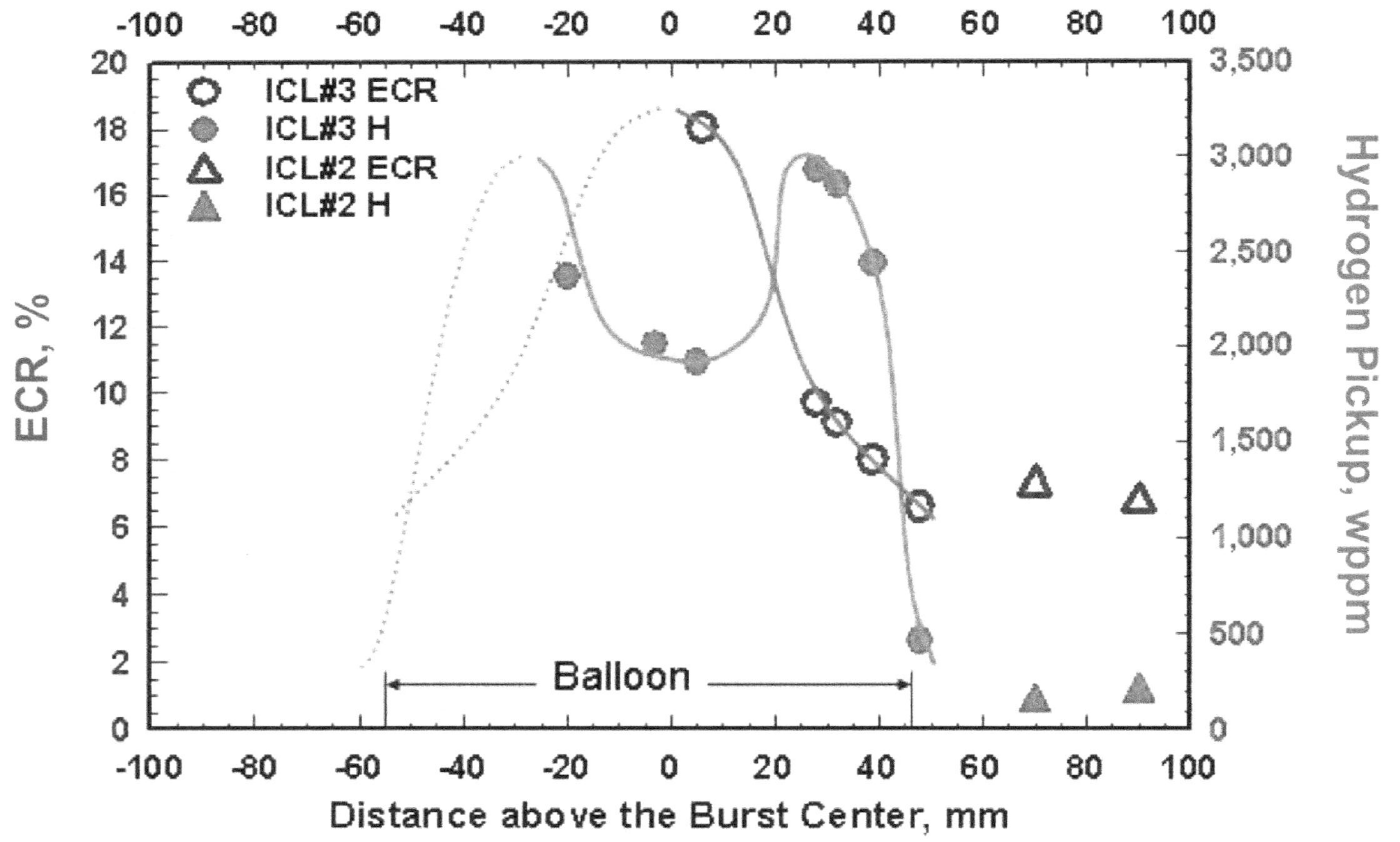

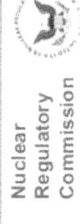

## Summary of In-Cell LOCA Integral Tests with High-Burnup Limerick BWR Fueled Specimens

| Parameter | ICL#1 | ICL#2 | ICL#3 | ICL#4 | OCL#11 |
|---|---|---|---|---|---|
| Hold Time, minutes | 0 | 5 | 5 | 5 | 5 |
| Max. Pressure, MPa | 8.96 | 8.87 | 9.0 | 8.86 | 8.61 |
| Burst Pressure, MPa | ≤8.61 | ≤8.01 | 8.6 | 8.0 | ≤7.93 |
| Burst Temperature, °C | ≈755 | ≈750 | ≈730 | ≈790 | ≈750 |
| Burst Shape | Oval | Oval | Oval | Oval | Dog Bone |
| Burst Length, mm | 13 | 14 | 11 | 15 | 11 |
| Max. Burst Width, mm | 3 | 3.5 | 4.6 | 5.1 | 1 |
| Length of Balloon, mm | ≈70 | ≈90 | ≈100 | ≈80 | ≈140 |
| $(\Delta D/Do)_{max}$, % | 38±9 | 39±10 | 43±9 | 36±9 | 43±10 |
| Max. Calculated ECR, % | 0 | ≈20 | ≈21 | ≈20 | ≈21 |
| Max. $\Delta H$, wppm | ... | >220 | ≥2900 | TBD | 3900 |

Balloon region: Diametral strain > ≈2%

Nuclear
Regulatory
Commission

# High-Burnup PWR (H.B. Robinson) LOCA Integral Test Plans

- **Data Required for High-burnup PWR LOCA Integral Test Planning**

  - Oxidation rate constant for ANL Equivalent Cladding Reacted (ECR) Model to determine ECR vs. test time

  - Hydrogen (pre-test H content and H pickup during steam oxidation) effects on post-quench ductility (PQD)

- **Issues Addressed in High-Burnup Test Planning**

  - Pre-transient ECR and hydrogen content: choose sample locations

  - Decrease in alpha-beta phase change temperature with hydrogen content: increase internal pressure to get burst in alpha phase

  - Embrittlement effects of hydrogen for samples oxidized at $\approx 1200°C$: requires a reduction in hold time (<300 s) to find ductile-to-brittle transition ECR for ballooned region and beyond-balloon region

  - Hold times and expectations

130

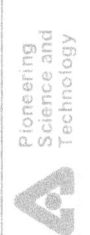

# High-burnup PWR Samples

## Sample from Grid Span #2

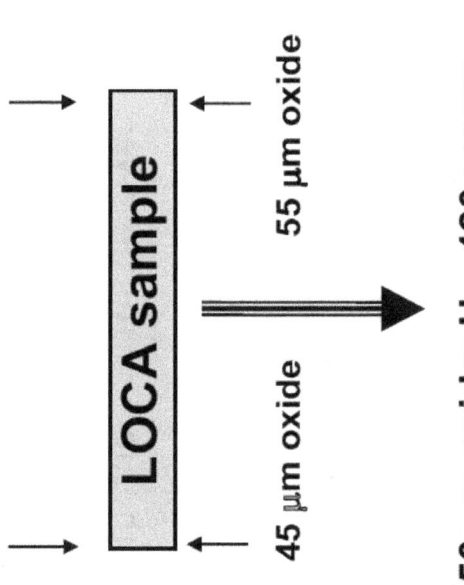

$\approx$ 380 wppm        $\approx$ 460 wppm

**LOCA sample**

45 μm oxide        55 μm oxide

50 μm oxide; H $\approx$ 420 wppm

Wall thickness: 760 – 50/1.75 $\approx$ 730 μm
Pre-transient $ECR_{ss}$ = 3.8%

## Sample from Grid Span #4

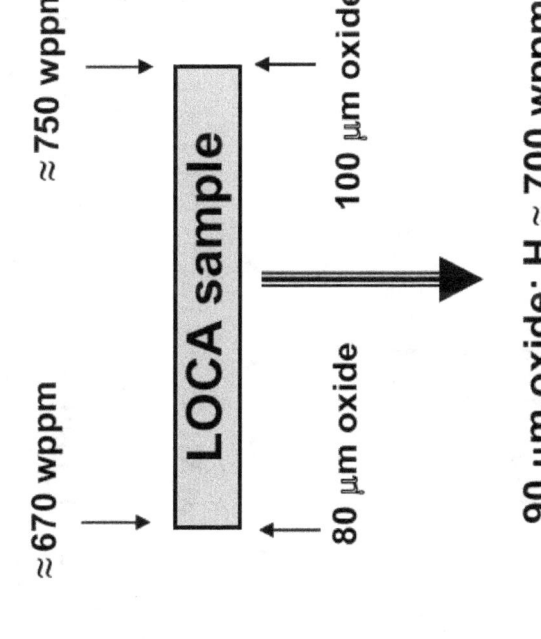

$\approx$ 670 wppm        $\approx$ 750 wppm

**LOCA sample**

80 μm oxide        100 μm oxide

90 μm oxide; H $\approx$ 700 wppm

Wall thickness: 760 – 90/1.75 $\approx$ 710 μm
Pre-transient $ECR_{ss}$ = 6.8%

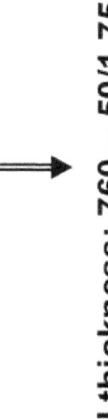

# *Baseline Data for PWR LOCA Integral Test Planning*

- **Baseline Data**

  - PQD tests with as-received 15x15 Zry-4: oxidized at 1204°C (completed)

  - PQD tests with pre-hydrided 15x15 Zry-4: Oxidized at 1204°C (completed)

  - PQD test with defueled high-burnup HBR Zry-4 cladding: (Dec. 2004)

- **Issues Addressed:**

  - Oxidation rate constant for Zry-4 cladding with thick corrosion layer

  - H effects on post-quench ductility

  - Possible ductile behavior beyond the balloon region

    - ➢ *Note: maintaining ductility in balloon region is unlikely because of primary and secondary hydrogen pick-up and oxygen-induced embrittlement*

Pioneering
Science and
Technology

132

# Offset Strain vs. ECR for As-received 15x15 Zry-4 Oxidized at 1204°C (H <25 wppm)

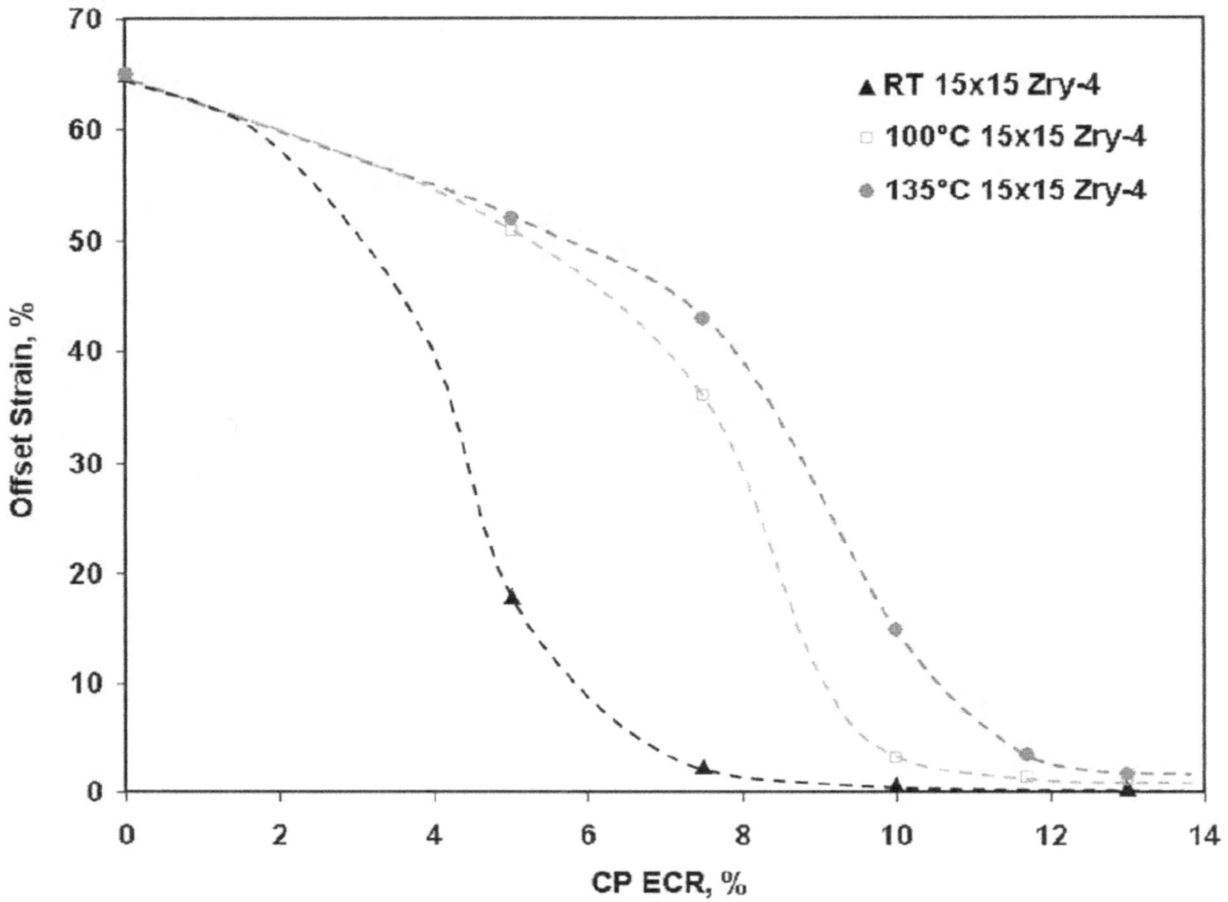

# *Effects of Hydrogen on Post-Quench Ductility at 135°C for Prehydrided 15x15 Zry-4 Oxidized at 1204°C*

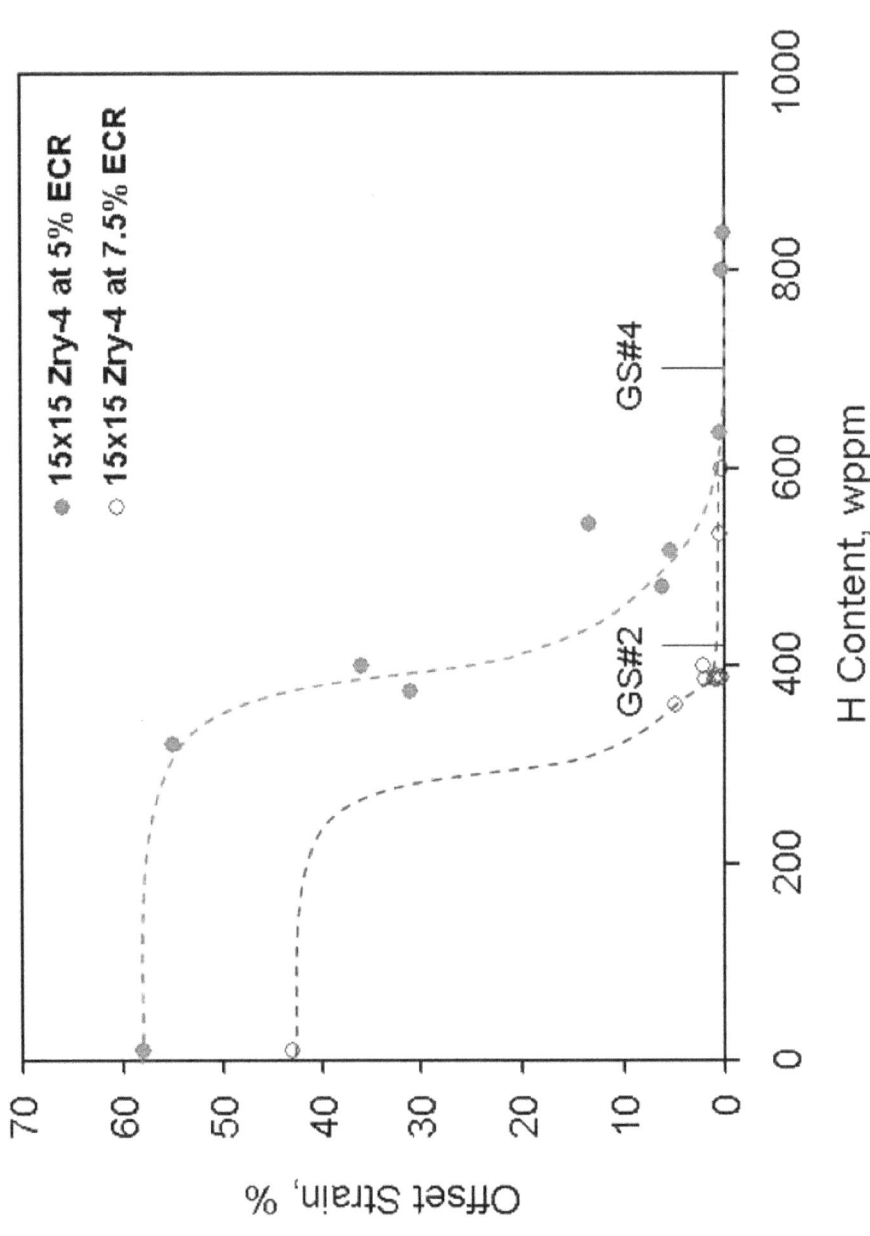

# *Two-sided oxidation in burst region vs. one-sided oxidation in beyond-neck region for a full LOCA sequence for GS#4*

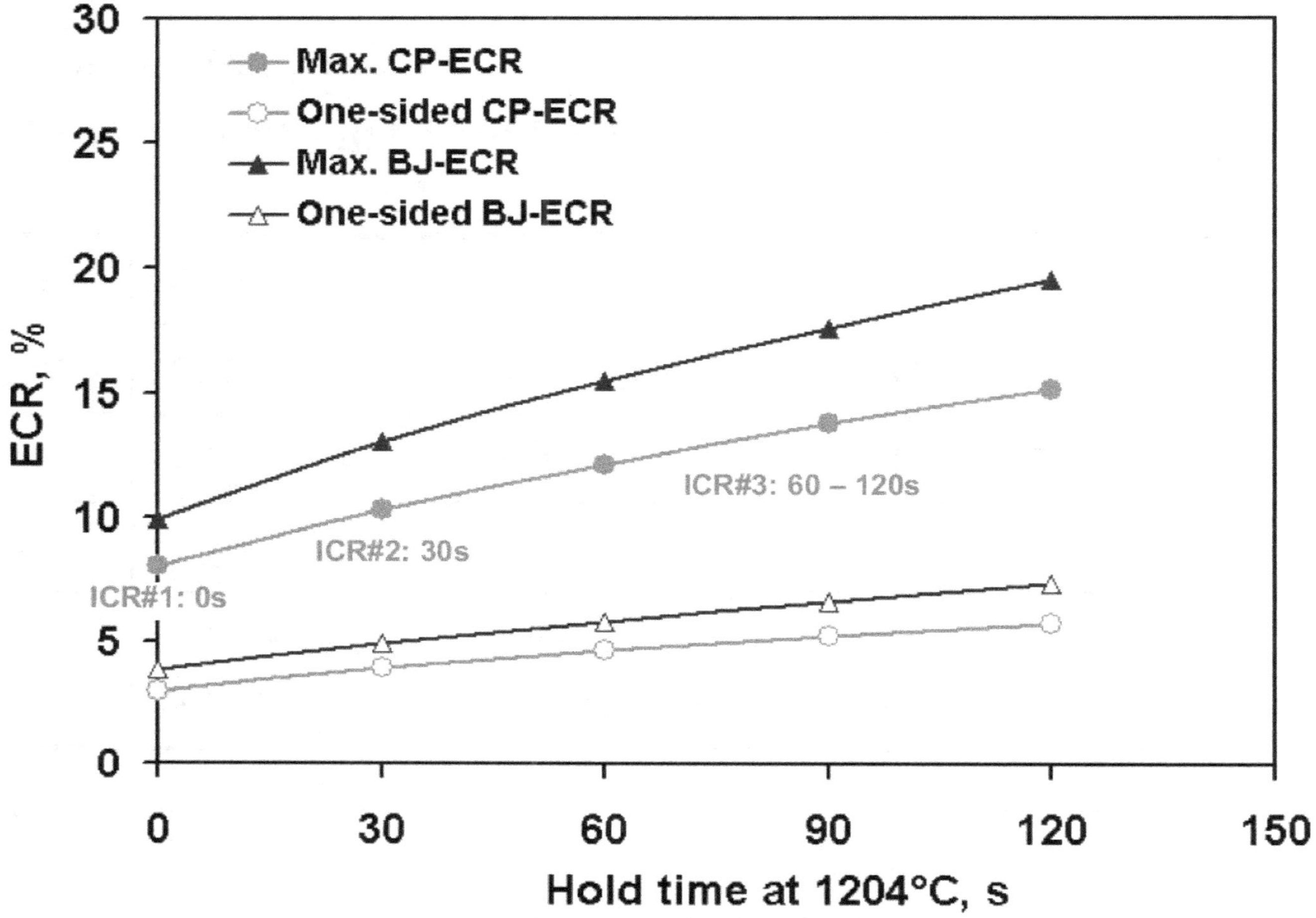

Pioneering
Science and
Technology

Nuclear
Regulatory
Commission

# Oxidation Kinetics at 1204°C and PQD Testing of High-Burnup PWR (H.B. Robinson) Cladding

| Test ID | H Content (wppm) | CP-ECR (%) | Ring Compression Test Temp. (°C) |
|---------|------------------|------------|----------------------------------|
| RO#1 | 550 - 600 | 5 | 100 and 135 |
| RO#2 | 550 - 600 | 7.5 | 100 and 135 |
| RO#3 | 550 - 600 | 10 | 100 and 135 |
| RO#4 | 550 - 600 | 13 | TBD |

\* Additional tests will be conducted later at 1000°C and 1100°C oxidation temperature

136

# *Summary of Plans for H.B. Robinson Integral Tests*

## Start with samples from GS#4

- ● **ICR#1: 0 s hold time**
  - – **Expect ductility along the length of the sample**
    - ✓ *Total BJ-ECR = 17% for GS#4 and 14% for GS#2*
    - ✓ *Transient CP-ECR in beyond-balloon region ≈ 3% for GS#4*
  - – **Determine pressure increase needed to burst in alpha phase**
  - – **Quantify gas communication, ballooning and burst shape, fuel relocation**

- ● **ICR#2: 30 s hold time**
  - – **Expect balloon region to be brittle and beyond-balloon region to be ductile**
    - ✓ *Total BJ-ECR = 20% for GS#4 and 17% for GS#2*
    - ✓ *Transient CP-ECR in beyond-balloon region ≈ 4% for GS#4*

- ● **ICR#3: 60 – 120 s hold time, depending on results of ICR#1 and #2**
  - – **Expect embrittlement in both regions**
    - ✓ *Total BJ-ECR in this region ≈ 23-27% for GS#4*
    - ✓ *Transient CP-ECR in beyond-balloon region = 4.6-5.7% for GS#4*

# LOCA Integral Test Results for High-Burnup BWR Fuel

Yong Yan, Michael C. Billone, Tatiana A. Burtseva, and Hee M. Chung

Argonne National Laboratory (ANL), Argonne, IL 60439

## Abstract

Results from loss-of-coolant-accident (LOCA) integral tests are reported for high-burnup fuel-rod specimens from a boiling water reactor (BWR) at Limerick. These results are compared to baseline data for nonirradiated Zry-2 cladding specimens filled with zirconia pellets and exposed to the same tests. Four LOCA integral tests have been conducted with specimens from Limerick BWR fuel rods at burnup of 56 GWd/MTU. The ICL#1 test specimen was heated to bursting in argon and slowly cooled. The ICL#2 specimen was exposed to the LOCA test sequence with the exception of quench. The ICL#3 specimen achieved partial quench (800°C to 470°C) before failure of the quartz chamber that surrounded the specimen. The full LOCA sequence in ICL#4 calls for heating in steam at 5°C/s to 1204°C, holding for 5 minutes at 1204°C ($\leq$ 20% CP best-estimate ECR), slow cooling at 3°C/s to 800°C, and bottom-flooding to quench the cladding from 800 to 100°C. Destructive examinations showed two-sided oxidation in the ballooned-and-burst region of both high-burnup and nonirradiated specimens. For nonirradiated specimens, the hydrogen pickup was low in the burst region and very high at 70-90 mm above and below the burst mid-plane. For high-burnup-fuel cladding, the hydrogen peak was toward the burst mid-plane. In addition, the effects of hydrogen on diametral-compression post-quench ductility (PQD) have been investigated with prehydrided (300-to-800 wppm hydrogen), nonirradiated 15x15 Zry-4 cladding rings after oxidation at 1204°C and quench. The baseline data from prehydrided cladding are being used to plan the LOCA test times for specimens from a pressurized water reactor (PWR) such that the embrittlement equivalent cladding reacted (ECR) can be determined effectively in the ballooned and non-ballooned regions.

## Introduction

The LOCA licensing criteria (10 CFR 50.46) limit peak cladding temperature to 2200°F (1204°C) and maximum oxidation (expressed as equivalent cladding reacted, ECR) to 17% to ensure cladding ductility during quenching from the emergency core cooling system and during possible post-LOCA events (e.g., seismic).  In formulating these criteria, it was assumed that knowledge of the detailed loading modes and magnitudes experienced by the cladding, beyond the thermal stresses induced by rapid cooling, are not well defined.  Cladding that retains some plastic ductility has more margin for surviving quench and post-quench loads without fragmenting  compared with brittle cladding. Based on Appendix K of this regulation, the Baker-Just (BJ) correlation is to be used to calculate the metal-water (i.e., steam) reaction.  Regulatory Guide 1.157 (1989) allows the use of a best-estimate correlation, such as the Cathcart-Pawel (CP) model, to calculate the oxidation rate in steam for T > 1900°F (1038°C).  At 1204°C, the ratio of the BJ-to-CP prediction is ≈1.3.  To compensate for the possible effects of high burnup operation (e.g., hydrogen pickup), NRC Information Notice 98-29 (1998) defines total oxidation to include in-reactor corrosion ($ECR_{ss}$), as well as transient steam oxidation ($ECR_t$).

The LOCA integral tests at ANL, using high burnup fuel-rod segments from boiling and pressurized water reactors, are designed to address the adequacy of the embrittlement criteria, of the correlations (CP vs. BJ) used to calculate oxidation, and of the decrease in allowable transient oxidation ($ECR_t \leq 17\%$ - $ECR_{ss}$) to compensate for the embrittling effects of hydrogen.  In addition to this confirmatory aspect of the research, the fundamental behavior of high-burnup fuel and cladding, exposed to a LOCA transient, is investigated and characterized.

The LOCA-relevant research at Argonne National Laboratory (ANL) includes high-temperature steam oxidation studies of cladding [1], LOCA integral testing of fueled segments [2], post-quench ductility testing of LOCA integral specimens, and post-quench ductility testing of nonirradiated zirconium-based cladding alloys [3].  Four LOCA integral tests with high-burnup BWR samples (from Limerick fuel rods at 56 GWd/MTU) have been completed. The ICL#1 test specimen was heated to bursting in argon and slowly cooled.  The ICL#2 specimen was exposed to the LOCA test sequence with the exception of quench. The ICL#3 specimen achieved partial quench (800°C to 470°C) before failure of the quartz chamber that surrounded the specimen. A full LOCA sequence in ICL#4 was completed in March 2004.

This paper presents the results of the post-test examinations of the ICL#1-4 samples: fuel morphology; cladding inner- and outer-surface oxidation within the ballooned region; post-test gamma scanning; fractography; and hydrogen pickup in the neck and ballooned regions. After completion of the high-burnup BWR test matrix, high-burnup PWR rods (from the H. B. Robinson reactor) will be subjected to the LOCA test sequence indicated in Fig. 1. Post-quench ductility (PQD) baseline data with prehydrided 15x15 Zry-4 cladding are also presented for the planning of the H.B. Robinson LOCA integral tests.

## Experimental Procedure

LOCA integral tests consist of rapidly heating (5 °C/s) a 300-mm-long fuel segment under internal pressure in a steam environment, holding it at 1200°C for ≤5 minutes, cooling it (3 °C/s) to 800°C, and then quenching it with room-temperature water. A schematic illustration of the LOCA sequence is given in Fig. 1. The LOCA integral apparatus, as shown schematically in Fig. 2, contains the following features: radiant furnace, argon purge system, high-pressure system to internally pressurize the 300-mm-long test sample, steam supply system, and quench system.

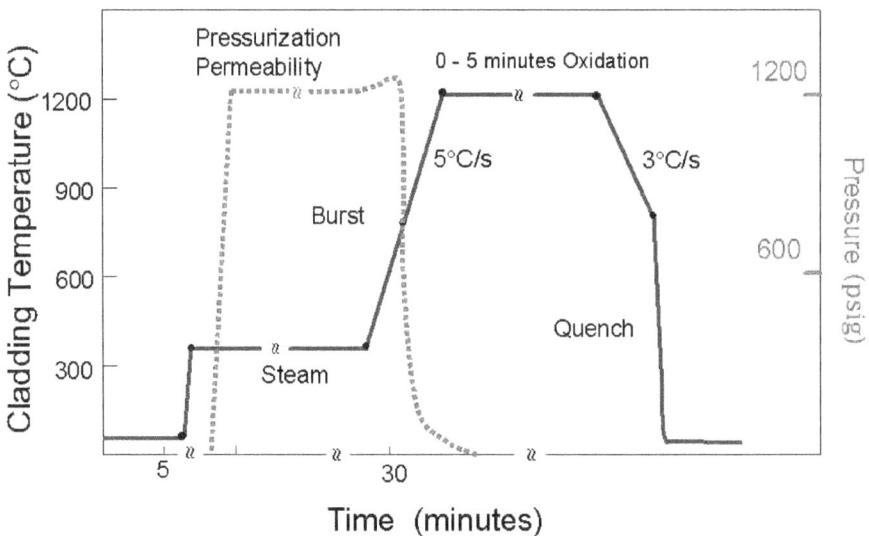

Fig. 1. Temperature and pressure histories for full LOCA integral test sequence, including quench from 800°C to 100°C.

A series of out-of-cell LOCA integral tests was conducted with unirradiated Zry-2 cladding at 1204°C in steam. The out-of-cell LOCA samples are used to benchmark the testing methods and to obtain data for post-quench ductility, as well as properties data for weight gain, oxide-alpha-beta layer thicknesses, and hydrogen and oxygen distributions. This data set serves as a baseline for in-cell LOCA tests with high-burnup-fuel samples.

Results for the OCL#11 (an out-of cell LOCA test with unirradiated 9x9 Zr-2 for 5 minutes at 1204°C), companion test to ICL#2, are briefly discussed here. The concentrations of oxygen and hydrogen were measured by the LECO method after the OCL#11 test. These concentrations are referenced to the weight of the oxidized samples. The data need to be converted to concentrations referenced to the pre-oxidized sample weight in order to determine pickup values during the transient. The algorithms for calculating oxygen and hydrogen pickup from the LECO data were given in our previous work [3]. In addition, quantitative metallography (see Fig. 3), along with the CP models for interface oxygen concentrations and diffusion within the oxide, alpha, and beta layers, was used to determine the weight gain per unit surface area and the corresponding ECR.

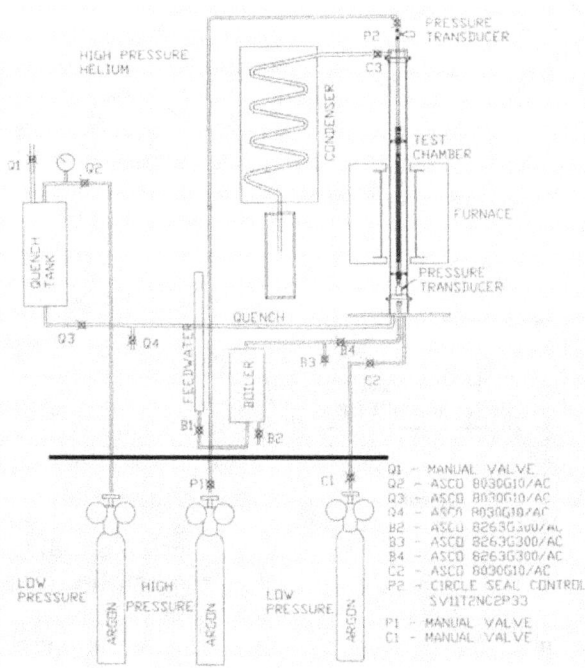

Fig. 2.  Schematic illustration of LOCA system.

Fig. 3.   Cladding metallographic results for OCL#11 specimen.

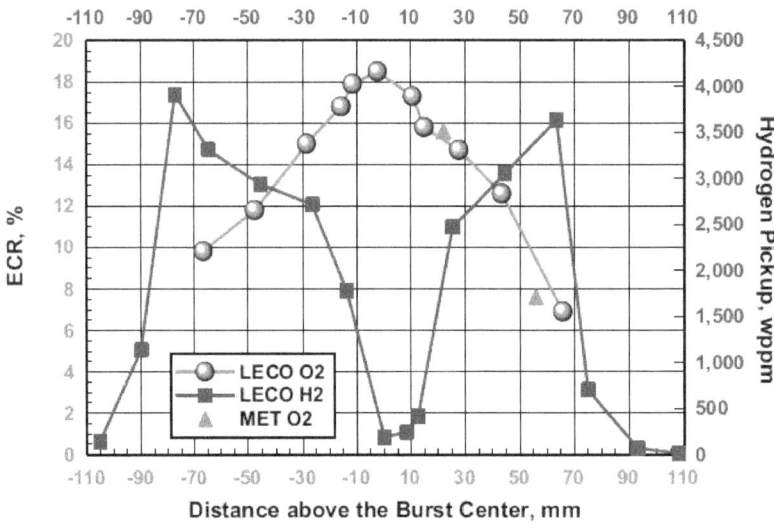

Fig. 4.   Axial distributions of hydrogen pickup and ECR for out-of-cell test OCL#11 with nonirradiated
Zry-2 cladding.  The average values do not give information regarding the local concentrations
of oxygen and hydrogen across the wall of the cladding.  Oxygen concentrations in the oxide and
alpha layers are much higher than the average value, while hydrogen concentration in the prior-
beta layer is much higher than the average concentration.

The axial distributions of ECR and hydrogen pickup of the OCL#11 sample are shown in Fig. 4. As expected, the oxygen pickup and ECR peak at the center of the burst region where the cladding is thinnest and the oxidation is fully two-sided. The ECR decreases away from the burst center as the cladding wall thickness increases and the degree of inner-surface oxidation decreases. Also, as expected, the hydrogen pickup, due to secondary hydriding, peaks near the balloon neck regions. The magnitude of these hydrogen peaks, however, is larger than previously reported [4] and may depend on ballooning strain profile, burst opening, diameter of pellets (zirconia) inside the cladding, heating method (internal vs. external vs. direct-electrical), and cladding type (lined Zry-2 vs. Zry-4). As these hydrogen peaks, as well as the hydrogen within the balloon region, are potentially embrittling, it is important to determine the magnitude of such effects for high-burnup cladding. Additional nondestructive examination results and some destructive results for the OCL#11 sample were reported in Ref. 3.

**LOCA In-cell Integral Test Results**

Limerick Rod J4 was selected for the in-cell LOCA tests ICL#3 and ICL#4. Based on our gamma scan data (see Fig. 5), there appear to be no unusual features in the pre-test specimens. Figure 6 shows the temperature and pressure histories for the ICL#4 test at an average hold temperature of $\approx 1204°C$ for 5 minutes in steam. Two thermocouples (180° apart) were strapped at 2-inches above the mid-plane, and the temperature difference between them was less than 14°C.

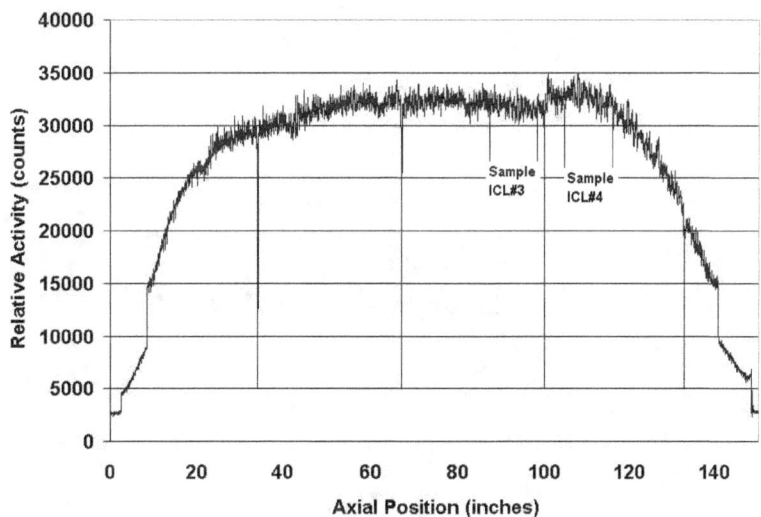

Fig.5.   Gamma scan profile of Limerick fuel rod J4.

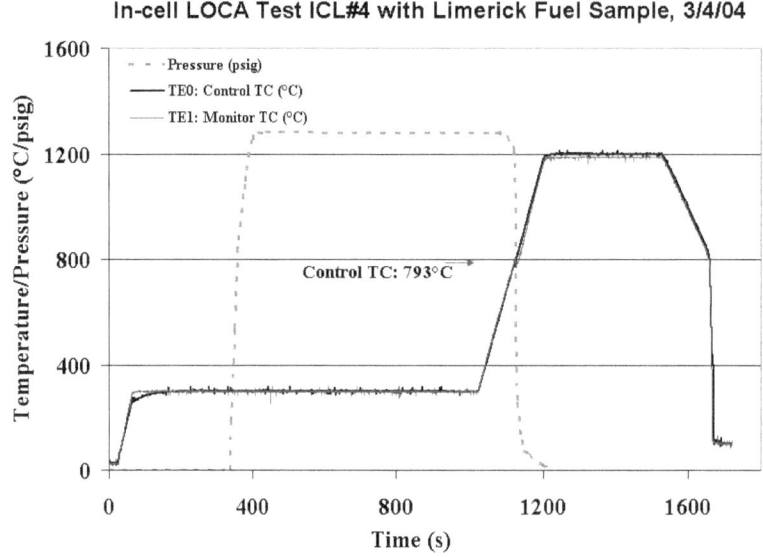

Fig.6.    Temperature and pressure histories of in-cell LOCA test ICL#4.

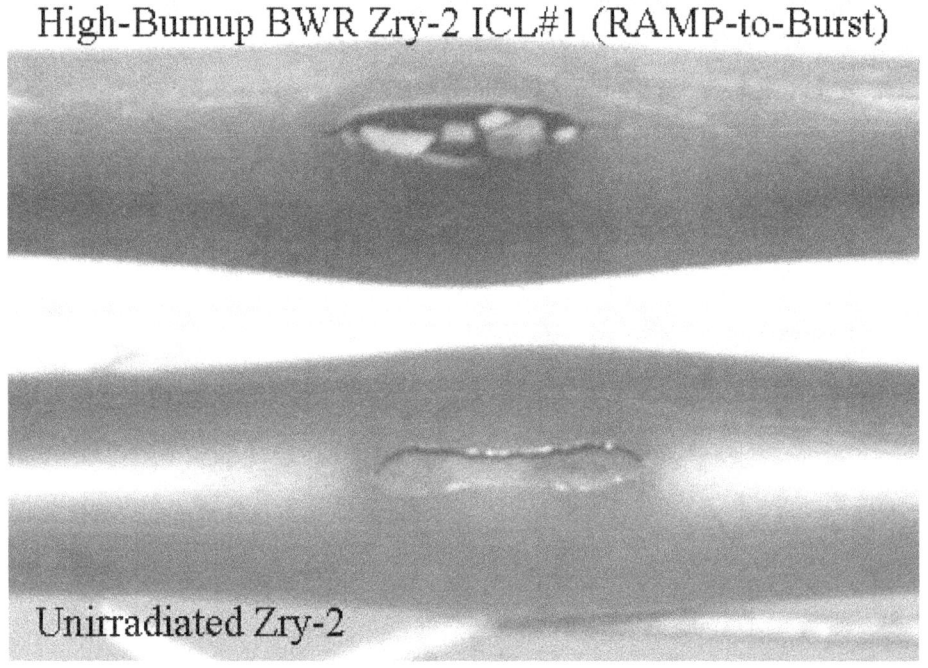

Fig. 7.    High magnification micrographs of the burst opening for ramp-to-burst tests.

144

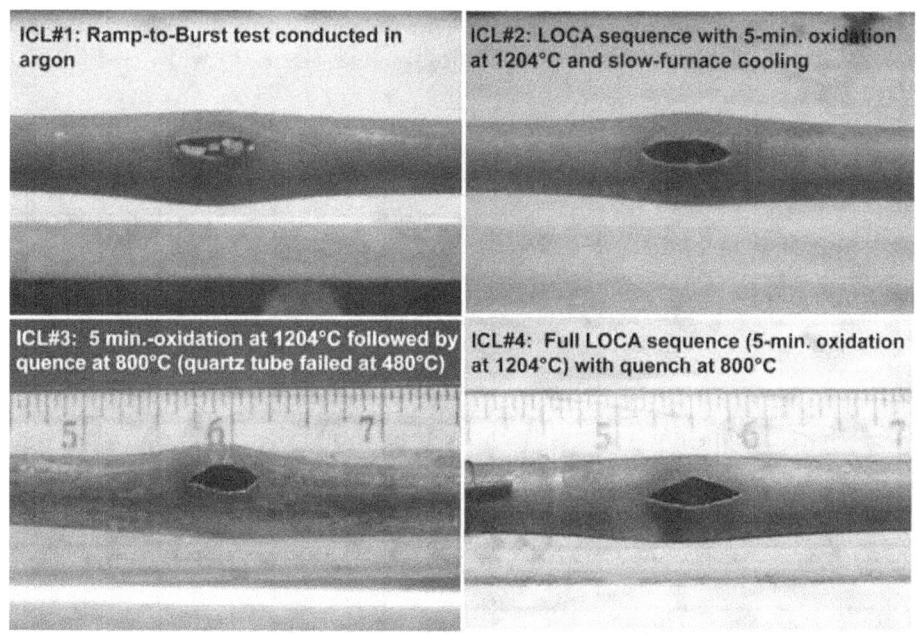

Fig. 8.   Micrographs showing the balloon and burst regions for ICL#1 – #4 specimens.

<u>Nondestructive Characterizations</u>

Nondestructive characterizations for the ICL#1 - #4 samples were completed shortly after the tests. The results are summarized in Table 1.   Unlike the dog-bone burst shape of the unirradiated OCL#11 sample (see Fig. 7), an oval burst shape was observed for all LOCA in-cell samples (see Fig. 8). The burst temperatures for the LOCA in-cell tests range from 730 to 790°C, and their burst lengths are in the range of 11-15 mm. In addition, the balloon lengths (defined by OD strains ≈2%) of the LOCA in-cell samples are shorter than that of the LOCA out-of-cell sample. No significant difference was found for the maximum OD strain between the unirradiated samples and the high-burnup samples.

<u>Axial Locations of Specimens for Destructive Evaluation</u>

Figure 9 shows the axial locations at which sample ICL#3 was broken (locations A, B, and C) during the sample handling, before the sectioning was performed at location D. Metallographic examinations were conducted at location B, and scanning electron microscopy (SEM) examinations were conducted at locations A and C. The hydrogen and oxygen analysis samples were further sectioned from the specimens A-B and C-D.

Table 1. Summary of in-cell LOCA integral tests (ICL) with high-burnup fueled cladding specimens from Limerick BWR. Also shown are the results of out-of-cell test OCL#11 with non-irradiated Zry-2 cladding.

| Parameter | ICL#1 | ICL#2 | ICL#3 | ICL#4 | OCL#11 |
|---|---|---|---|---|---|
| Hold Time, minutes | 0 | 5 | 5 | 5 | 5 |
| Max. Pressure, MPa | 8.96 | 8.87 | 9.0 | 8.86 | 8.61 |
| Burst Pressure, MPa | ≤8.61 | ≤8.01 | 8.6 | 8.0 | ≤7.93 |
| Burst Temperature, °C | ≈755 | ≈750 | ≈730 | ≈790 | ≈ 750 |
| Burst Shape | Oval | Oval | Oval | Oval | Dog Bone |
| Burst Length, mm | 13 | 14 | 11 | 15 | 11 |
| Max. Burst Width, mm | 3 | 3.5 | 4.6 | 5.1 | 1 |
| Length of Balloon, mm | ≈ 70 | ≈ 90 | ≈ 100 | ≈ 80 | ≈ 140 |
| $(\Delta D/D_o)max$, % | 38±9 | 39±10 | 43±9 | 36±9 | 43±10 |
| Max. Calculated ECR, % | 0 | ≈20 | ≈21 | ≈20 | ≈21 |
| Max. $\Delta H$, wppm | … | > 220 | ≥ 2900 | TBD | 3900 |

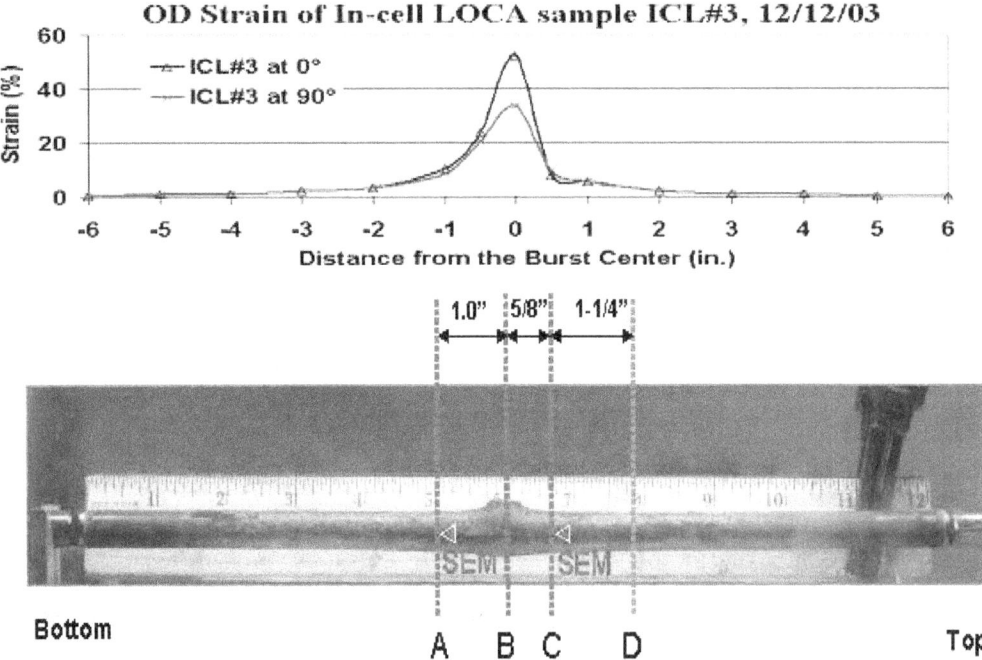

Fig. 9   Post-test characterization for ICL#3 sample. The ICL#3 sample was broken at axial positions A, B, and C during the sample handling.

146

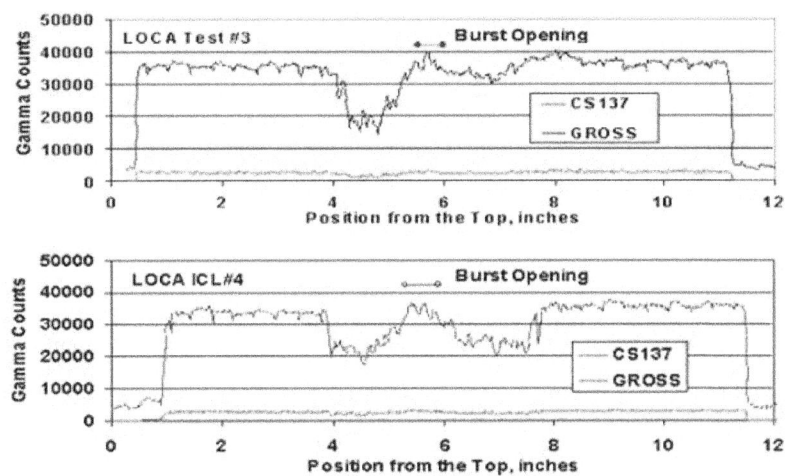

Fig. 10.   Gamma scan profiles for ICL#3 and ICL#4 specimens.

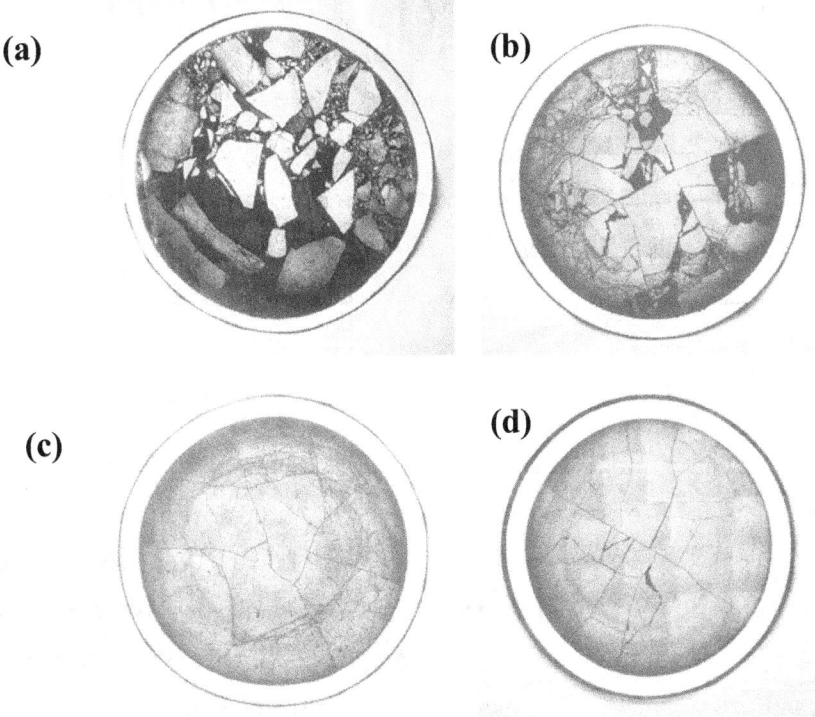

Fig. 11.   Low-magnification images of the post-LOCA test ICL#2 fuel samples at ≈12 mm above the burst center (a); ≈50 mm above the burst center (b); ≈130 mm below the burst center (c); and the Limerick fuel prior to LOCA testing (180 mm from the LOCA sample) (d).   Cladding diametral strains are 2-4% for the Fig. 11b cross section and 15-25% for the Fig. 11a cross-section.

Fuel Relocation

There is considerable interest in the behavior of high-burnup fuel during a LOCA transient. Prior to the transient, the fuel is tightly bonded to the cladding. During ballooning, the cladding pulls away from the fuel. This allows space for fuel particles (macro-cracked, micro-cracked, and very small particles from the rim layer) to fall into the balloon region. If such movement were to result in a local increase in fuel per unit length, then the higher decay heat per unit length would increase cladding oxidation temperature and maximum ECR in the burst region. Also, if the fuel-cladding bond material moves with the cladding, such a layer could slow down the initial steam oxidation rate and could protect the cladding from the large hydrogen absorption observed in tests with bare-wall, nonirradiated cladding (see Fig. 4). As methods that could be used to freeze the fuel particles in place (e.g., epoxy) conflict with cladding characterization, the ANL program is more focused on the details of cladding oxidation, hydriding, and ductility than on fuel behavior. This focus was certainly the case for the ICL#2 sample, as no attempt was made to prevent fuel fallout during handling. For the ICL#3 and ICL#4 samples, the burst areas were taped following the test to minimize fuel fallout and the samples were gamma-scanned – prior to other nondestructive characterizations – to determine the axial distribution of fuel in and beyond the balloon region. Figure 10 shows the gamma scan profiles for the ICL#3 and ICL#4 specimens. For the axial locations with little-to-no permanent strain, gamma counts received from the fuel most likely represent the condition of the fuel at the end of the LOCA test. For the ICL#3 and ICL#4 balloon regions, some redistribution of fuel particles likely took place between the end of the LOCA test and the gamma scan due to the sample handling and transfer.

Figure 11 shows low-magnification images of the fuel structure of the ICL#2 sample at axial locations: (a) ≈12 mm above the burst center, (b) ≈50 mm above the burst, and (c) ≈130 mm below the burst center (45 mm above the bottom end-cap). Also shown is (d) the fuel structure of the as-received Limerick fuel. The structures of Figs. 11c and d are similar, except that the post-LOCA fuel shows a ring of circumferential tearing about mid-radius. This tearing may have occurred as the cladding tried to move a small distance (0.1 mm) away from the fuel and/or because the fission-product gases affected the fuel (see dark ring near mid-radius for the pre-LOCA fuel in Fig. 11d). At ≈50 mm above the burst, the circumferential tearing is enhanced as compared to the ≈130-mm location, most likely due to the larger cladding strain. Some fuel fallout may have occurred during cutting, although this region of the fuel column was embedded in a soft epoxy prior to cutting. Smaller fuel particles are also observed. In Fig.

11a, a wide range of fuel particles is observed, although these particles and fuel chunks are not co-planar. The particles and chunks are held in place by soft epoxy. Because this photograph was taken after extensive handling of the sample, resulting in axial redistribution of particles and fuel fallout through the burst opening, it does not represent the fuel condition near the burst center during the LOCA test or after the quench. The most that one can glean from such a picture is that the wide range of fuel-particle sizes would allow some fuel to fall from <50 mm above the burst center to the burst region.

Cladding Metallography

Low-magnification photographs were taken at 16 circumferential locations of the burst midplane of the ICL#3 sample and pieced together (see Fig. 12a) to obtain an image of the metal (oxygen-stabilized alpha and prior-beta) thickness vs. circumferential location, as compared to the cladding structure of the unirradiated material (see Fig. 12b). The inner and outer oxide layers are not visible in Fig. 12a. Also, the burst tips, which are very thin and heavily oxidized, are likely lost in the cutting process.

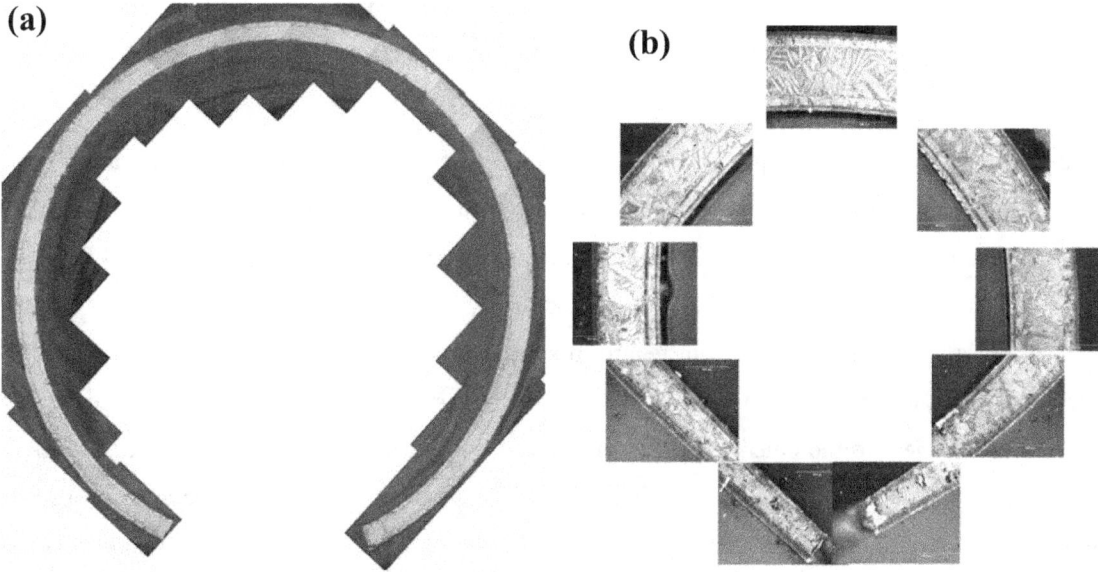

**(a)**          **(b)**

Fig. 12. Composites of the cladding cross section at the burst midplane of the in-cell LOCA sample ICL#3 (a) and the out-of-cell LOCA sample OCL#11 (b). The tips of samples were lost during the post-test sample handling.

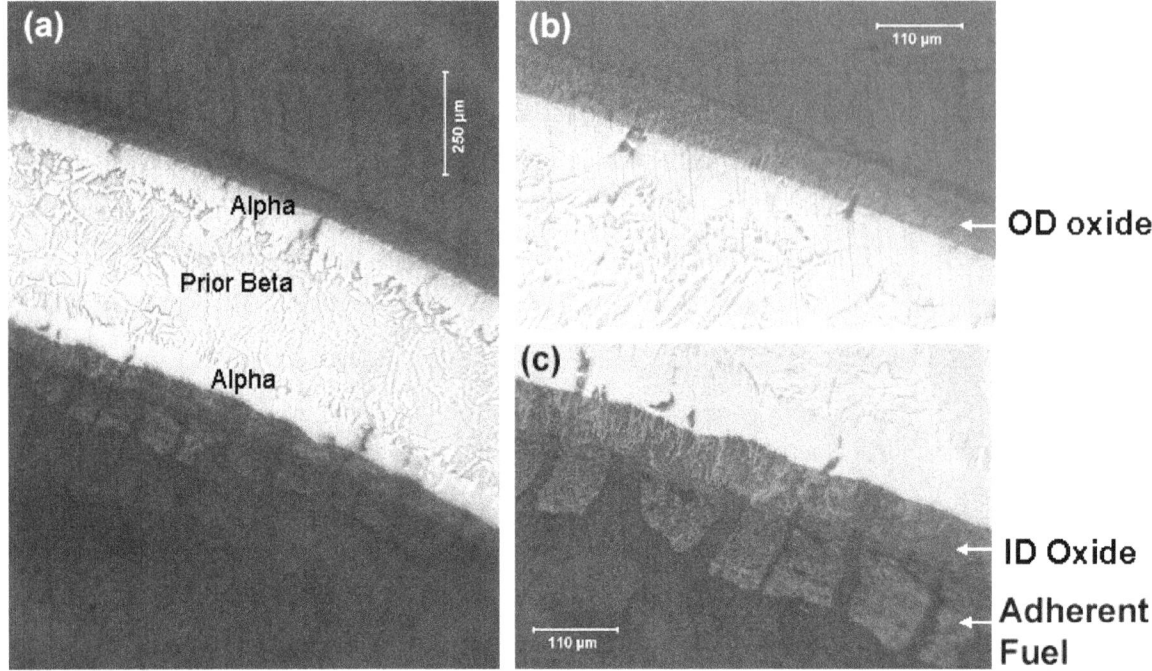

Fig. 13. Metallographic images of the ICL#3 specimen at the burst midplane. Micrograph (a) shows well-defined alpha layers that formed at 1200°C and large prior-beta grains surrounded by oxygen-stabilized alpha layers formed during slow cooling to 800°C; and micrographs(b) and (c) show good definition of the outer- and inner-surface oxide layers.

High-magnification micrographs were obtained of the inner and outer cladding regions away from the burst tips. These are shown in Figs. 13b (outer surface) and 13c (inner surface). The thickness of the outer-surface oxide layer is consistent with the CP-calculated oxide thickness and with the results from oxidation tests conducted on undeformed cladding samples [2]. The results demonstrate that the $\approx$10-$\mu$m-thick corrosion layer is not protective with regard to steam oxidation. The inner-surface oxide layer is wavy in appearance, which is likely due to the high hydrogen-to-steam ratio within the burst region. It is comparable in thickness to the outer-surface layer, which suggests that the fuel-cladding bond layer is also not protective with regard to steam oxidation. Alpha incursions into the prior-beta layer are observed at this location, just as they were observed in the oxidation tests. These most likely formed during the 3°C/s cooling from 1200°C to 800°C. They represent regions with higher oxygen content than the remaining prior-beta material and lower oxygen content than the alpha layer that was formed at 1200°C. Adherent fuel was observed on the inner surface of the ICL#3 burst region, as shown in Fig. 13c.

More detailed metallography was obtained for the ICL#2 specimen at the axial position 12 mm above the burst midplane. Double-sided oxidation is evident at this location, with the thickness of the inner-surface oxide layer greater than that of the outer-surface. Quantitative metallography was performed to determine the distribution of cladding thickness and outer-surface oxide-layer thickness (inner and outer) with respect to circumferential orientation. These results are compared to the baseline results obtained for the nonirradiated Zry-2 sample used in the OCL#11 test. To focus on the transient oxidation of the high-burnup LOCA sample (ICL#2), 10 μm was subtracted from the total outer-surface oxide-layer thickness to generate the transient oxidation data. For the OCL#11 sample, the weight gain and ECR at this location were determined to be 11.4 mg/cm$^2$ and 15.7%, respectively, while the prior-beta-layer thickness was measured to be 398 μm. The high-burnup sample differs somewhat in that there is more circumferential variation in the inner-surface oxide-layer thickness. The weight gain and ECR were determined to be 10.5 mg/cm$^2$ and 14.9%, respectively, while the prior-beta-layer thickness was measured to be 435 μm for the ICL#2 sample at this location. Although the axial locations with respect to the burst center are slightly different (18 mm above for OCL#11 and 12 mm above for ICL#2), the values for oxide-layer thickness, weight gain, ECR, and prior-beta-layer thickness are remarkably close. These results indicate that close to the burst region the steam oxidation of both nonirradiated and high-burnup Zry-2 samples is essentially the same. No significant high-burnup effects were observed.

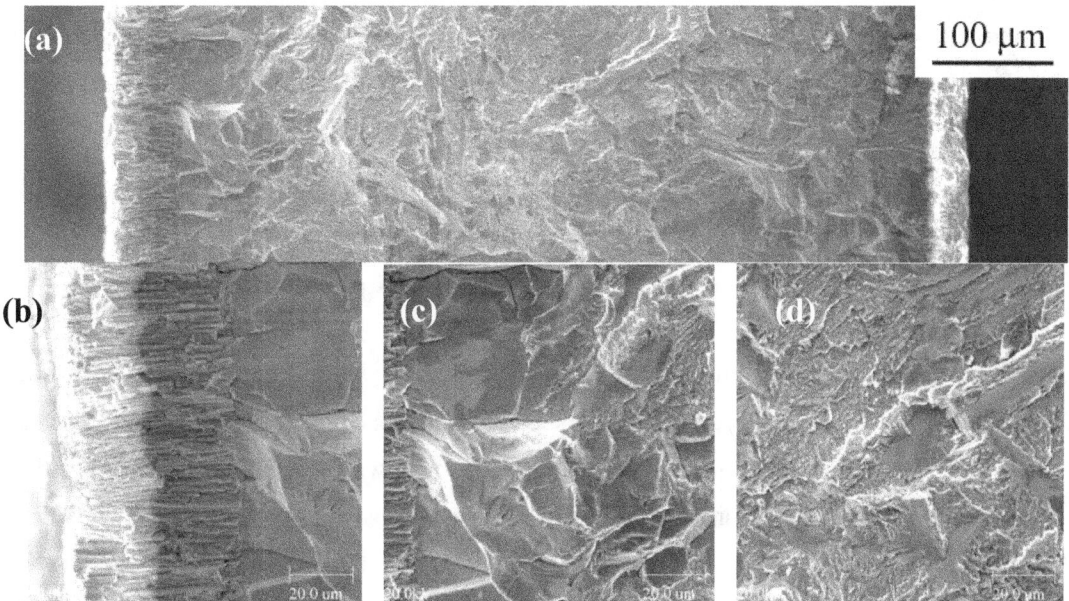

Fig. 14.    Fractography of ICL#3 sample at the location of 20 mm above the burst midplane.

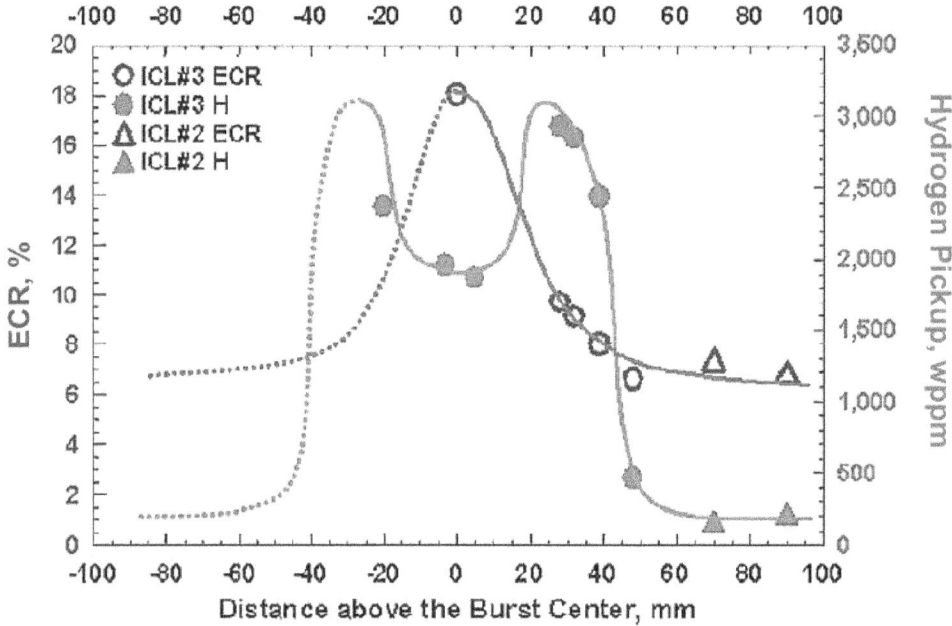

Fig. 15. Axial distributions of hydrogen and ECR for in-cell LOCA tests ICL#2 and ICL#3 with high-burnup fueled BWR samples.

During the post-test sample handling, the ICL#3 sample was broken three times. Cross-sectional SEM examinations were conducted for fractographic analysis at locations A and C of Fig. 9. Figure 14 shows typical fractographs of the oxide, the alpha phase, and the prior-beta-phase layers for the sample at location A. The failure region appears to be nearly brittle even in the prior-beta-phase area. No significant differences are observed between locations A and C.

Secondary Hydriding

Hydrogen is released during inner-surface oxidation within the balloon region, particularly near the burst region. Because of the resistance to flow through the small burst opening, a relatively high fraction of this hydrogen remains within the sample and migrates up and down the sample toward the burst neck-and-beyond regions. For the unirradiated samples, there is little resistance to this migration, and the bare cladding inner surface absorbs a large amount of hydrogen (see Fig. 4). Qualitatively, the same behavior would be expected for fresh and low-burnup fuel cladding. For high-burnup fuel, the axial extent of hydrogen that could come in contact with the cladding would be limited by the presence of the fuel, and the local hydrogen absorption would be limited by the fuel-cladding bond layer. This layer has been shown to not be protective with regard to steam oxidation within the ballooned region. It is of great interest to determine whether or not the layer is protective with respect to hydrogen diffusion into the cladding. Hydrogen and oxygen concentrations were measured for ICL#2 and ICL#3 samples sectioned from different axial positions below and above the burst center. The raw data give oxygen and hydrogen concentrations. The hydrogen pickup and ECR can be determined by applying Eqs. 1 and 2 given in reference 3, the results of which are shown in Fig. 15. These results are compared to the peak hydrogen-content locations and values for OCL#11. This shows a significant difference in the post-LOCA behavior between the axial distribution of hydrogen in high-burnup cladding vs. nonirradiated cladding. The peaks of the hydrogen pickup in the high-burnup specimens are shifted toward the burst center, compared with the unirradiated samples. However, the maximum hydrogen pickups are nearly at the same level for both high burnup ($\approx$3000 wppm) and unirradiated (3500 – 4000 wppm) specimens. More data are to be provided to map out the axial distribution of hydrogen pickup in high-burnup cladding subjected to the LOCA transient.

## Effects of Hydrogen on Post-quench-ductility for Prehydrided 15x15 Zry-4

While extensive literature data are available for unirradiated Zry-4, relatively little data have been published for high-burnup PWR samples, which have a higher hydrogen content than BWR during normal operation in the reactor. The high-temperature steam oxidation tests and PQD tests with prehydrided Zry-4 provide some guidance for planning high-burnup PWR LOCA integral tests. In this program, all non-irradiated samples (as-received 15x15 Zry-4 and prehydrided 15x15 Zry-4) are oxidized at the same heatup rates, hold times, and cooling rates (slowly cooled to 800°C and water-quenched). The 25-mm-long samples are exposed to two-sided steam oxidation prior to cooling. Also, the samples are compressed in the same Instron machine, and the load-displacement data are analyzed by a common method to determine ductility.

The 15x15 Zry-4 materials provided by Framatome ANP have an outer diameter of 10.75 mm and a wall thickness of 0.76 mm. Following steam oxidation and quench, 8-mm rings are cut from the 25-mm-long samples. Ring compression tests are performed at a cross-head displacement rate of 2 mm/min and room temperature, 100°C, and 135°C. The load-displacement curves are analyzed by the traditional offset-displacement method. The offset displacement, which is a measure of permanent displacement, is normalized to the outer diameter (10.75 mm) to give a nominal plastic hoop strain. Samples that exhibit offset strains ≥3% are considered to be ductile. However, for samples with <3% offset strain, another method is used to better determine permanent deformation and ductility. For this second method, the sample is unloaded after the first significant load drop, indicating through-wall failure along the length of the sample. The post-test diameter along the loading direction is measured directly and compared to the pre-test diameter to give a direct measure of permanent strain. For these low-offset-strain samples, the permanent diameter change in the loading direction provides a direct measure of ductility. Rings that exhibit ≥1% permanent diameter change are considered to be ductile.

Oxidation and quench have been completed for as-received and prehydrided Zry-4 samples oxidized at 1204°C. Weight gains were recorded for each sample, normalized to the oxidation surface area, and compared to CP predictions for weight gain (in mg/cm$^2$). For prehydrided samples hydrogen-content analyses have been performed before and after the steam oxidation tests. This characterization is performed to allow correlation between the ductility observed in the ring compression test and the microstructure (e.g., prior-beta-layer thickness, extent of alpha incursions into this layer, and the hydrogen content).

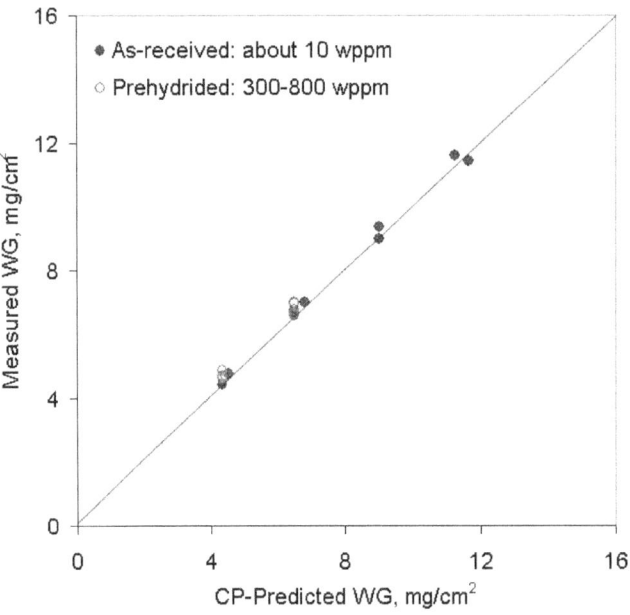

Fig. 16. Measured vs. predicted weight gain comparison of as-received and pre-hydrided samples for 15x15 Zry-4 after steam oxidation at ≈1204°C.

154

Weight Gain Kinetics

The weight gain data for both as-received and prehydrided 15x15 Zry-4, oxidized at 1204°C, were in excellent agreement with the CP-predicted weight gain values (see Fig. 16). At the longest test time, measured values were 11.45 mg/cm$^2$, as compared to the predicted value of 11.67 mg/cm$^2$. Using the measured wall thickness for each alloy, these weight-gain values convert to the "measured" ECR values of 13.2% in our tests. No significant hydrogen influence was found on the oxidation kinetics.

Post-quench-ductility Results

The load-displacement curves from the ring compression tests were analyzed by the offset method. Tests were stopped very shortly after the first significant load drop (≈30–50%), which indicated a through-wall crack along the length of the ring. The offset displacement prior to the first through-wall crack is determined mathematically by unloading the specimen at the elastic loading stiffness. Normalizing this offset displacement to the as-fabricated outer diameter gives an offset strain.

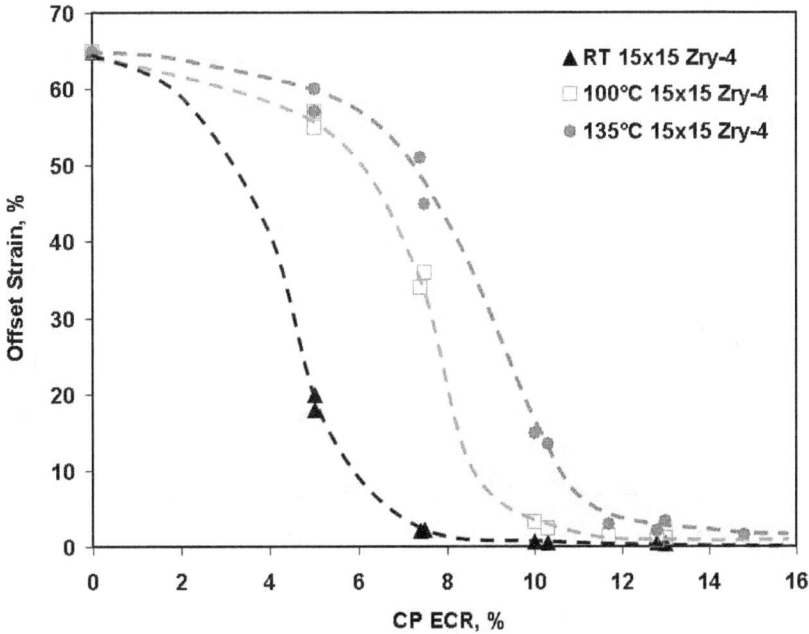

Fig. 17.   Offset strain vs. CP ECR for as-received 15x15 Zry-4 oxidized at 1204°C (H < 25 wppm)

To evaluate the temperature influence on PQD, 25-mm-long samples were oxidized at 1204°C up to 13% ECR. Each sample was then cut into three 8-mm-long rings and tested at room temperature (RT), 100°C, and 135°C. It was found that the threshold of PQD shifts toward higher ECR values as the test temperature increases. For oxidized, as-received 15x15 Zry-4 samples (H < 25 wppm), embrittlement occurred at 8-9% ECR at RT; 11-12% ECR at 100°C; and 13-14% ECR at 135°C. Figure 17 shows the offset strain vs. ECR for as-received 15x15 Zry-4 oxidized at 1204°C.

It is well known that the post-LOCA test samples can be very brittle due to hydriding effects (pre-test H content and the H pickup during the steam oxidation). The pre-test hydrogen content during normal operation in reactors for high-burnup PWR samples is quite high (400-800 wppm), compared to high-burnup BWR samples. Thus, steam oxidation tests with prehydrided 15x15 Zry-4 were performed at 1204°C, followed by the ring compression tests, to study the effects of hydrogen on post-quench ductility. The hydrogen charging was performed at 400°C in the flowing mixture gas 4% $H_2$-Ar, and the range of H-charging was carefully chosen to be close to the pre-test H content of the high-burnup PWR samples. Hydrogen analyses were conducted with the LECO hydrogen determinator on pre- and post-test samples.

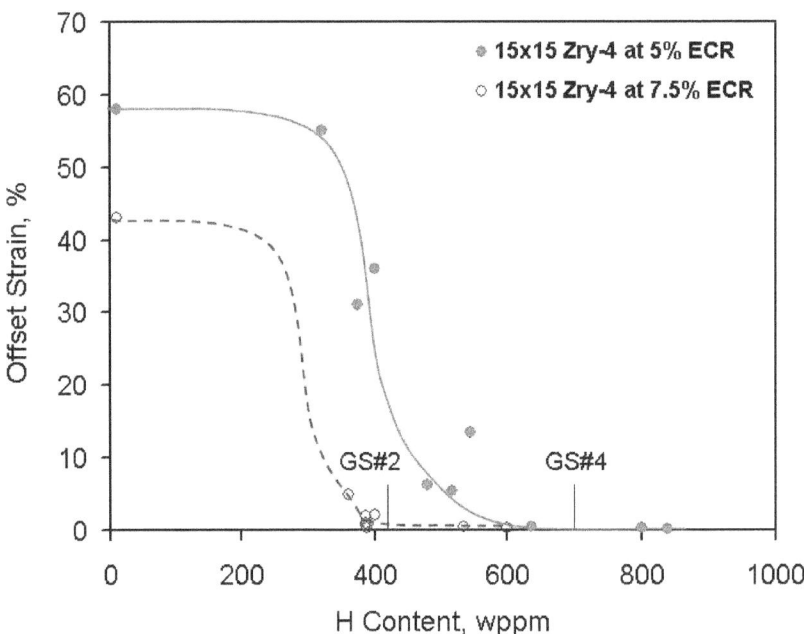

Fig. 18   Effects of hydrogen on post-quench ductility at 135°C for prehydrided 15x15 Zry-4 oxidized at 1204°C.

For samples oxidized at 1204°C for 5% CP-ECR, the samples remained ductile until the hydrogen content increased > 550 wppm. For the 7.5% CP-ECR samples, the samples became brittle when the hydrogen content reached 400 wppm. The plot in Fig. 18 shows the effects of hydrogen on PQD at 135°C for prehydrided 15x15 Zry-4 oxidized at 1204°C.

## Conclusions and Future Work

LOCA integral test results are reported for high-burnup BWR fuel specimens. These results are compared to baseline data for nonirradiated Zry-2 cladding specimens filled with zirconia pellets and exposed to the same tests. Four LOCA integral tests have been conducted with specimens from Limerick BWR fuel rods at 56 GWd/MTU, among which the ICL#4 specimen was exposed to the full LOCA sequence by heating in steam at 5°C/s to 1204°C, holding for 5 minutes at 1204°C (20% CP-ECR), slow-cooling at 3°C/s to 800°C, and bottom-flooding to quench the cladding from 800 to 100°C. The specimens were internally pressurized with helium to a gauge pressure of ≈8.3 MPa at 300°C. During heating in steam at 5°C/s, the internal pressure rose to ≤9 MPa prior to burst at ≈750°C.

Nondestructive examinations included photography and profilometry for all four specimens and post-test gamma scanning for the ICL#3 and #4 specimens. Destructive examinations were performed on ICL#2 and #3 specimens to determine oxide layer thickness, fuel morphology, and axial profiles of hydrogen and oxygen concentration. Based on measurements of cladding outer- and inner-surface oxide thickness at several axial locations, it appears that the presence of ≈10 μm of corrosion does not inhibit or slow down outer-surface oxidation and the presence of a fuel and fuel-cladding bond does not retard inner-surface steam oxidation. Oxide-layer thickness and oxygen-content results indicate two-sided oxidation in the ballooned-and-burst region of both high-burnup and nonirradiated specimens. With regard to steam oxidation, high-burnup Zry-2 behaved very similarly to nonirradiated Zry-2 during the LOCA transient. The major post-LOCA difference observed between high-burnup fuel cladding and nonirradiated cladding was the degree of secondary hydriding in the balloon neck region. For nonirradiated specimens, the hydrogen pickup was low in the burst region and very high (≈3900 wppm) at 70-90 mm above and below the burst mid-plane. For high-burnup fuel cladding, the hydrogen peak (≈3000 wppm) was toward the burst mid-plane. Because of the large secondary hydriding from the cladding inner surface, significant degradation of PQD is expected for the ICL#4 ballooned region, even at 100-135°C.

In the uniform burnup region (within grid spans 2-5), the high-burnup PWR cladding for the next set of LOCA tests differs from BWR cladding in terms of corrosion layer thickness ($\approx$40 to 100 $\mu$m) and hydrogen content ($\approx$400 to 800 wppm). For test planning purposes, the separate effects of hydrogen on diametral-compression PQD have been investigated with prehydrided, nonirradiated 15x15 Zry-4 cladding rings after oxidation at 1204°C and quench. For as-received ($\approx$10 wppm H) cladding that was oxidized at 1204°C, the ductile-to-brittle-transition CP-ECR was 8-9% at room temperature, 11-12% at 100°C, and 13-14% at 135°C. In contrast, cladding with 400-to-800 wppm hydrogen exhibited significant embrittlement, even after moderate oxidation at 1204°C. Samples prehydrided to 400-800 wppm and oxidized at 1204°C to 8% CP-ECR exhibited no ductility. With anticipated secondary hydrogen uptake from the cladding inner surface, the embrittlement ECR in the balloon region is expected to be $\ll$17% for high-burnup PWR specimens subjected to LOCA integral tests at 1204°C, even if the ECR is determined by the sum of the corrosion layer and the BJ-calculated transient ECR. The baseline data from prehydrided cladding are being used to plan the PWR hold times at 1204°C such that the embrittlement ECR can be determined effectively in the ballooned and non-ballooned regions of the PWR specimens from LOCA integral tests.

**References**

1. Y. Yan, R. V. Strain, T. S. Bray, and M. C. Billone, "High Temperature Oxidation of Irradiated Limerick BWR Cladding," Proceedings of the Nuclear Safety Research Conference (NSRC-2001), Washington, DC, October 22-24, 2001, NUREG/CP-0176 (2002) 353-372.

2. Y. Yan, R. V. Strain, and M. C. Billone, "LOCA Research Results for High-Burnup BWR Fuel," Proc. Nuclear Safety Research Conference (NSRC-2002), Washington, DC, October 28-30, 2002, NUREG/CP-0180 (2003) 127-155.

3. Y. Yan, T. Burtseva, and M. C. Billone, "LOCA Results for Advanced-alloy and High-burnup Zircaloy Cladding," Proc. Nuclear Safety Research Conference (NSRC-2003), Washington, DC, October 25-27

4. G. Hache and H. M. Chung, "The History of LOCA Embrittlement Criteria," Proc. 28th Water Reactor Safety Meeting, Bethesda, MD, October 23-25, 2000, NUREG/CP-0172 (2001) 205-237

# LOCA Testing at Halden

## E. Kolstad, W. Wiesenack, V. Grismanovs
### (OECD Halden Reactor Project)

The safety criteria for loss of coolant accidents are defined to ensure that the core will remain coolable. Since the time of LOCA experiments in the '70s, which were largely conducted with fresh fuel, changes in fuel design, the introduction of new cladding materials and in particular the move to high burnup have generated a need to re examine these criteria and to verify their continued validity. Hot cell programmes concentrating on embrittlement and mechanical properties of high burnup cladding have been initiated in some countries.

The Halden reactor is suited for integral in pile tests on fuel behaviour under LOCA conditions. It is intended to utilise fuel rods irradiated in commercial reactors to burnup levels >50 MWd/kg with a thorough characterisation regarding the state of the cladding and the bonding with the fuel. Participating organisations have supplied both PWR and BWR fuel with desired characteristics. It is the intention to include medium burnup (40 45 MWd/kg) fuel in the test series in order to assess the difference between medium burnup and very high burnup fuel (>60 MWd/kg.

The Halden experiments are single pin tests and will focus on effects that are different from those studied in out of reactor tests. A prototypical bounding LOCA transient does not exist, and it was recommended that the test conditions be selected to meet the following primary objectives:

to maximise the balloon size to promote fuel relocation, and to evaluate its possible effect on cladding temperature and oxidation

to investigate the extent (if any) of "secondary transient hydriding" on the inner side of the cladding around the burst region.

Target peak clad temperatures (PCT) for the pre irradiated rods have been set at 800°C and 1100°C for high and medium burnups.

The first LOCA trial runs were carried out in the Halden reactor in May 2003, using a fresh, unpressurised PWR rod with Zr 4 cladding. The main objective was practicing, to determine how to run the later experiments with pre irradiated segments. PCTs in the range 800°C 1100°C for the initial six transients were successfully achieved.

The rig with the fuel rod was located in a high pressure flask connected to an ex reactor high pressure loop incorporating a blow down system. The cladding temperature transients can be controlled by rod power and an annular heater surrounding the rod. A spray system at the top of the rig is used to supply steam for the oxidation and hydriding processes. The extensive rig/rod instrumentation enables power calibration and neutron flux monitoring, and includes a fuel centre thermocouple (first rod only), three cladding thermocouples (at two elevations), rod pressure sensor, cladding extensometer, heater thermocouples etc. The geometry of the test section is shown in Fig. 1.

The second trial LOCA test run was successfully carried out on 28 May this year. The test was performed with a fresh, pressurised PWR rod and consisted of a blowdown phase, heat up, hold at target PCT and termination by reactor scram (Fig. 2). The main objective was to achieve ballooning and cladding failure to find out how to run later experiments with pre irradiated rodlets.

The target cladding temperature of 1050°C was achieved, and rod rupture occurred at 800°C, as evidenced by rod pressure and elongation measurements (Fig. 3) as well as the gamma monitor on the blowdown line to the dump tank. The hold time above 900°C was 390 s and the average temperature increase rate between 600 and 800°C was ~7°C/s. The azimuthal temperature variation was small prior to cladding failure, within ± 2 3°C, and the tensile hoop stress ~55 MPa. The spray was applied intermittently during the high temperature period and the test was terminated by a reactor scram. The rod with its capsule will undergo gamma scanning at Halden before it is shipped to Kjeller hotcells for detailed PIE.

Pre test calculations were carried out by VTT using the FRAPTRAN/GENFLO code. The code predicted the maximum cladding temperature with good accuracy. Also the timing and temperature of the rod failure was well predicted. Further calculations will be performed in preparation of the next test (the first with a pre irradiated PWR segment).

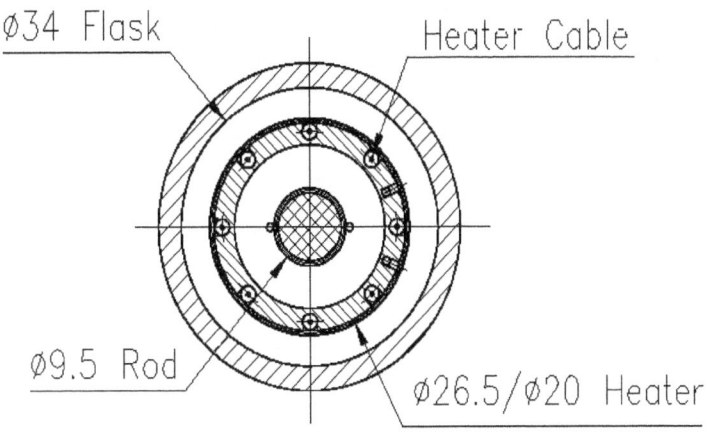

*Fig. 1. The geometry of the test section*

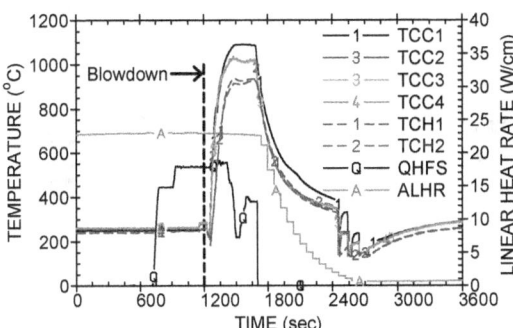

*Fig. 2. Cladding, heater and loop inlet and outlet temperatures during the LOCA test. Also fuel and heater power is shown. Nomenclature: A = fuel power, Q = heater power, I = inlet loop temperature, dashed 1 2 = heater temperature*

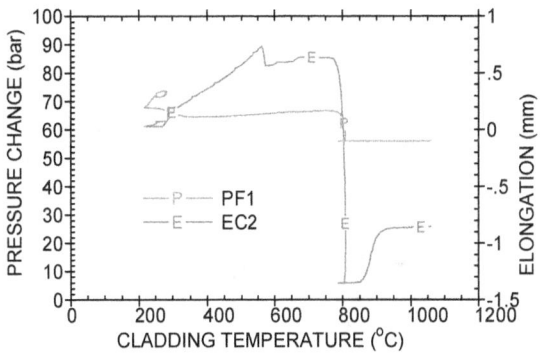

*Fig. 3. Measurements of rod pressure and cladding elongation as function cladding temperature during the LOCA test (Rod rupture at 800°C)*

Institutt for energiteknikk
**OECD HALDEN REACTOR PROJECT**

# LOCA TESTING AT HALDEN

Second in-pile test in IFA-650.2

*E. Kolstad, W. Wiesenack, V.Grišmanovs*

*OECD Halden Reactor Project*

*Nuclear Safety Research Conference*

*WASHINGTON DC, October 25th-27th, 2004*

File    *NSRC-2004-VGr*

**Institutt for energiteknikk**
**OECD HALDEN REACTOR PROJECT**

# CONTENT OF PRESENTATION

- Background and objectives
- Test facility and instrumentation
    - Test rig and test rod
    - Outer loop
- Pre-test code calculations
- Test execution and results
- Summary
- Further plans and test objects

**Institutt for energiteknikk**
**OECD HALDEN REACTOR PROJECT**

# NEED FOR LOCA STUDIES

The move to high burnup and the introduction of new cladding materials have generated a need to re-examine the safety criteria for LOCA and to verify their continued validity.

The in-pile tests in the Halden reactor will address LOCA issues using ex-LWR high burnup fuel segments.

# RESEARCH ON THERMAL-HYDRAULICS, CRITICAL HEAT TRANSFER AND LOCA AT HALDEN REACTOR

**1963-68**　Experiments on natural convection flow instabilities and dry-out limit

**1965-72**　Dry-out experiments in natural convection flow channels

**1979-83**　Safety-related tests:　Blow-down, heatup & quench behaviour of nuclear rods and electric simulators

　　　　　IFA-511:　Thermal response studies

**1982-85**　Safety-related tests:　Blow-down, heatup & quench behaviour of nuclear rods and electric simulators

　　　　　IFA-54x:　Ballooning and rod-to-rod interaction studies

**1996-98**　IFA-613:　Short-term dry-out test series

165

Institutt for energiteknikk
**OECD HALDEN REACTOR PROJECT**

# BASIC CONCEPTS OF LOCA

- **3 Phases:**

  - Blowdown (fuel/core uncovered), de-pressurisation

  - Refill (ECCS systems start)

  - Reflood (water level above core top)

- **Timing:** uncovery, quenching, long-term cooling

- **After-effects:**

  fuel temperature rises - - > cladding oxidation and hydriding - - >
  embrittlement (melting) of cladding - - > fuel fragmentation

  → **Safety criteria:**

  - Peak Clad Temperature (PCT)    < 2200 F (1204 C)
  - Oxidation of cladding          < 17% of its thickness

- Rod swelling/ballooning may alter core geometry/coolability

  → **Safety requirement:**

  calculated geometry changes must still warrant core cooling

File    *NSRC-2004-VGr*

166

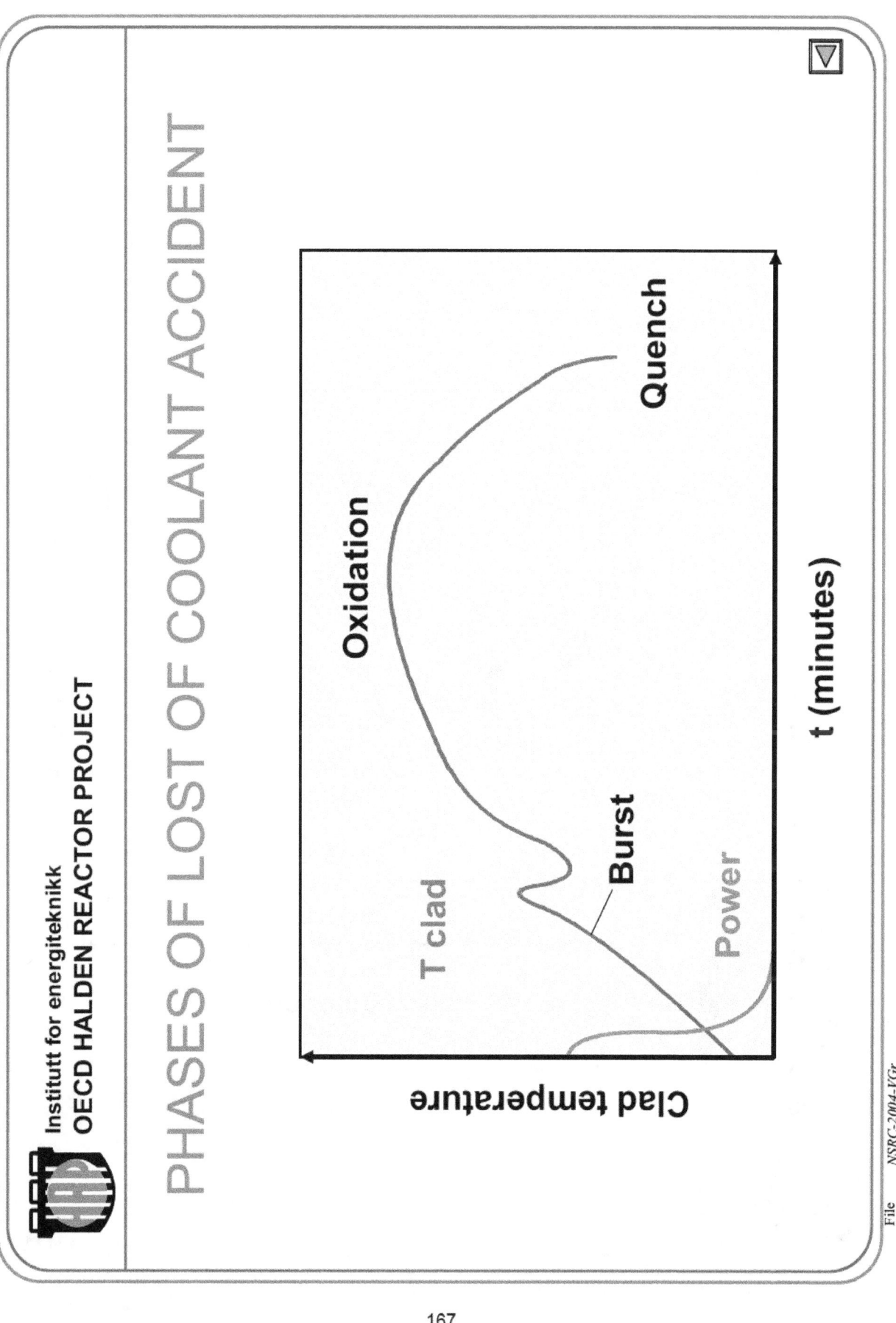

# PHASES OF LOST OF COOLANT ACCIDENT

File    NSRC-2004-VGr

# CREEP OF ZIRCALOY (MATPRO)

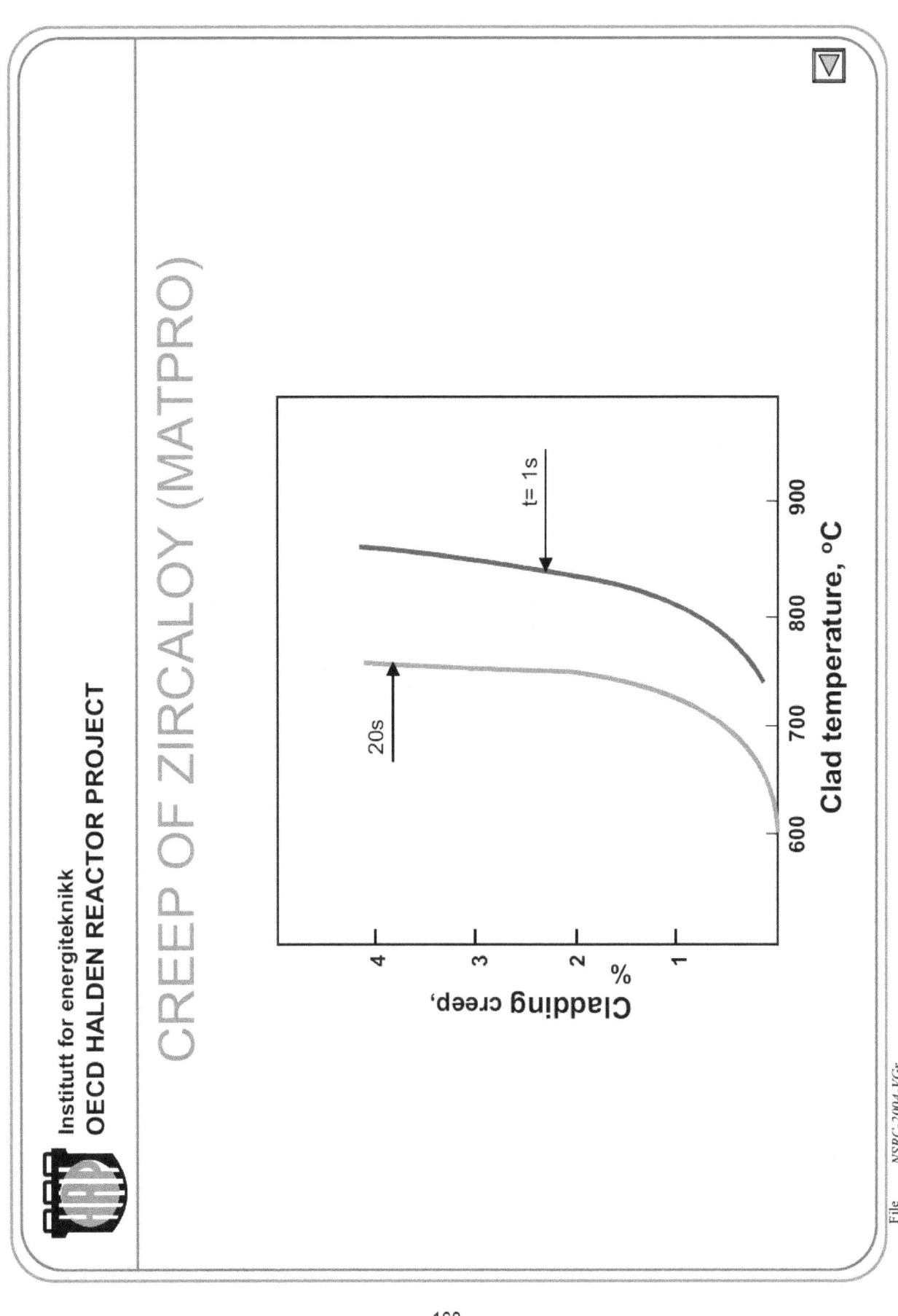

# OXIDATION OF ZIRCALOY IN STEAM (MATPRO)

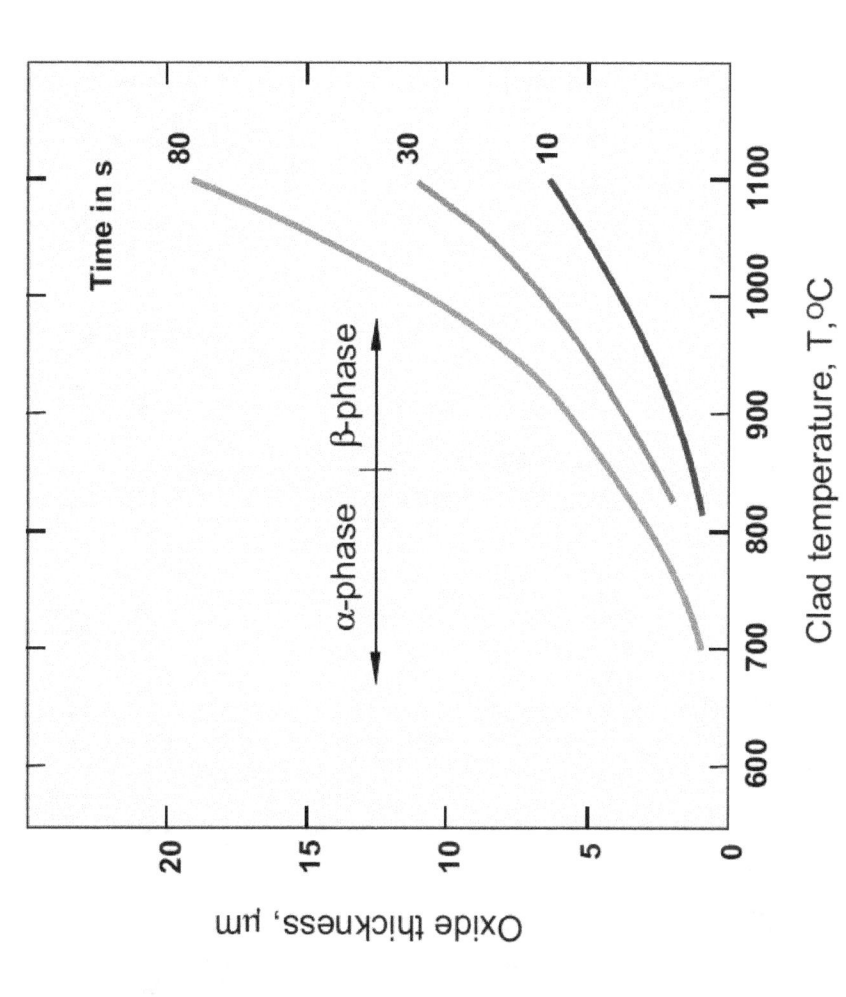

Institutt for energiteknikk
**OECD HALDEN REACTOR PROJECT**

# HALDEN LOCA EXPERIMENT (IFA-650.2)
## (USNRC/EPRI/IRSN/EDF/FRAMATOM-ANP/GNF)

## Primary objectives

- Measure the extent of fuel (fragment) relocation into the ballooned region and evaluate its possible effect on cladding temperature and oxidation

- Investigate the extent of – "secondary transient hydriding" - on the inner side of the cladding above and below the burst region

**Institutt for energiteknikk**
**OECD HALDEN REACTOR PROJECT**

# SETUP OF THE LOCA TEST

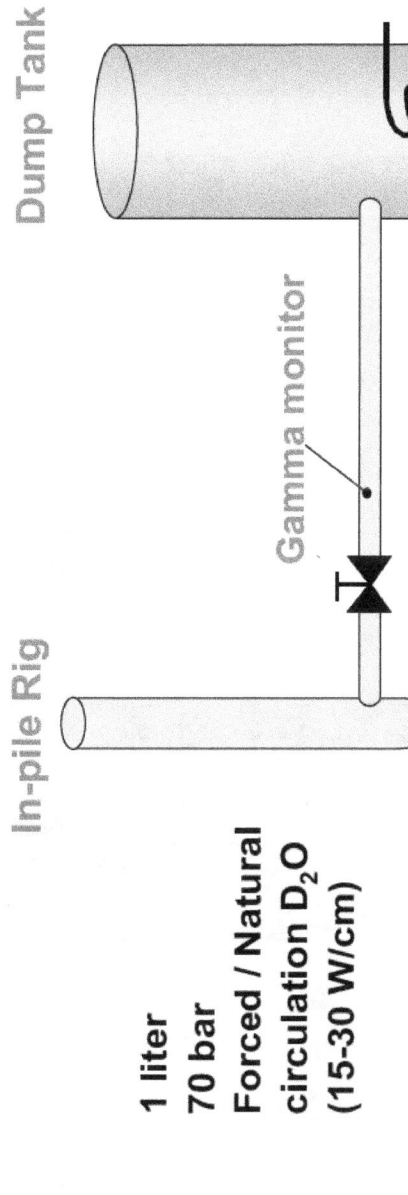

In-pile Rig

Dump Tank

Gamma monitor

1 liter
70 bar
Forced / Natural
circulation D$_2$O
(15-30 W/cm)

**Blow-down line**

° 16 m long
° 6-9 mm ID
° Gamma monitor
° Blow-down rate control

° Shielded tank
° Cooled
° Slight over-
  pressure (2-3 bar)
° Level gauge

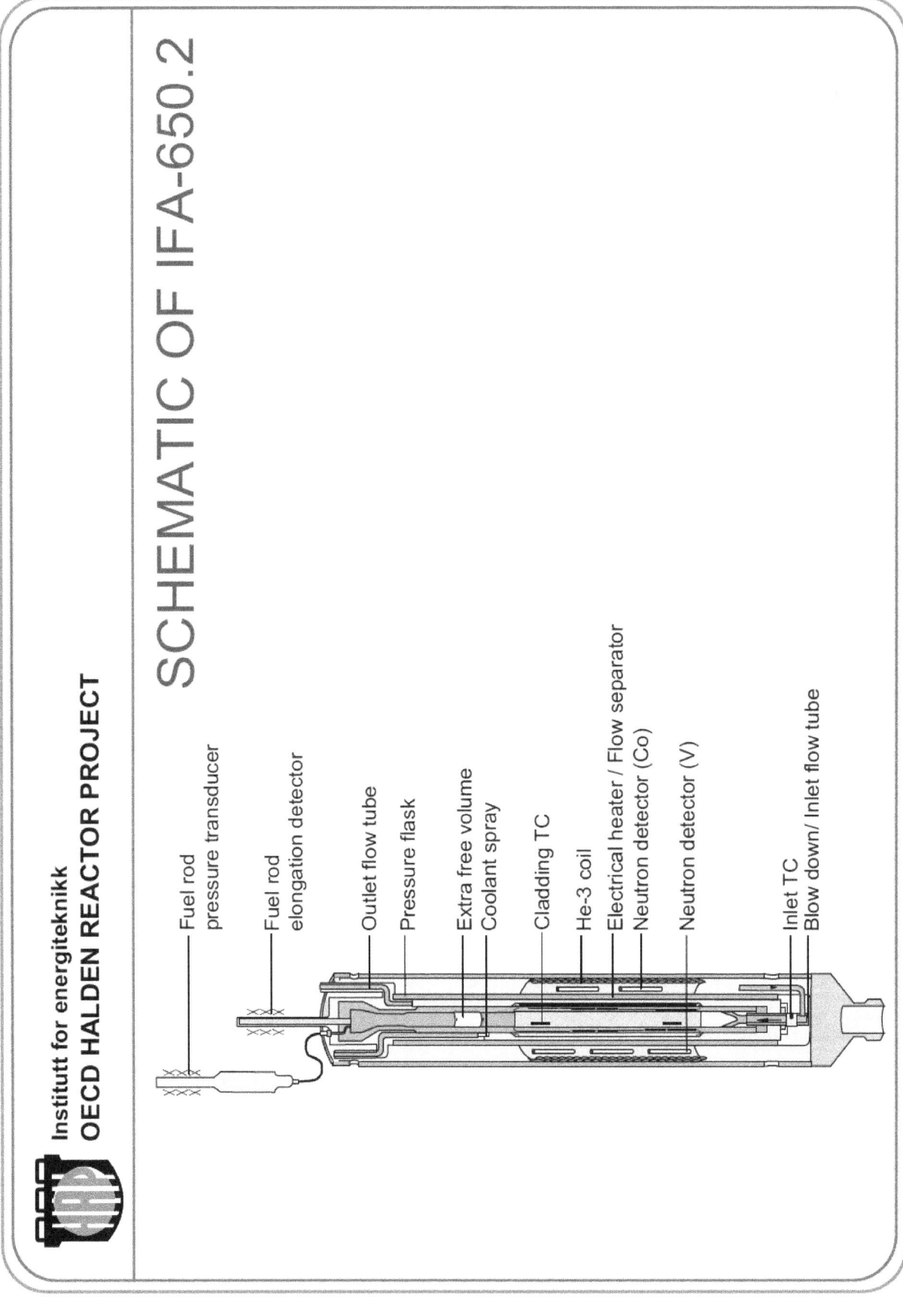

Institutt for energiteknikk
**OECD HALDEN REACTOR PROJECT**

# SCHEMATIC OF IFA-650.2

Fuel rod pressure transducer

Fuel rod elongation detector

Outlet flow tube

Pressure flask

Extra free volume

Coolant spray

Cladding TC

He-3 coil

Electrical heater / Flow separator

Neutron detector (Co)

Neutron detector (V)

Inlet TC

Blow down/ Inlet flow tube

File    NSRC-2004-VGr

Institutt for energiteknikk
**OECD HALDEN REACTOR PROJECT**

# CROSS SECTION OF IFA-650.2

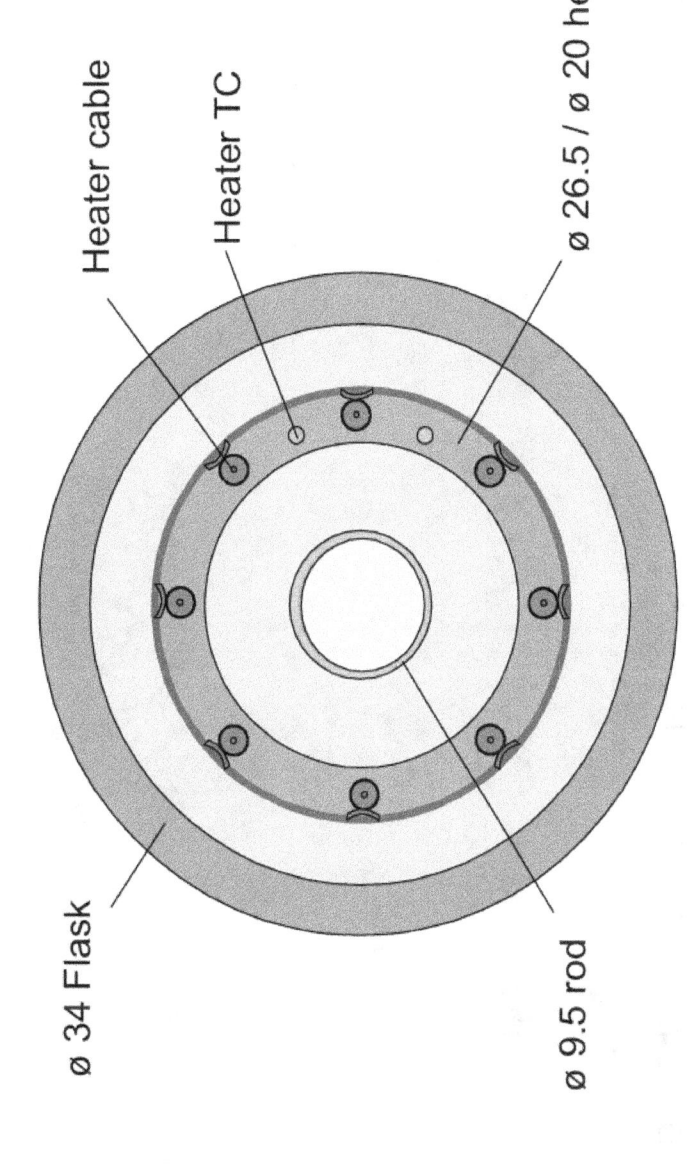

Heater cable

Heater TC

ø 26.5 / ø 20 heater

ø 34 Flask

ø 9.5 rod

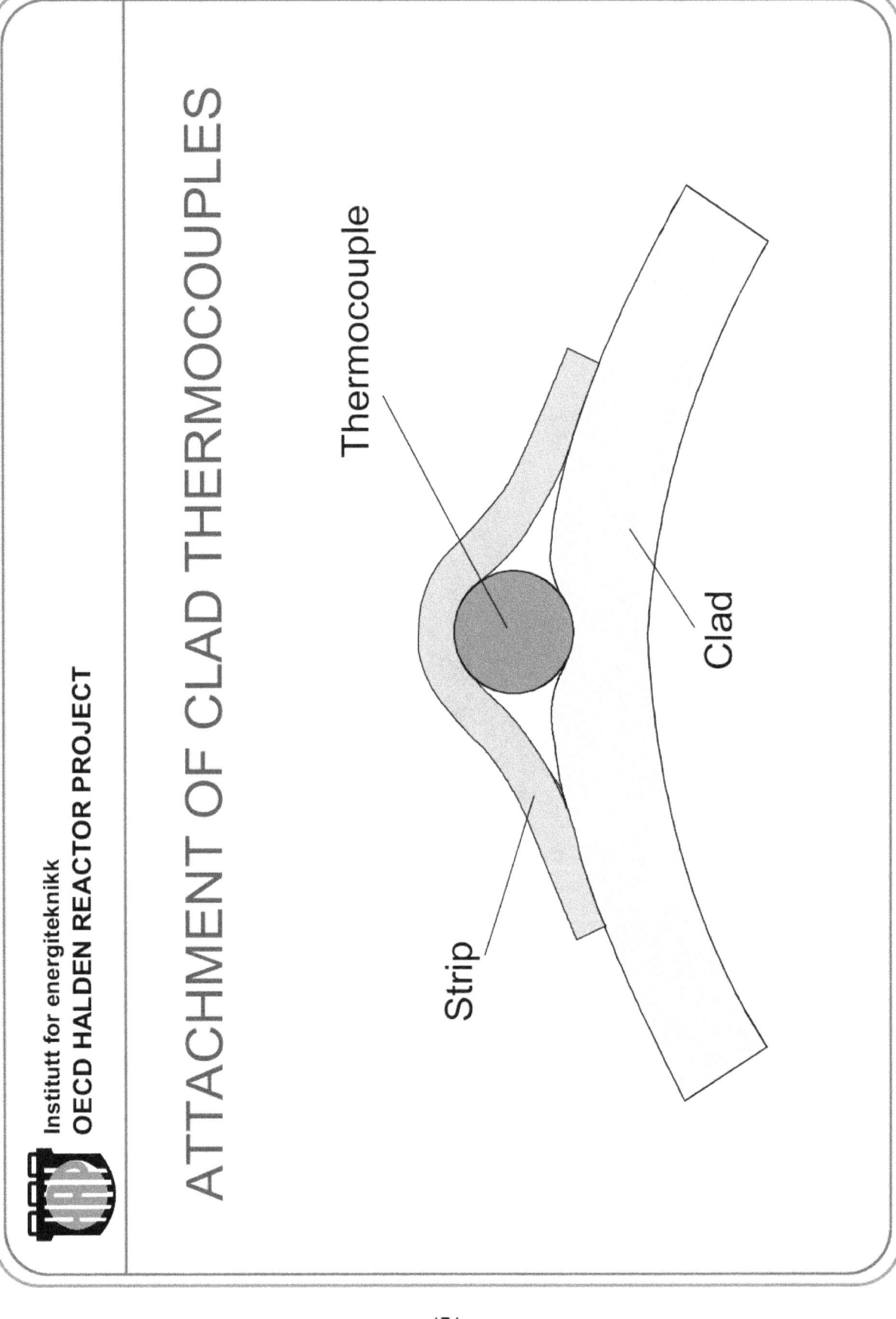

**Institutt for energiteknikk**
**OECD HALDEN REACTOR PROJECT**

# PRE-TEST CODE CALCULATIONS

- TRAC-BF1 code (PSI)

- FRAPTRAN – GENFLOW (VTT)

- SCTEMP and ALGOR (Halden)

# LOCA TRIAL RUNS: IFA-650.1

## Test rod and instrumentation

### Rod (Zry-4)

| | |
|---|---|
| Length: | 50 cm |
| OD / ID: | 9.50/8.36 mm |
| Gap size: | 70 µm |
| Enrichment: | 4 wt% U-235 |
| Fill pressure: | 2 bar He |
| Free volume: | 15 cm$^3$ |

Dished pellets

### Instrumentation

3 Clad thermocouples

1 Clad extensometer

1 Fuel thermocouple

1 Rod pressure sensor

2 Heater thermocouples

Institutt for energiteknikk
**OECD HALDEN REACTOR PROJECT**

# LOCA TRIAL RUNS: IFA-650.1

- Blow-down tests at zero power
- Power calibrations (reactor power ~18 MW)
- Trial runs (reactor power 5-6 MW)

<u>Six test runs:</u>

- 14 W/cm (rod) + 6 W/cm ( heater)  (~ 830°C)
- 14 W/cm (rod) + 6 W/cm ( heater)  (~ 830°C)
- 14 W/cm (rod) + 12 W/cm ( heater)  (~ 900°C)
- 25 W/cm (rod) + 6 W/cm ( heater)  (~ 930°C)
- 25 W/cm (rod) + 18 W/cm ( heater)  (~1030°C)
- 30 W/cm (rod) + 20 W/cm ( heater)  (~1120°C)

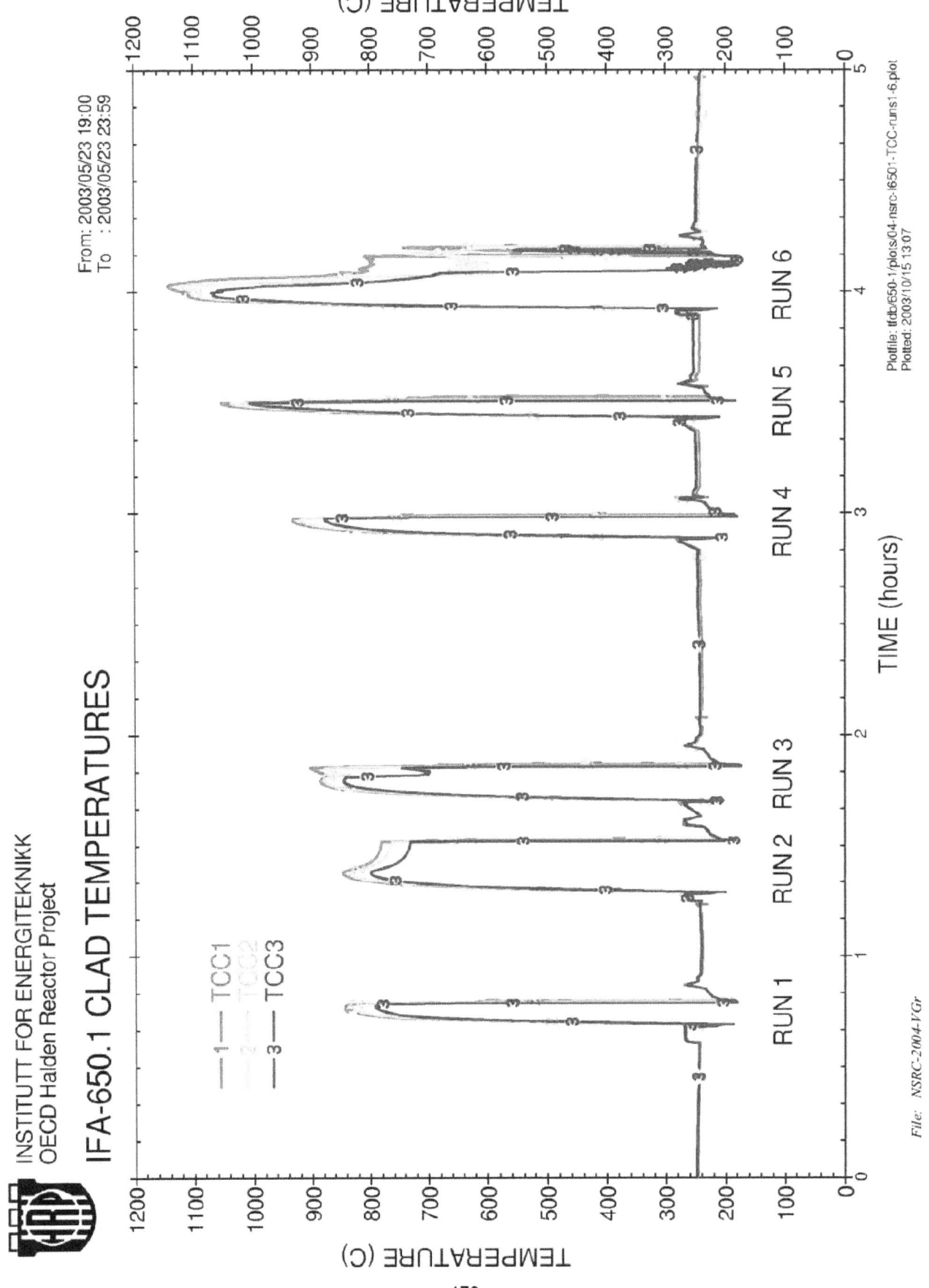

**Institutt for energiteknikk**
**OECD HALDEN REACTOR PROJECT**

# LOCA TRIAL RUN: IFA-650.2

## Test rod and instrumentation

### Rod (Zry-4)

| | |
|---|---|
| Length: | 50 cm |
| OD / ID: | 9.50/8.36 mm |
| Gap size: | 70 µm |
| Enrichment: | 2 wt% U-235 |
| Fill pressure: | 40 bar He (RT) |
| Free volume: | 15 cm³ |

Dished pellets

### Instrumentation

4 Clad thermocouples

1 Clad extensometer

1 Rod pressure sensor

2 Heater thermocouples

Institutt for energiteknikk
**OECD HALDEN REACTOR PROJECT**

# LOCA TRIAL RUN: IFA-650.2

- Blow-down tests at zero power
- Power calibration (reactor power ~17 MW)
-                      (reactor power   8.6 MW)
- Trial run

<u>One run:</u>

22 W/cm (rod) + 17 W/cm ( heater)   (~1050°C)

Institutt for energiteknikk
**OECD HALDEN REACTOR PROJECT**

# IFA-650.2 Fuel rod power history

R - Reactor power
F - Fuel rod ALHR

LOCA test

REACTOR POWER (MW)

FUEL ROD ALHR (kW/m)

TIME (day)

File    NSRC-2004-VGr

181

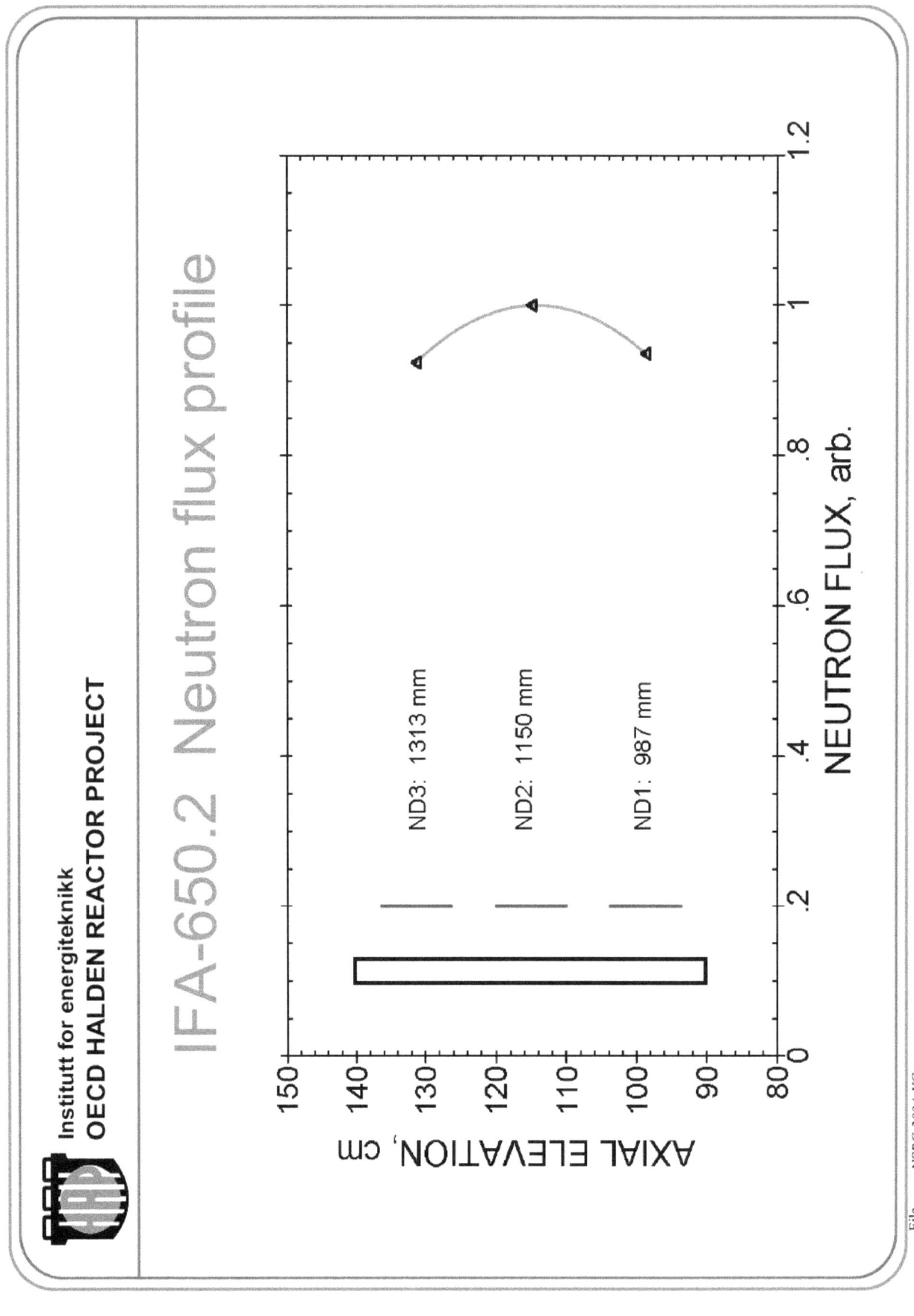

File    NSRC-2004-VGr

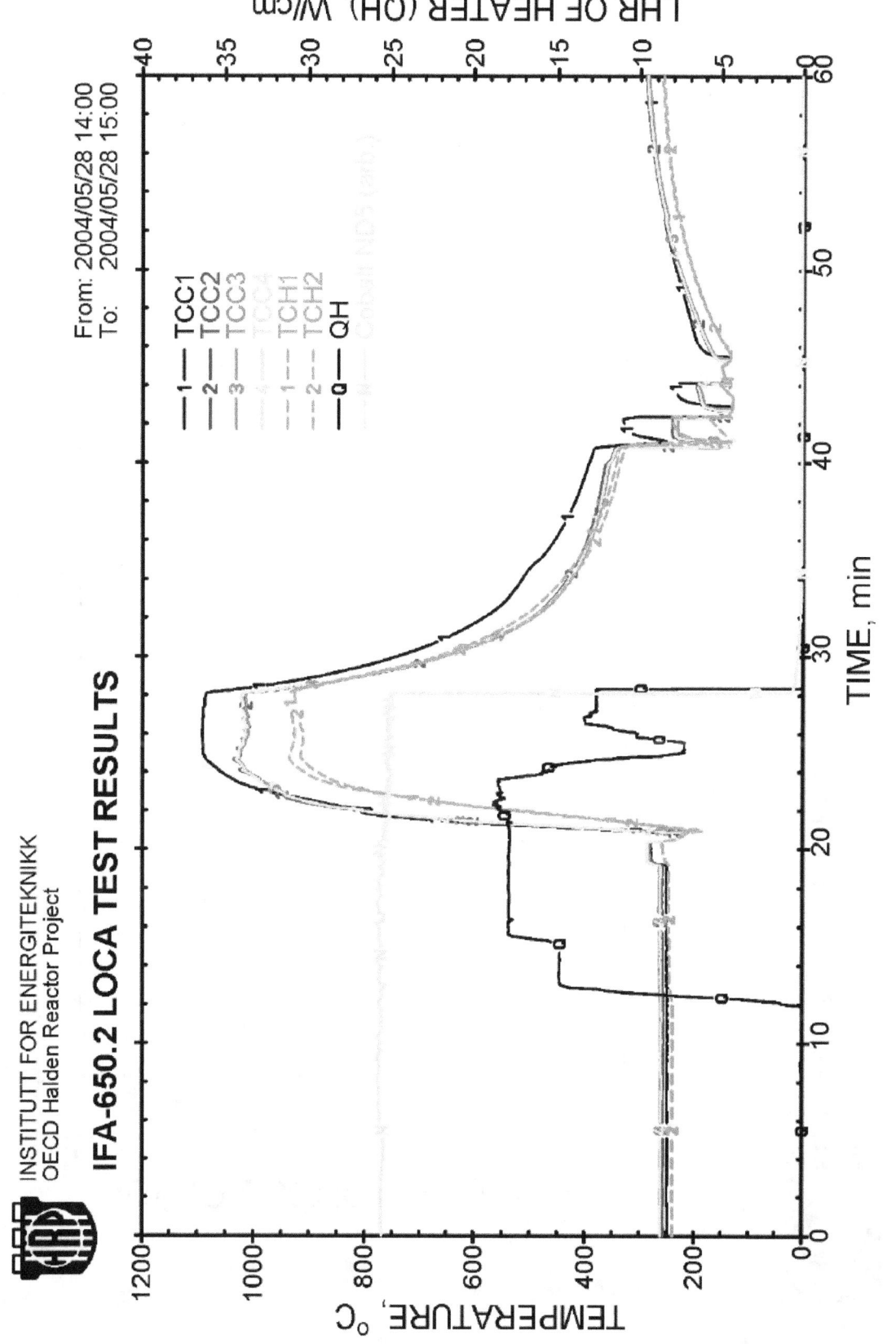

IFA-650.2 Rod pressure and clad elongation

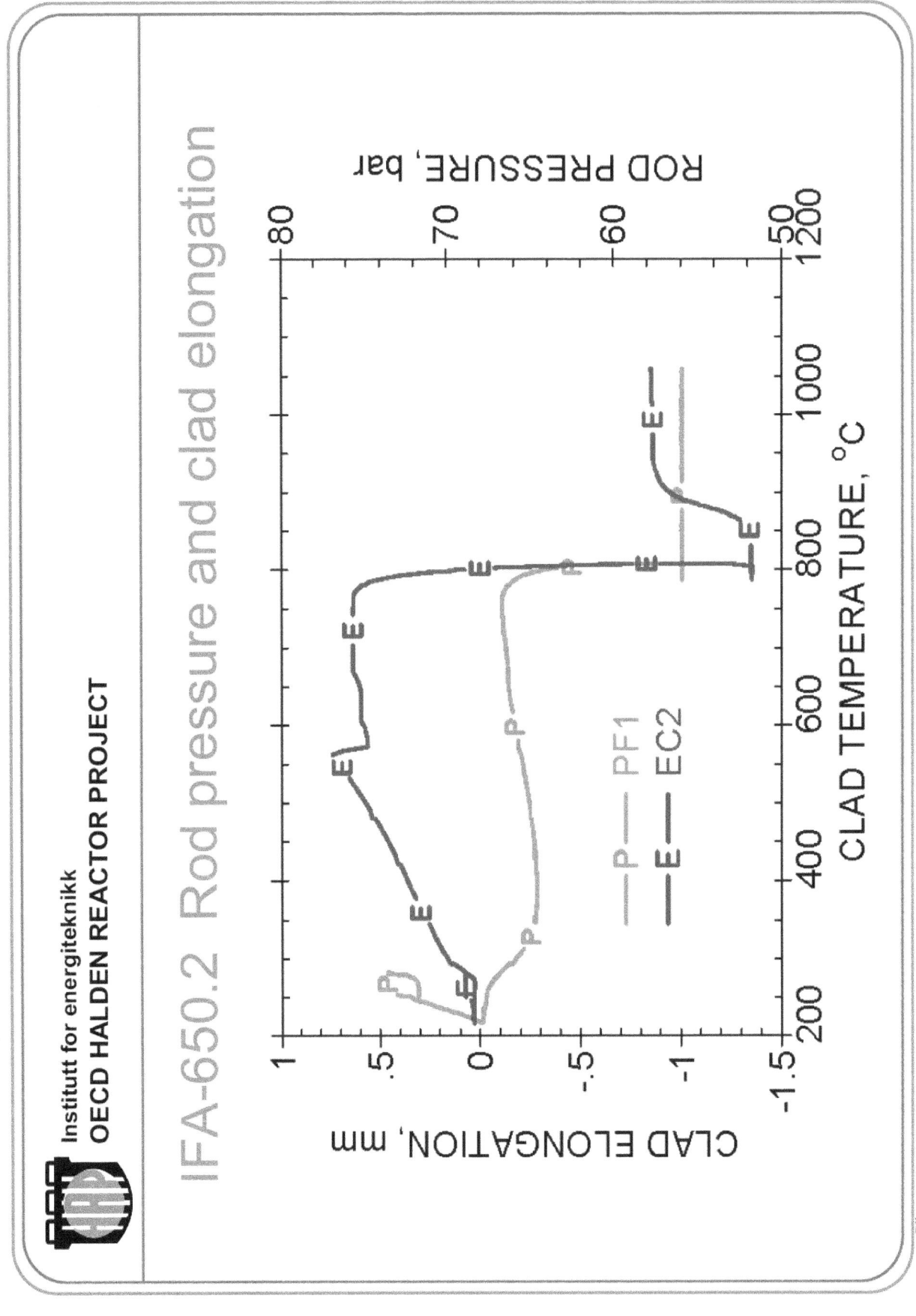

File    NSRC-2004-VGr

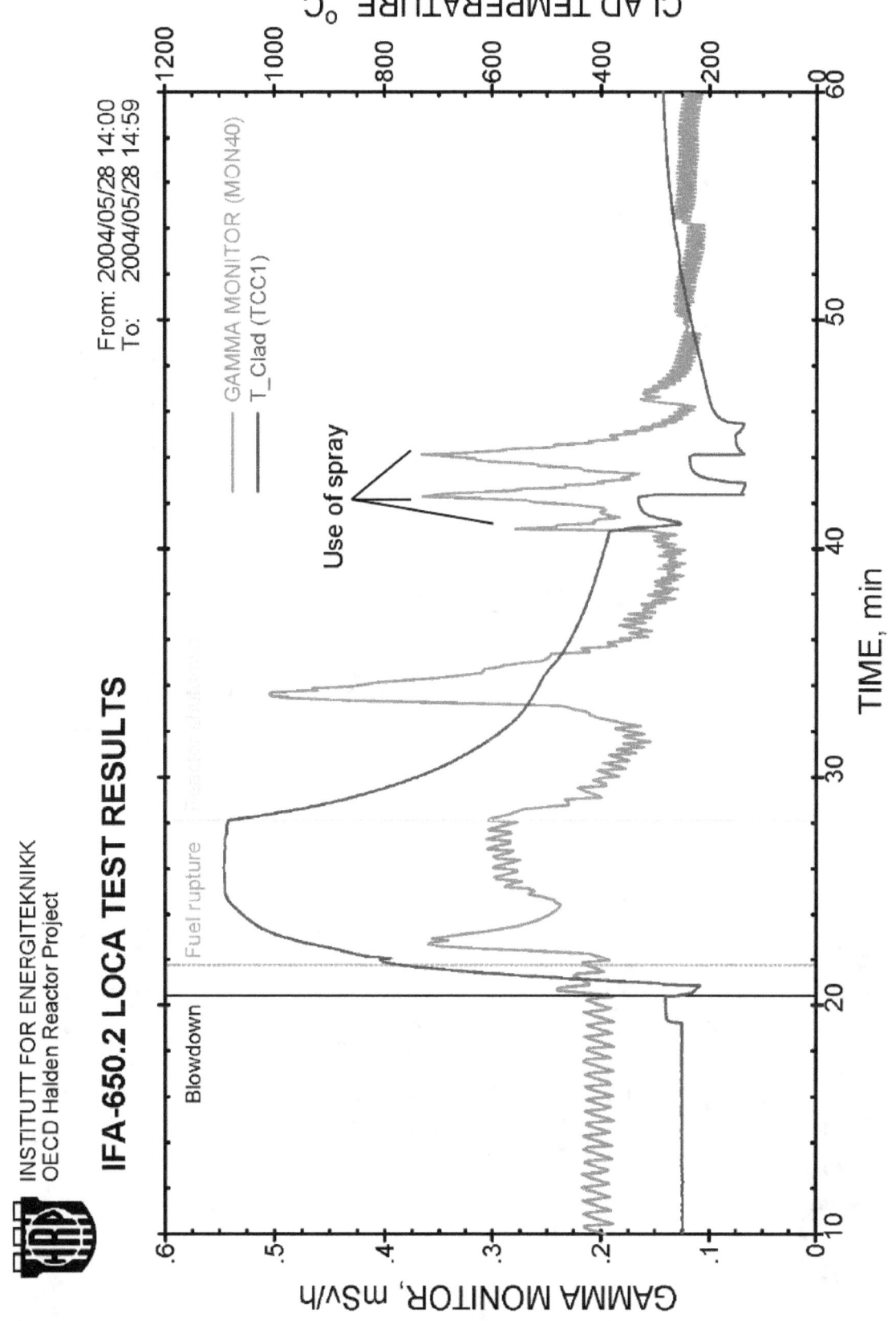

# Summary of 2$^{nd}$ in-pile test in Halden (1)

- Loop and rig, and rod instrumentation worked well

- Target PCT ~1050 °C was achieved

- Rod rupture was detected at 800 °C

- Hoop stress at failure: ~55 MPa

- Clad temperature increase rate: 7.8 °C/s

- Small azimuthally temperature variation: +/- 3 °C

- Holding time above 900 °C: 6.5 minutes

- Water spray was applied intermittently above 900 °C

- Test was terminated by reactor scram

Institutt for energiteknikk
**OECD HALDEN REACTOR PROJECT**

# Summary of 2<sup>nd</sup> in-pile test in Halden (2)

## Further actions

- Gamma-scanning at Halden

- Detail PIE at Kjeller hot cells

  - oxide thickness (axial distribution: inside/outside)

  - hydride content in balloon area and its vicinity

  - characterization of balloon (dimensions, shape)

# FURTHER PLANS AND TEST OBJECTS

- Pairs of rods (50 - 80 MWd/kg, 40-50 cm length) from commercial LWRs to be tested in PWR and BWR conditions, suitable to address possible effect of axial fuel fragment relocation
  - when does it occur (heat-up, quenching)?
  - does bonding prevent the movement of fragments?

- Include medium burnup fuel (~40 MWd/kg, less or no bonding) to bridge the gap between low and high burnup

- VVER fuel envisaged for testing at a later stage

# LOCA testing at Halden,
# Second in-pile test in IFA-650.2

**E.Kolstad, V.Grišmanovs, W.Wiesenack**
OECD Halden Reactor Project

## ABSTRACT

The safety criteria for loss-of-coolant accidents (LOCA) are defined to ensure that the core will remain coolable. Since the LOCA experiments that were performed in the 1970s largely with fresh fuel, changes in fuel design, the introduction of new cladding materials and in particular the move to high burn up have generated a need to re-examine these criteria and to verify their continued validity. The Halden reactor is suitable for integral in-pile tests on fuel behaviour under LOCA conditions. It is aimed to utilize BWR and PWR fuel rods irradiated in commercial reactors to burn up levels over 80 MWd/kg, with a thorough characterization of the cladding and its bonding with the fuel. There is an intention to include medium burnup fuel (40-45 MWd/kg) in the test series in order to bridge the gap between the low and high burn up fuels.

The second trial experiment on LOCA was successfully carried out in May 2004. The test was performed with a fresh pressurized PWR rod and consisted of a blow down phase, a heat-up, a hold at target peak clad temperature (PCT) and termination by a reactor scram. The main objective was to achieve ballooning and cladding failure to gain experience that will be used to run later experiments with pre-irradiated fuel rods. The PCT of 1050 °C was achieved and rod rupture occurred at 800 °C as evidenced by rod pressure and elongation measurements, as well as by gamma monitoring of the blow down line to the dump tank. The rod with its capsule will undergo gamma scanning at Halden and then be shipped to Kjeller hot cells for detailed post-irradiation examination (PIE).

Pre-test calculations were carried out by VTT using the FRAPTRAN/GENFLO code. The code predicted the maximum cladding temperature with good accuracy. Also the timing and temperature of the rod failure were well predicted.

# 1. INTRODUCTION

The move to high burnup fuels, new fuel designs and introduction of new cladding materials have generated a need to re-examine and verify the validity of the safety criteria for LOCA. The LOCA tests at Halden are integral, single pin, in-pile tests using high burnup fuel rods irradiated in commercial reactors. Participating organizations in the Halden Project have offered both PWR and BWR fuels with the desired high burnup. It has also been proposed to include medium burnup (40-45 MWd/kg) fuel in the test series in order to bridge the gap between the low and high burnup fuels. In future, tests with VVER-fuel are being considered.

The test conditions are planned to meet the following primary objectives:
- to maximize the ballooning size to promote fuel relocation and to evaluate its possible effect on cladding temperature and oxidation;
- to investigate the extent (if any) of "secondary transient hydriding" of the cladding around the burst region.

Target peak clad temperatures for the pre-irradiated rods have been set at 800°C and 1100°C for high and medium burnups, respectively. The lower temperature is used because high burnup fuel is not expected to reach higher clad temperatures in a LOCA. The higher temperature is quite close to the current LOCA temperature limit prescribed by regulations.

The first LOCA trial test runs were carried out in the Halden reactor in May 2003 using a fresh, tight-gap and unpressurised PWR rod with Zry-4 cladding. The main objective was to gain experience in the operation of the rig and to determine how to run the later experiments with pre-irradiated fuel rods. The PCTs were successfully achieved and the test gave a good basis for further experiments.

This paper deals with the second trial experiment on LOCA performed with a fresh pressurized PWR rod. The main objective was to achieve ballooning and cladding failure to gain experience on how to run future experiments with pre-irradiated fuel rods.

# 2. EXPERIMENTAL

## 2.1 Fuel and rig

In IFA-650.2, the 50 cm long fuel rod was located in the center of a high-pressure flask connected to a heavy water loop and a blow-down system. The latter is located outside the reactor; it has a level gauge and is cooled and shielded. The test rod had a Zry-4 cladding with an outer diameter of 9.50 mm and a wall thickness of 0.57 mm, and contained dished fuel pellets made of 2 wt% enriched $UO_2$. The diametral gap between the fuel and cladding was about 70 $\mu$m. The fuel rod was pressurized with helium at 40 bar (room temperature). A heated flow separator and the pressure flask surrounded the fuel rod. A schematic of the rig is presented in Figure 1. Figure 2 shows the cross-section of the test channel. The heating is provided from within the fuel rod and by the heater surrounding the fuel rod. The heater is used to simulate the thermal boundary conditions. The cladding temperature can be controlled by adjustment of the rod and heater powers.

The test rig instrumentation consists of two heater thermocouples, two inlet and outlet coolant thermocouples, a flow meter, three self-powered vanadium neutron detectors (NDs) and two fast response cobalt NDs. The two embedded heater thermocouples are located above and below the axial mid height of the fuel rod. The volume flow rate is measured in the external loop. The three vanadium NDs are placed at three elevations to measure the axial power distribution. Rapid power changes are monitored using the two cobalt NDs.

The rod instrumentation includes four cladding surface thermocouples, a cladding extensometer and a pressure sensor. Three cladding thermocouples are circumferentially located at the middle of the upper half of the fuel segment and one at the middle of the lower half of the fuel segment. The three circumferentially located thermocouples enable study of the azimuthal temperature distribution of the cladding. The cladding extensometer and the pressure sensor are located at the top of the rod.

## 2.2 Test execution and results

The second trial LOCA test in IFA-650.2 was carried out in May 2004. The irradiation history of the fuel rod is shown in Figure 3. Before the LOCA test, two power ramps were performed to pre-crack the fuel pellets. The fuel rod was irradiated for 1.5 days to accumulate fission products. The axial neutron flux profile just before the start of the LOCA test is shown in Figure 4. It can be seen that the flux profile is peaked at the center of the fuel rod, as desired.

The test consisted of a blow down phase, heat-up, hold at PCT and termination by a reactor scram. The target for peak cladding temperature was 1050 °C. In order to achieve this target, the heat rates of the fuel rod and heater were adjusted to 22 and 17 W/cm, respectively. These parameters were chosen on the basis of earlier experience obtained during the first trial tests in IFA-650.1.

Some results from the LOCA test are graphically shown in Figure 5. The cladding thermocouples (TCC2, TCC3 and TCC4) show that the target PCT of 1050 °C was successfully achieved. The peripheral temperature variation was negligible, i.e. within ±3 °C as measured by the upper three thermocouples. The observed average increase rate of the cladding temperature was 7.8 °C/s up to the time of rod failure. Failure of the fuel rod was detected at a cladding temperature of ~ 800 °C by means of cladding elongation and rod pressure measurements (see Figure 6). The rod pressure rapidly dropped to 58 bar. This is due to the pressure monitoring technique - the bellows expansion is stopped mechanically. In reality the rod pressure will soon reach the rig pressure (2-3 bar) after ballooning and burst. The hoop stress at failure was calculated to be ~ 55 MPa on the basis of measured rod pressure. The rod failure was also detected, after some delay, by the gamma monitor mounted on the blow down line (see Figure 7). Several activity peaks of released fission products were also observed shortly after the reactor scram and use of the spray system.

The hold time above 900 °C was about 6.5 minutes. In order to provide conditions for oxidation of the cladding, the water spay was intermittently applied during the hold at PCT. Spraying was started above 900 °C and the readings of the clad upper thermocouples (TCC2, TCC3 and TCC4) began to deviate from that of thermocouple TCC1 at the lower end. The duration of each spraying pulse was about 1 second and the intervals between the pulses varied from 20 to 40 seconds. When the spray was used, temperatures of TCC2-4 started to decrease slightly. However, at the same time the heater power was reduced from about 18 W/cm to 7-8 W/cm in order to keep its temperature below 950 °C and hence this is the main contributor to the decrease in cladding temperature as recorded by TCC2-4.

# 3. CODE ANALYSIS

Pre-test calculations were performed by outside laboratories using the codes FRAPTRAN/GENFLO (VTT) and TRAC-BF1 (PSI). FRAPTRAN/GENFLO is a coupled code where FRAPTRAN calculates fuel performance and GENFLO thermal-hydraulics. TRAC-BF1 is the LOCA code developed for BWR simulations. The predicted thermal response was in good accordance with the measurements. The code calculations will be continued and carried out also by other organisations after the test.

# 4. SUMMARY

The second LOCA trial test was successfully performed in the Halden reactor in May 2004 using a fresh pressurized PWR rod with Zry-4 cladding. The outcome of the test is summarized below:
- Loop and rig, and rod instrumentation worked well;
- Target PCT ~1050 $^{\circ}$C was achieved;
- Rod rupture was detected at 800 $^{\circ}$C;
- Hoop stress at failure: ~55 MPa;
- Clad temperature increase rate up to the rod failure: 7.8 $^{\circ}$C /s;
- Small peripheral temperature variation: +/- 3 $^{\circ}$C;
- Hold time above 900 $^{\circ}$C: 6.5 minutes;
- Termination by reactor scram.

Pre-test calculations were performed by outside laboratories using the codes FRAPTRAN/GENFLO (VTT) and TRAC-BF1 (PSI). The predicted thermal response was in good agreement with the measurements. The code calculations will be continued and carried out also by other organisations after the test.

The following post-irradiation examination will be performed on the fuel rod:
- Gamma scanning;
- Characterization of balloon (dimensions and shape);
- Estimation of oxide thickness (axial distribution at inside/outside of ballooned region);
- Analysis of hydride content in ballooned area and its vicinity.

The latter three items will be carried out at Kjeller hot cells, while the gamma scanning will be performed at Halden.

A meeting to discuss the outcome of the second test will be arranged before proceeding with testing of the pre-irradiated rods. The test scheme for the pre-irradiated rods will be decided on the basis of the inspection and PIE results as well as further code calculations.

## REFERENCES

1. V.Lestinen, E.Kolstad, W.Wiesenack: *"LOCA testing at Halden, Trial runs in IFA-650"*, Proceedings of NSRC-2003, Washington DC, October 2003.
2. E. Kolstad, W.Wiesenack: *"High burnup and safety relevant experiments at Halden"*, Fuel Safety Research Meeting, Tokyo, March 2004.

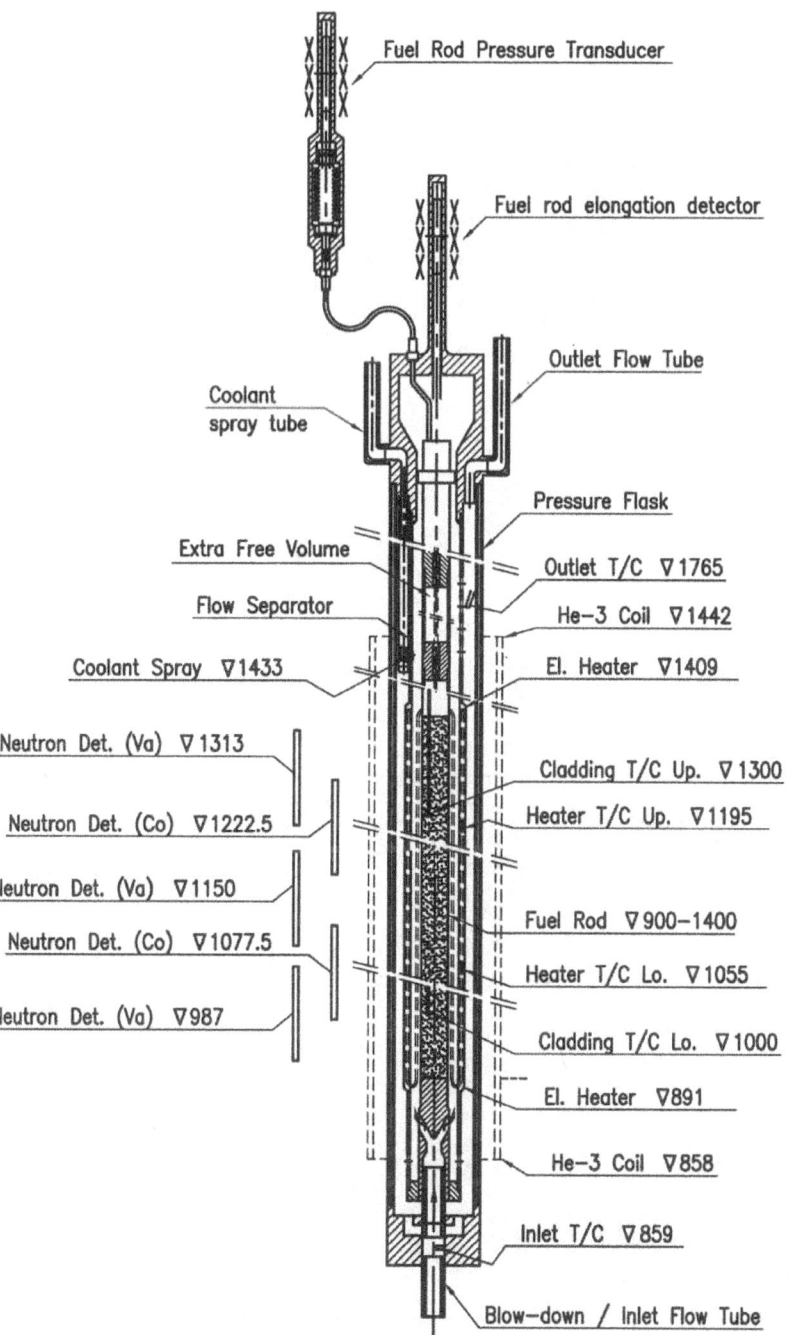

Figure 1. Schematic of test rig for LOCA experiments.

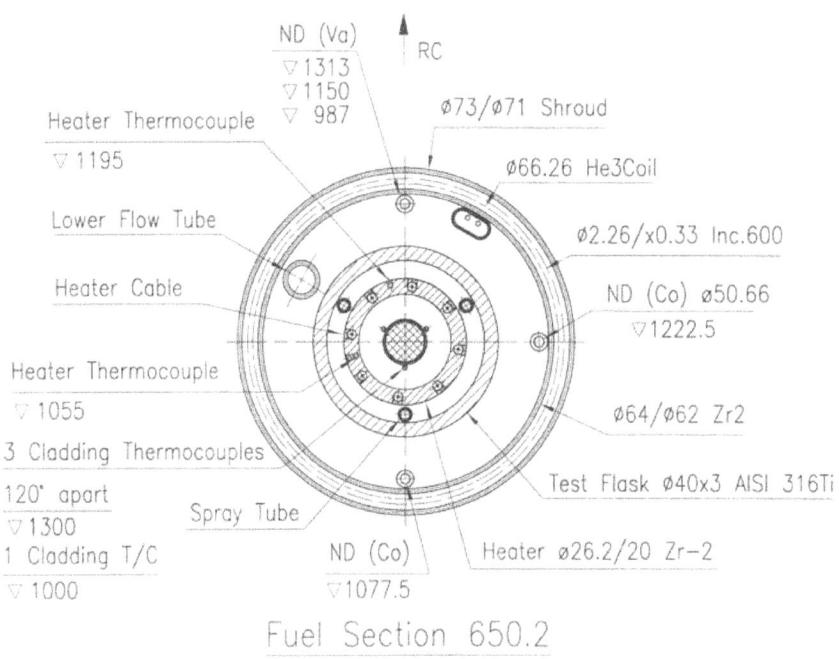

Figure 2. Cross section geometry of the LOCA test rig.

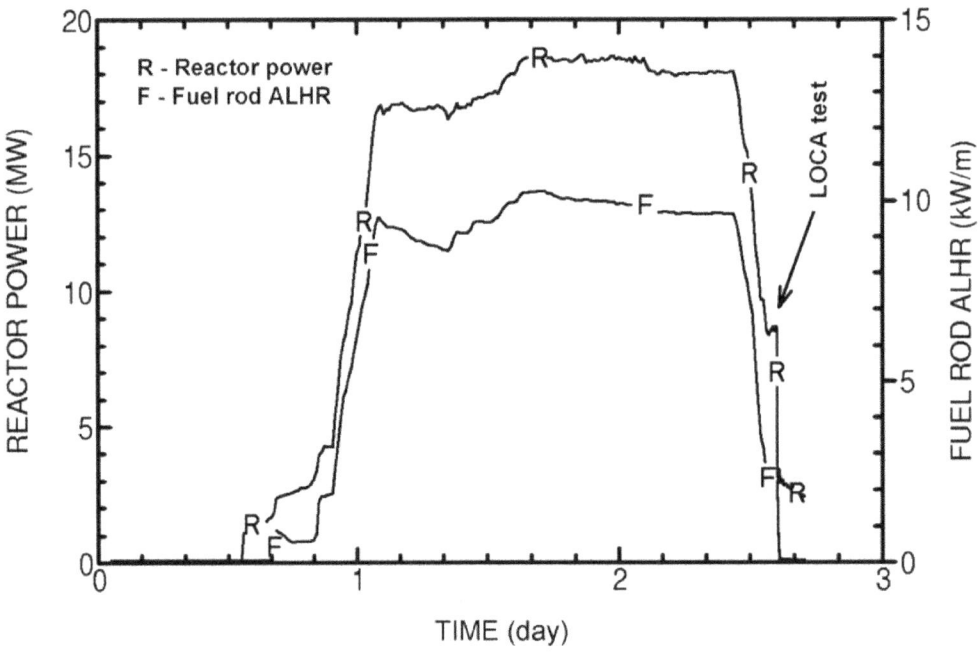

Figure 3. The reactor power and the average linear heat rate of the test fuel rod.

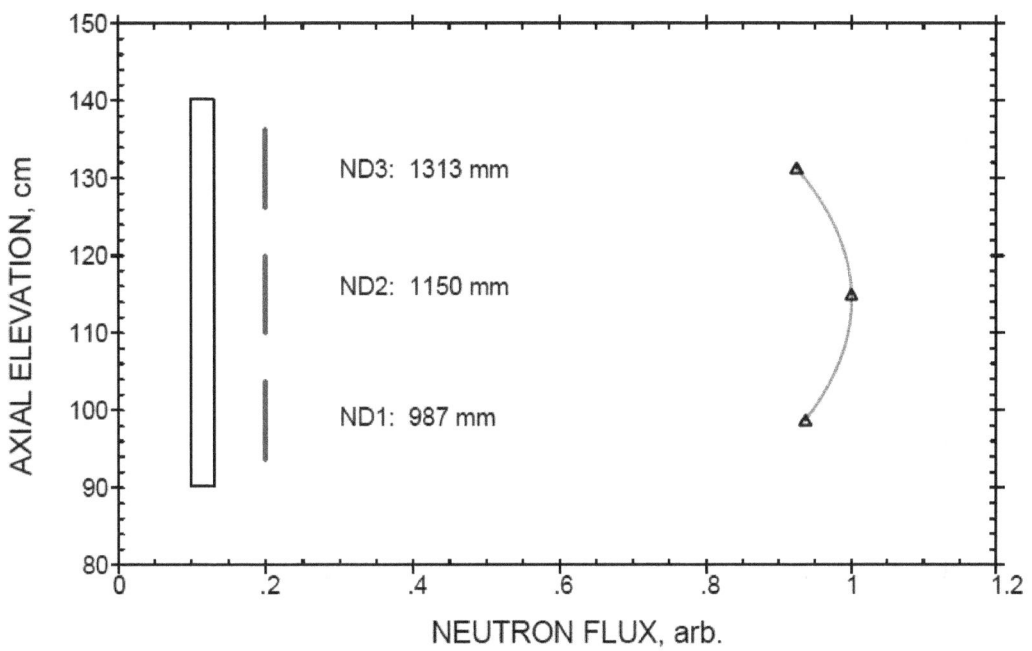

Figure 4. Measured neutron flux profile along the fuel rod just before the LOCA test.

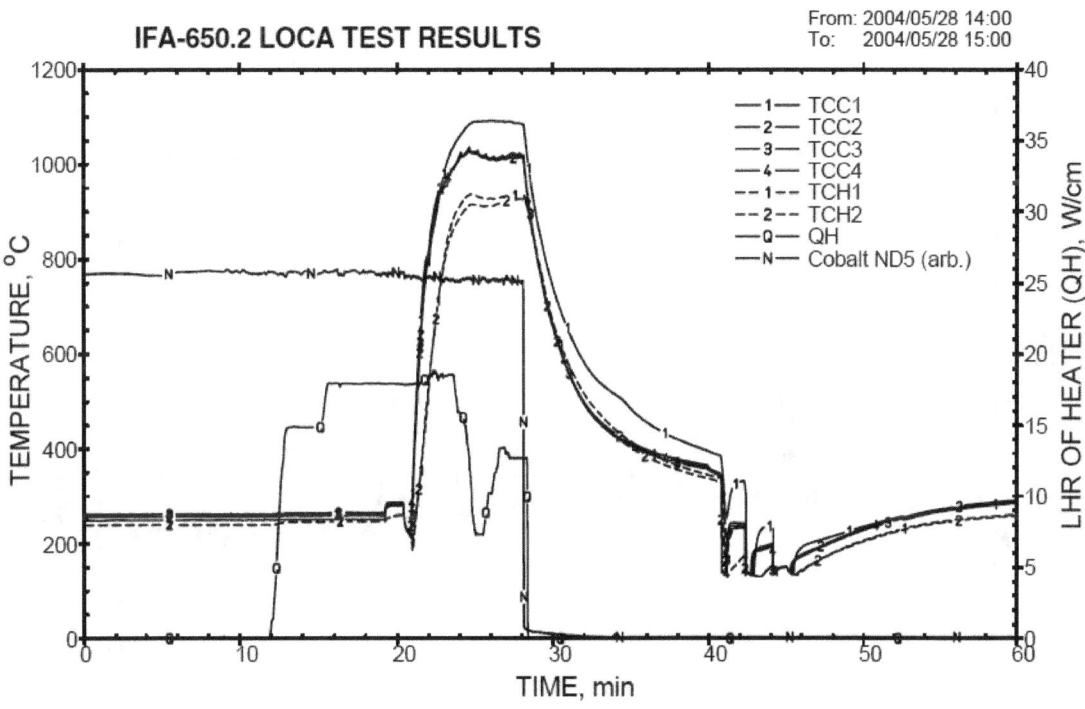

Figure 5. Measured clad (TTC1-4) and heater (TCH1-2) temperatures, linear heat rate of heater (QH) and readings of Co neutron detector (Cobalt ND5) during the LOCA test run.

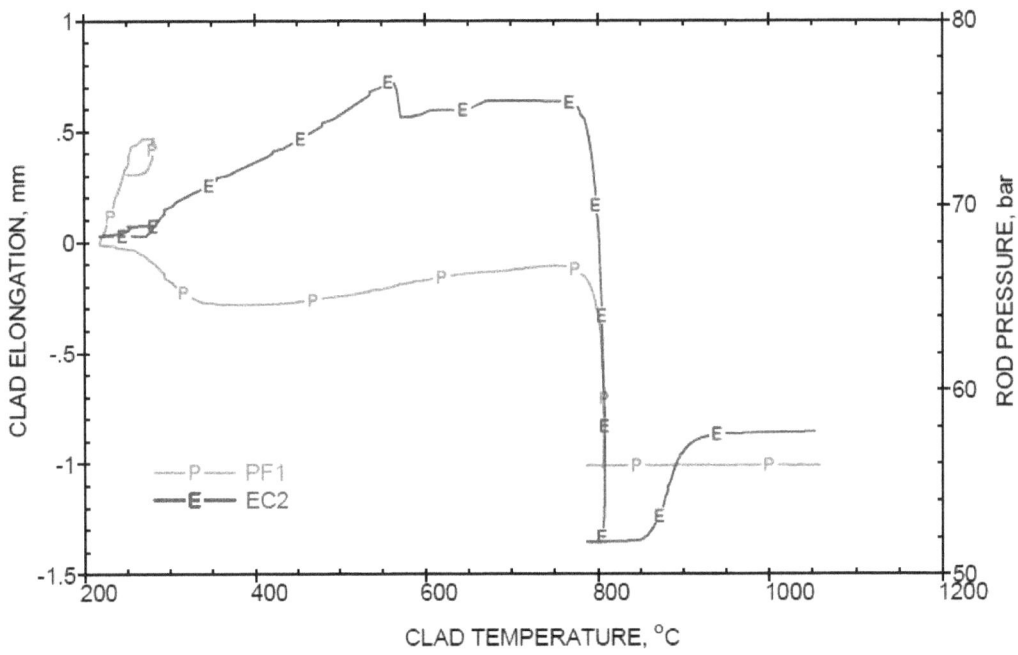

Figure 6. Cladding elongation and rod pressure versus clad temperature.

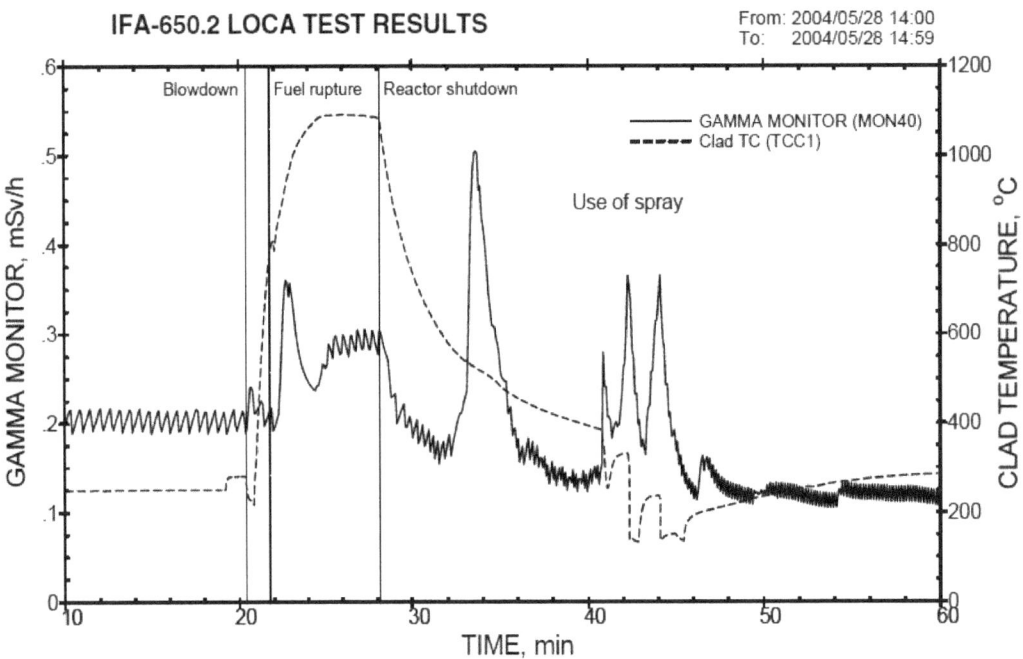

Figure 7. Readings of gamma monitor during LOCA test.

# RESULTS FROM STUDIES ON HIGH BURN-UP FUEL BEHAVIOR UNDER LOCA CONDITIONS

Fumihisa NAGASE and Toyoshi FUKETA
Department of Reactor Safety Research
Japan Atomic Energy Research Institute

## SUMMARY

To promote a better understanding of high burnup fuel behavior under loss-of-coolant accidents (LOCAs), a research program is being conducted at the Japan Atomic Energy Research Institute (JAERI). The program consists of integral thermal shock test and other separate tests for oxidation rate and mechanical property of fuel claddings.

1. Integral thermal shock test

In the test, short rods are heated up, burst, oxidized in steam and quenched by flooding water in order to evaluate fracture-bearing capability of oxidized fuel claddings under the simulated LOCA condition. The Japanese LOCA criteria for cladding embrittlement are based on results from the integral thermal shock tests. Tests were performed with non-irradiated cladding tubes, which were mechanically thinned and pre-hydrided to evaluate the separate effects. Claddings fractured into two pieces during the quench, depending primarily on the oxidation. The threshold of the fracture in terms of the oxidation decreases with the higher concentration of initially-absorbed hydrogen and with the larger load in axial constraint.

Two PWR fuel rods, irradiated to 39 and 44GWd/t (rod average) at Takahama unit-3 reactor, are subjected subsequently to the tests. The cladding material is low-Sn (1.3%Sn) Zircaloy-4 and thickness of oxide layer formed in the reactor ranged 18 to 25 μm. Before the tests, fuel pellets were removed from 190mm-long segments, and alumina dummy pellets were loaded in the defueled claddings. Zircaloy end-plugs were welded at the both ends of the claddings and the fabricated test rods were pressurized to about 5MPa with argon gas. Six test rods were quenched after rupture at about 800 deg C and isothermal oxidation at 1030 to 1192 deg C. Two claddings which were oxidized to about 26 to 30%ECR* fractured during quench. The fracture of the irradiated claddings agrees with the failure criteria for non-irradiated claddings containing similar hydrogen concentrations. Four claddings oxidized to about 16 and 25%ECR survived the quench. These indicate that fracture/no-fracture threshold is not reduced so significantly by irradiation to the examined burnup level.

2. Mechanical tests of oxidized and quenched cladding

Besides the integral thermal shock tests, mechanical tests are performed to develop methodology for predicting cladding fracture on quenching as well as to examine mechanism of cladding embrittlement. Ring-tensile and ring-compression tests were performed on non-irradiated Zircaloy-4 claddings which were pre-hydrided to 400 and 800 ppm, oxidized at 1000 to 1250 deg C, and finally quenched.

- Ductility reduction observed in the ring-tensile tests was not remarkable for the oxidation between 10 and 20%. Uniform tensile stress in the circumferential direction is applied to the cladding in the ring tensile test, and this stress state is quite different from that is applied during quench. This suggests that test methods should be carefully selected in order to estimate cladding embrittlement under LOCA condition.

- The ring-compression tests detected the sudden ductility drop above 15% oxidation for claddings which were oxidized at 1200 deg C without hydriding. The significant ductility reduction occurred at lower oxidation level in the pre-hydride claddings (400 and 800 ppm). This indicates that pre-hydriding enhances cladding embrittlement of oxidized cladding and agrees with the results of the integral thermal shock tests.

- It has been generally considered that slow cooling after high-temperature oxidation enhances oxygen diffusion from oxide layer into metallic prior-$\beta$ phase, and consequently microstructure and ductility of metallic prior-$\beta$ phase changes depending on the cooling rate. To confirm that, ring compression tests were performed with claddings which were cooled at different rates after oxidation at 1100 and 1200 deg C. It is shown that the influence of slow cooling differs depending on oxidation temperature and oxidation amount. The influence becomes greater in the cladding oxidized at 1200 deg C and at the lower oxidation level.

Acknowledgment

The integral thermal shock test with irradiated PWR fuel claddings has been performed as a corporative research program between JAERI and Japanese PWR utilities.

\* ECR is estimated by the Baker-Just equation, taking account of double sided oxidation and wall thinning by ballooning. The 'initial' cladding thickness used in the estimation is metallic thickness after corrosion during the reactor operation.

# Results from Studies on High Burn-up Fuel Behavior under LOCA Conditions

Fumihisa Nagase and Toyoshi Fuketa

*Japan Atomic Energy Research Institute*

JAERI

# Main Tasks

- To clarify fuel behavior in more detail

- To clarify loading condition during quench including restraint condition

- To evaluate influence of high burnup effect on fuel behavior
  Pre-oxidation, pre-hydriding, irradiation, new alloy, etc.

- To improve and develop test & evaluation methods

→ Regulatory judgment and criteria revision for high burnup fuel

# Experimental Program at JAERI

- Integral test with short test rod
  - simulating the whole LOCA sequence

- Separate-effect tests with small specimen
  - on oxidation rate
  - on cladding embrittlement

JAERI

201

*JAERI*

## Japanese ECCS acceptance criteria

- Maximum cladding temperature of **1200°C** (1473K, 2192 F) and oxidation of **15%** are defined in order to avoid significant cladding embrittlement and to maintain coolable geometry of reactor core.

## The "15%" is based on

- A threshold of rod fracture during quenching under the simulated LOCA condition

# Test Apparatus and Test Rod

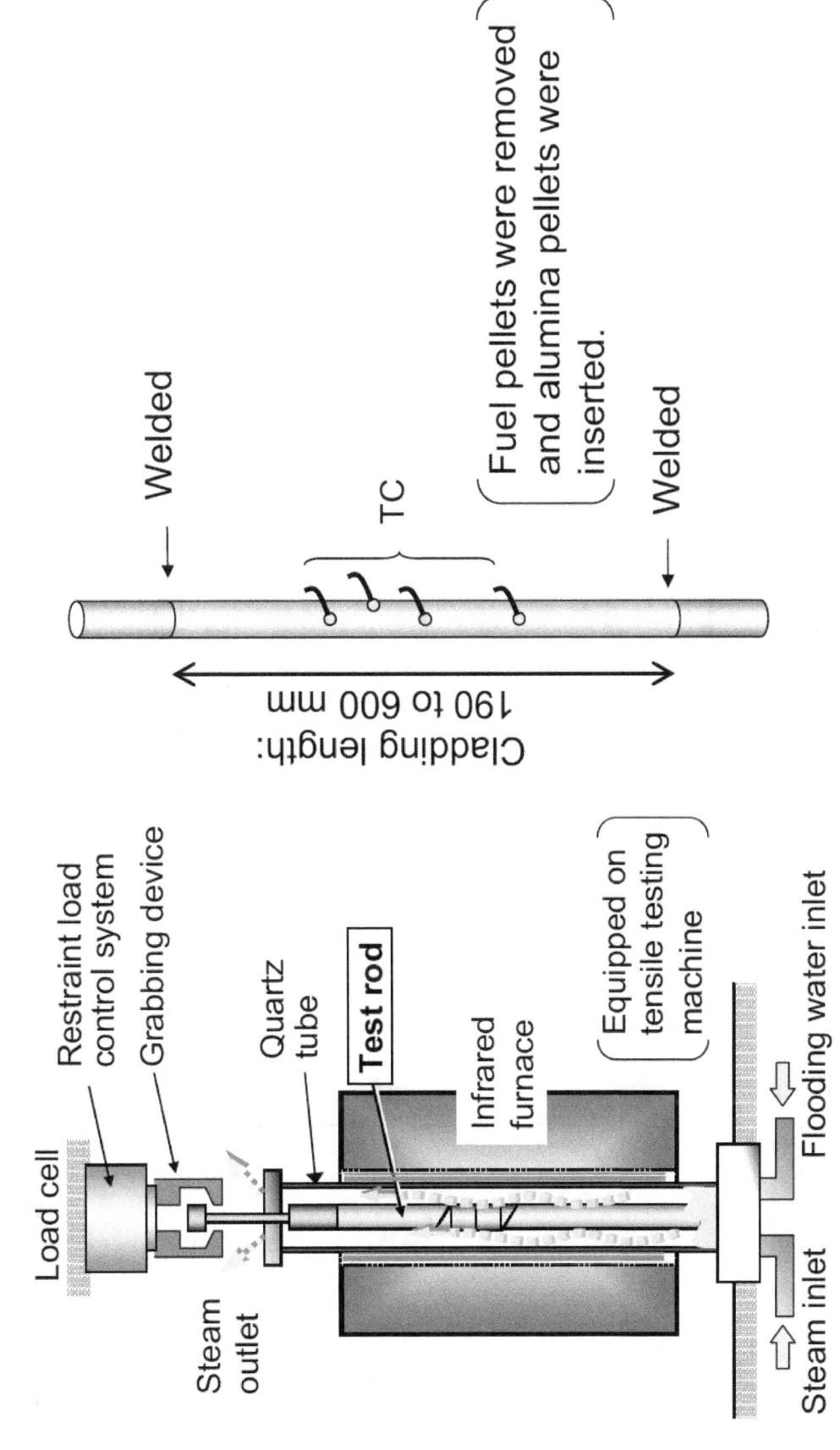

Welded

TC

Fuel pellets were removed and alumina pellets were inserted.

Welded

Cladding length: 190 to 600 mm

Load cell

Restraint load control system

Grabbing device

Quartz tube

**Test rod**

Infrared furnace

Equipped on tensile testing machine

Steam outlet

Flooding water inlet

Steam inlet

# Temperature and Axial Load Histories

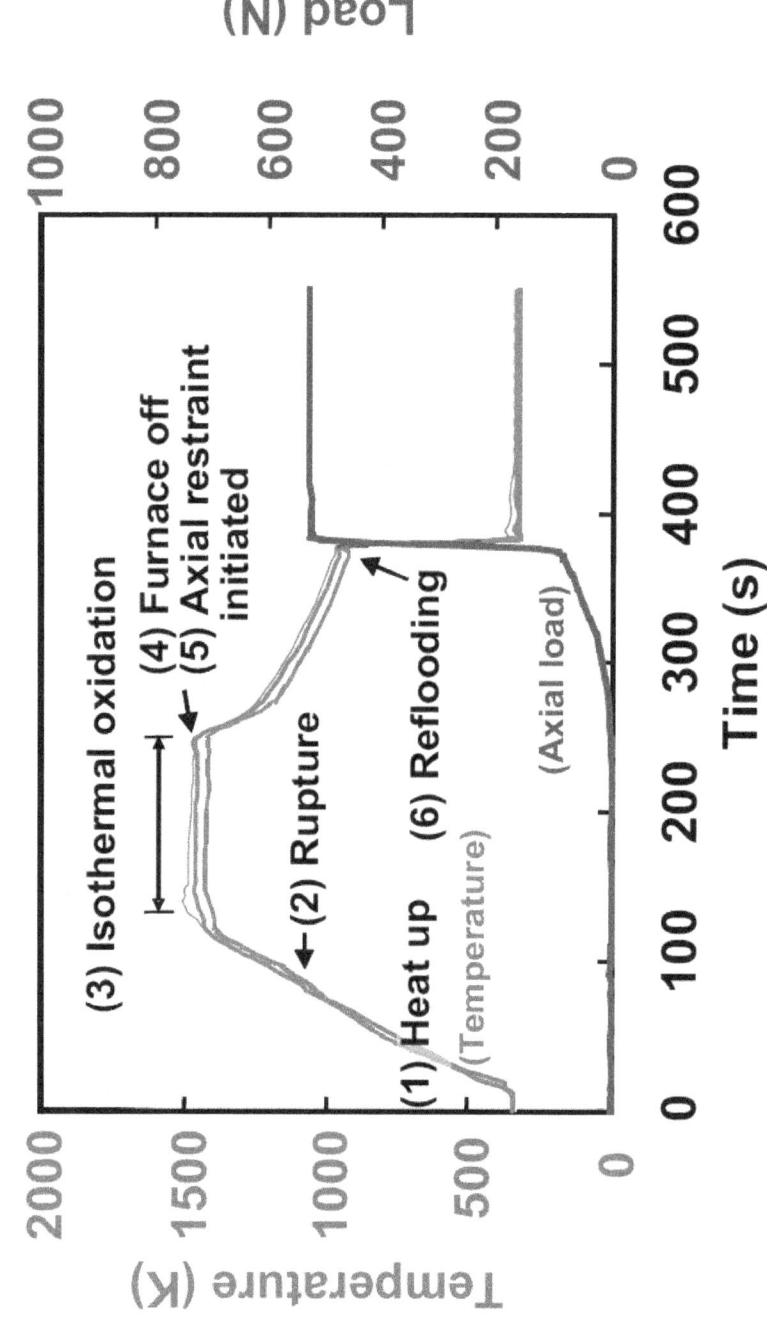

- Test rod is quenched with axial restraint to represent a possible condition of fuel rods between grid positions.

*JAERI*

# Fracture / No-fracture Threshold
## Relevant to ECR and Initial Hydrogen Concentration
### From tests with non-irradiated cladding

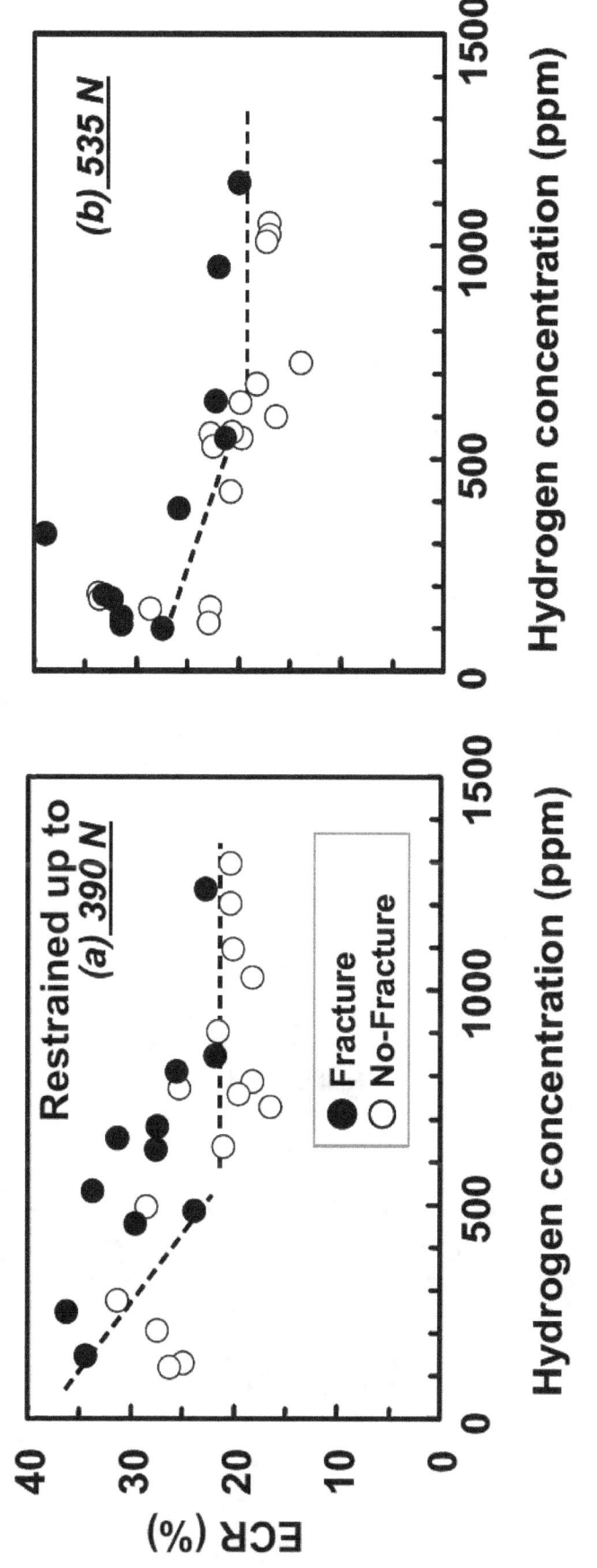

ECR was calculated with the Baker-Just equation, based on reduced cladding thickness after ballooning measure at rupture position

# Fracture / No-fracture threshold Relevant to ECR and Restraint Load
## From tests with non-irradiated cladding

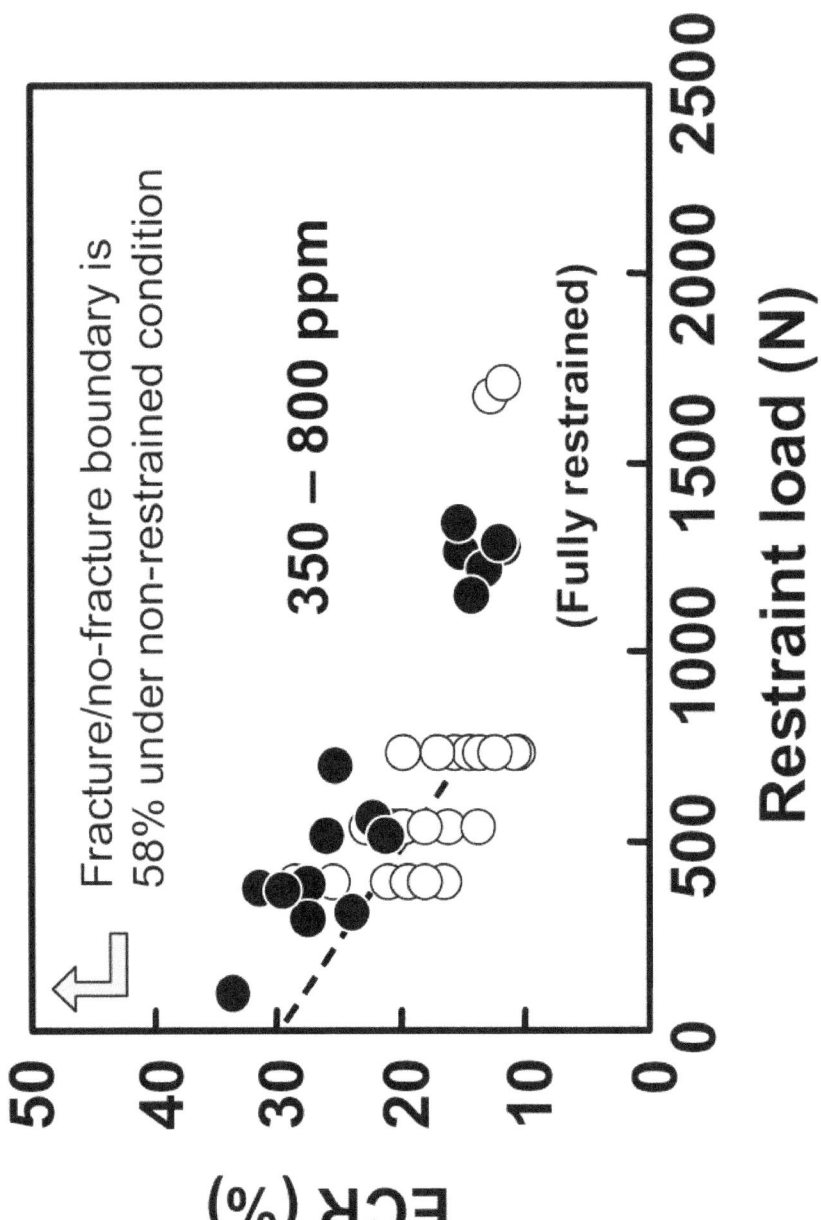

Fracture/no-fracture boundary is 58% under non-restrained condition

350 – 800 ppm

(Fully restrained)

Fracture condition is sufficiently higher than 20% ECR under non-restraint and intermediate restraint conditions.

# High burnup PWR Fuel Cladding

This study has been performed as a cooperative research program with Japanese PWR utilities.

| | |
|---|---|
| Fuel type | 17×17 |
| Rod average burnup (GWd/t) | 39 to 44 |
| Cladding material | Low tin Zircaloy-4 |
| Initial oxide layer (μm) | 18 to 25 |

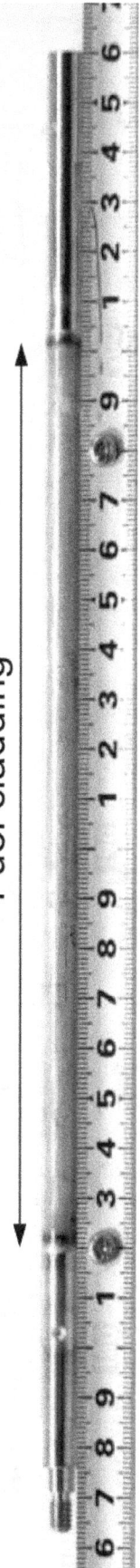

Fuel cladding

Fabricated test rod

| Test No. | 1 | 2 | 3 | 4 | 5 | 6 |
|---|---|---|---|---|---|---|
| Sample No. | A 3-1 | A1-2 | B L-3 | B I-3 | B I-5 | B L-7 |
| Burn-up (GWd/t) | 43.9 | 43.9 | 39.1 | 40.9 | 40.9 | 39.1 |
| Initial Oxide ($\mu$m) | 20 | 25 | 18 | 18 | 15 | 15 |
| Estimated initial hydrogen (ppm) | 170 | 210 | 140 | 140 | 120 | 120 |
| Rupture temperature (C) | 800 | 751 | 820 | 785 | 755 | 804 |
| Rupture strain (%) | 14.1 | 27.7 | 24.3 | --- | --- | --- |
| Oxidation temperature (C) | 1176 | 1178 | 1154 | 1172 | 1030 | 1177 |
| Oxidation time (s) | 486 | 120 | 200 | 363 | 2195 | 543 |
| ECR | 29.3 | 16.6 | 16.0 | 21.0* | 22.0* | 26.4* |
| Fractured / Survived | F | S | S | S | S | F |
| Load at fracture (N) | 498 | --- | --- | --- | --- | 39.3 |
| Maximum restraint load (N) | --- | 540 | 540 | 540 | 540 | --- |

ECR: Evaluated using Baker-Just equation, based on cladding thickness reduced by ballooning

*: Assumed 20% reduction of cladding thickness by ballooning

# Fracture / No-fracture threshold
## Relevant to ECR and Initial Hydrogen Concentration

ECR based on
reduced metal thickness after rupture

ECR based on initial metal thickness

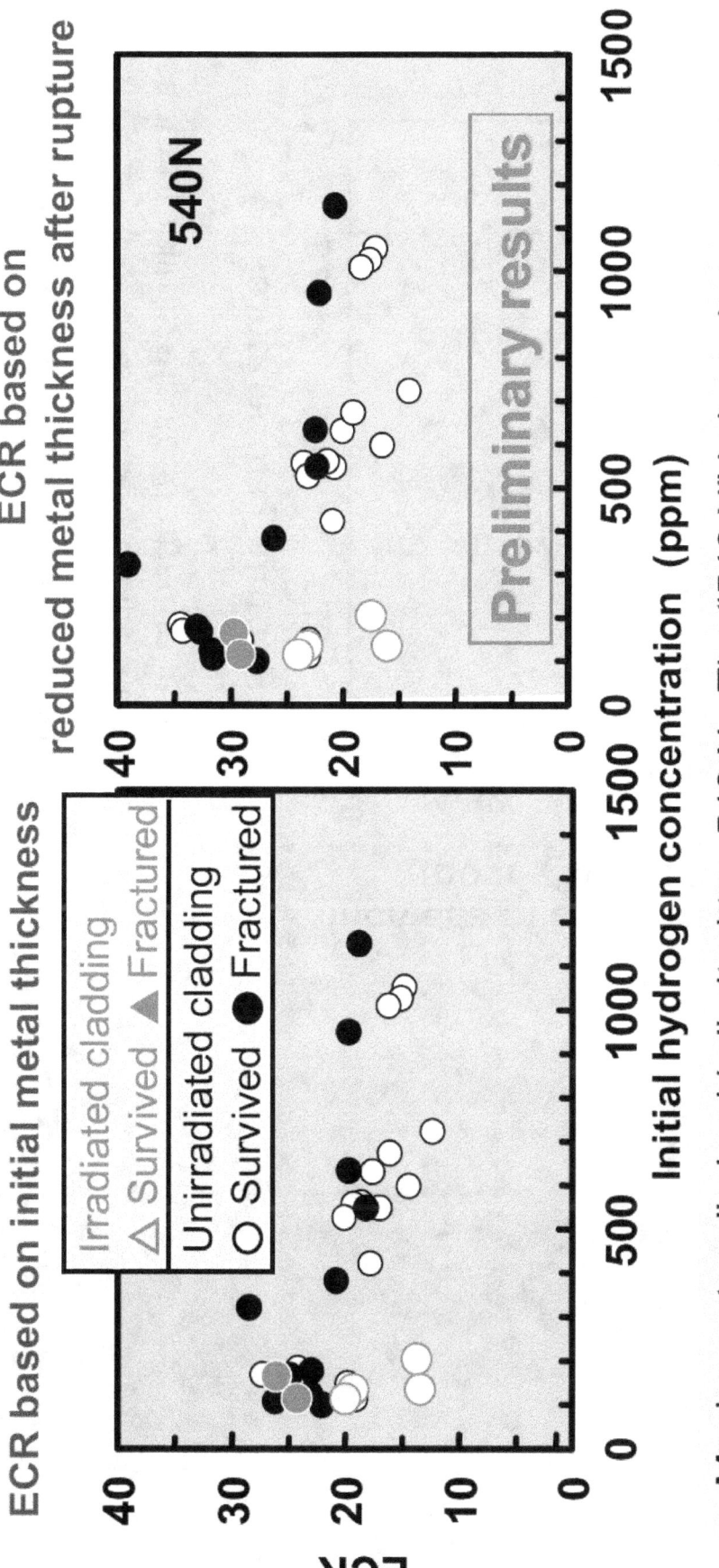

**Initial hydrogen concentration (ppm)**

- Maximum tensile load is limited to ~540 N. The "540 N" is based on measurement of resistant load between deformed or chemically interacted cladding and spacer grid (K. Homma et al., 2001).

- Rupture strain and reduced cladding thickness after ballooning and rupture have not yet been evaluated for irradiated claddings.

# Rupture Behavior

*JAERI*

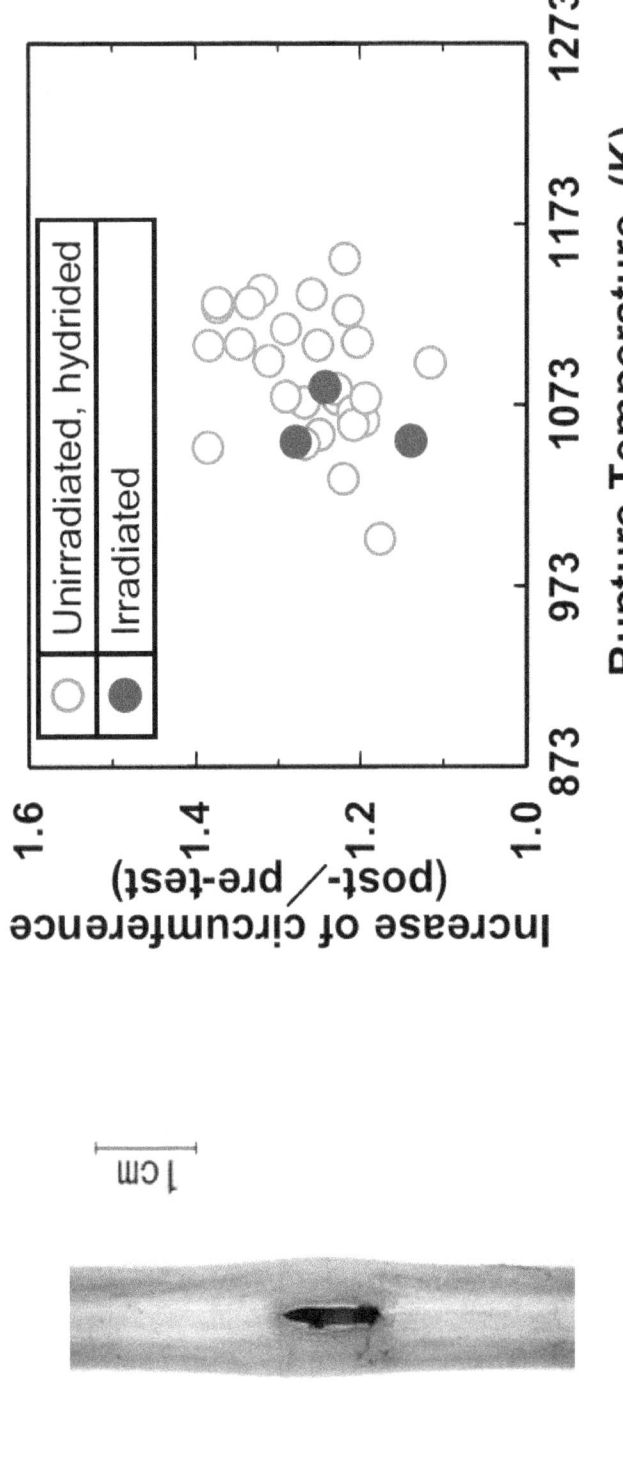

- Visual appearances are similar to those in unirradiated specimens.

- Opening size may be a little smaller.

- Increase of circumference depends on rupture temperature. Results from irradiated specimens are equivalent to those from unirradiated samples.

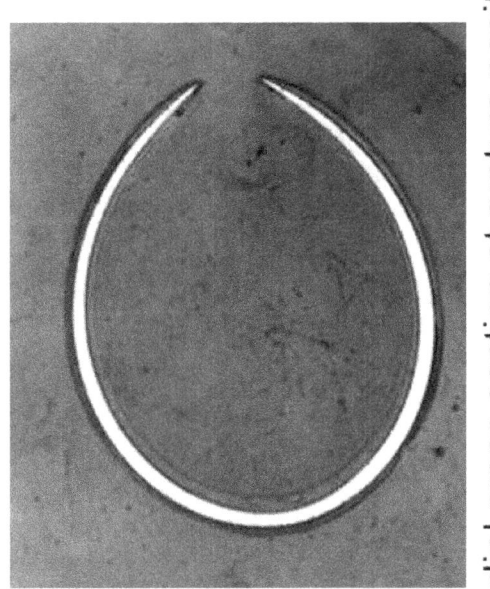

Radial cross section at rupture position

210

# Hydrogen Effect on Rupture Behavior
## Results from tests with unirradiated cladding

- Data from tests with unirradiated specimens in two different hydrogen ranges.

- The increase depends on rupture temperature, hence, on phase structure at rupture. Phase transformation temperatures change with hydrogen concentration.

- Rupture behavior, increase of circumference, depends on hydrogen concentration.

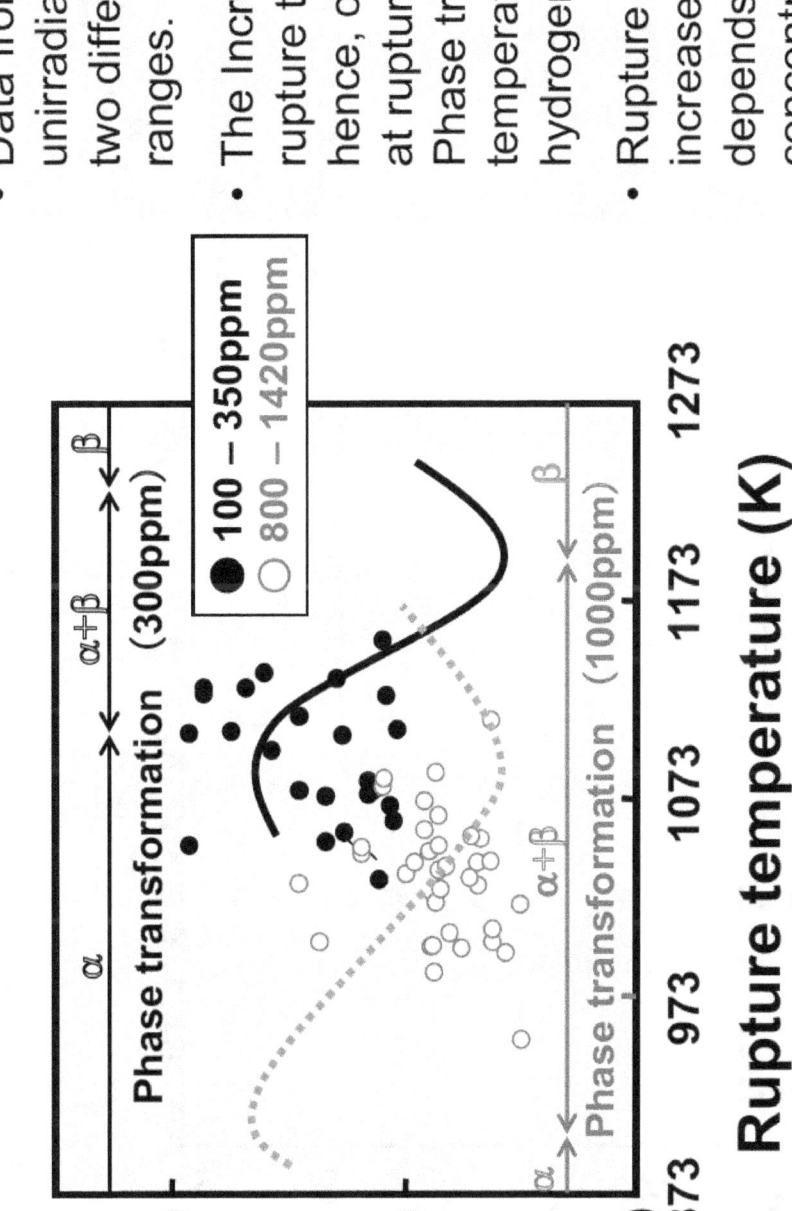

# Oxidation

Irradiated

1465K, 120sec, 17.9%ECR

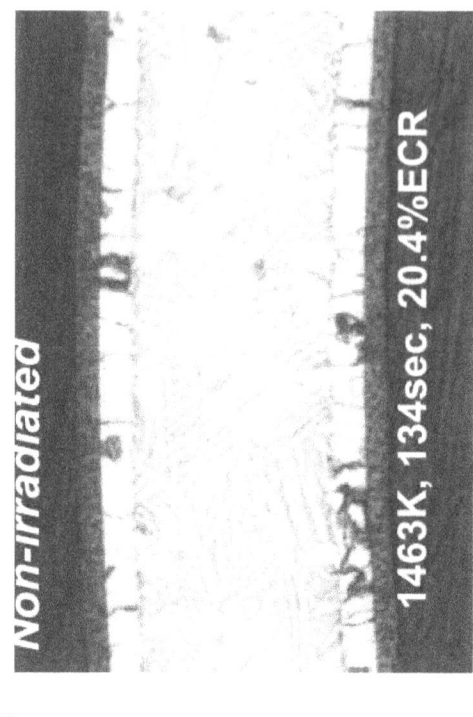

Non-irradiated

1463K, 134sec, 20.4%ECR

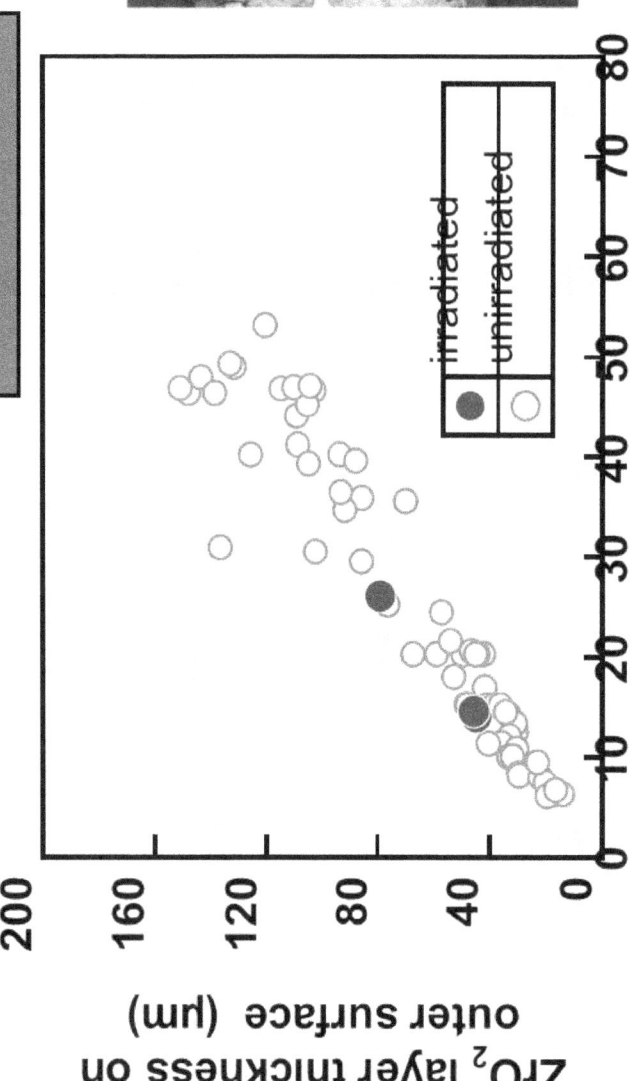

ZrO$_2$ layer thickness on outer surface (μm)

ECR (%)

- irradiated
- unirradiated

- Oxide thicknesses are in the same range in unirradiated and irradiated specimens. Effect of oxide layer pre-formed in PWR was not observed.

- Small cracks can be seen in the pre-formed oxide layer. Due to the cracks, oxidation in the high temperature LOCA conditions became same level in irradiated and unirradiated samples.

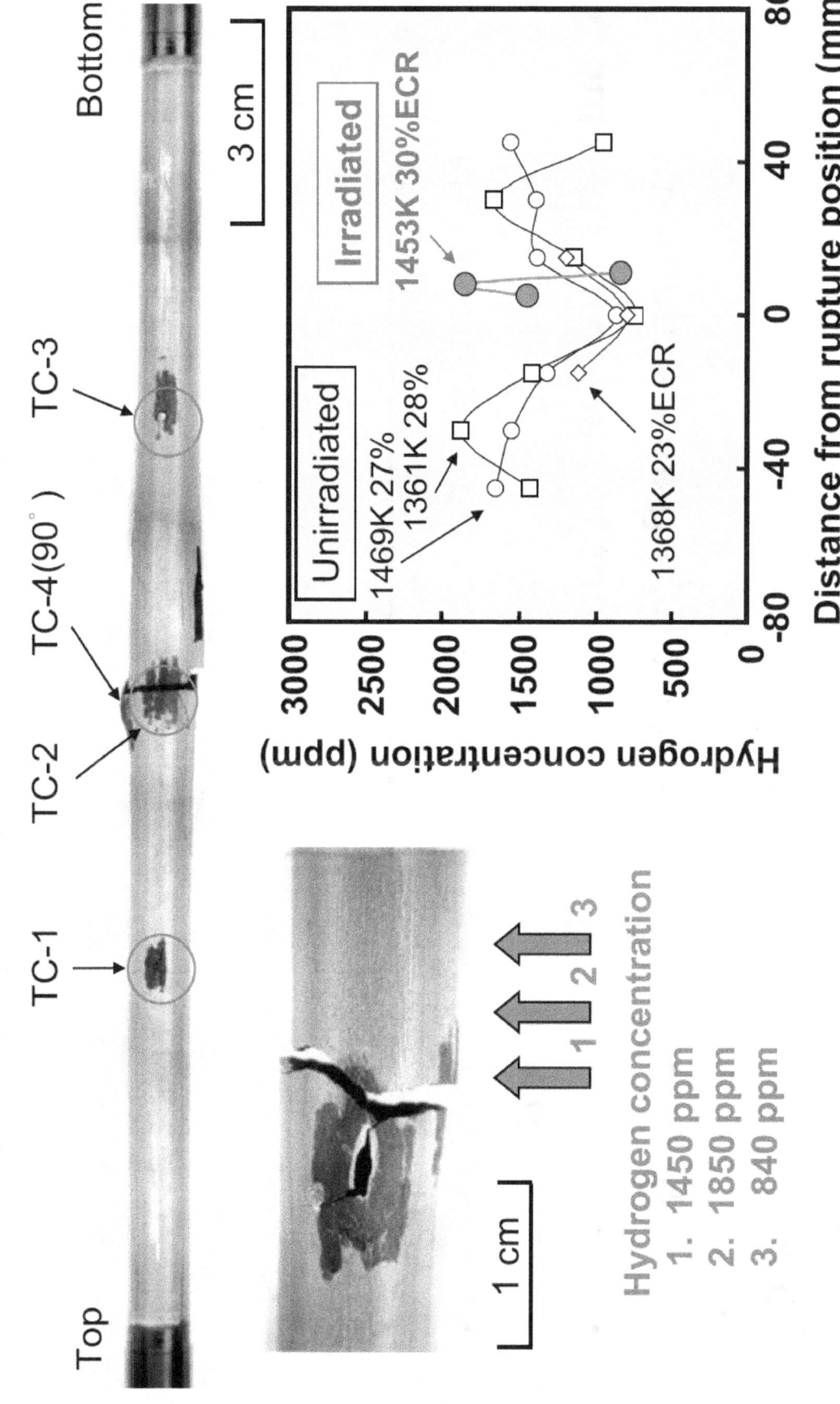

# Fracture Morphology and Secondary Hydriding

*JAERI*

**Hydrogen concentration**
1. 1450 ppm
2. 1850 ppm
3. 840 ppm

- The transient secondary hydriding is localized more towards the burst location in the irradiated sample. This is consistent with results from fueled rod tests in ANL.

213

# Effect of Cooling Conditions on Cladding Embrittlement

- It has been believed that;

  Quench temperature and cooling rate

  before quench affect cladding embrittlement.

- Those effects are very important to perform experiments under appropriate conditions.

- However, those effects have not been sufficiently confirmed.

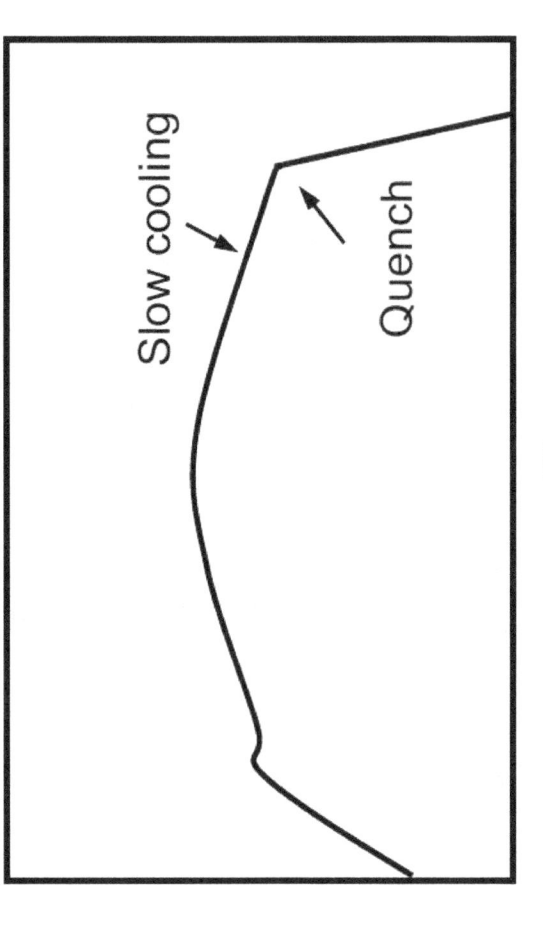

# Separate Tests on Cooling-Rate Effect

- **Oxidation temperature:**
  **1100, 1200 C**

- **Quench temperature:**
  **800, 900, 1000, 1100 C**

- **Cooling rate in slow cooling stage:**
  **2, 7 K/s and without slow cooling**

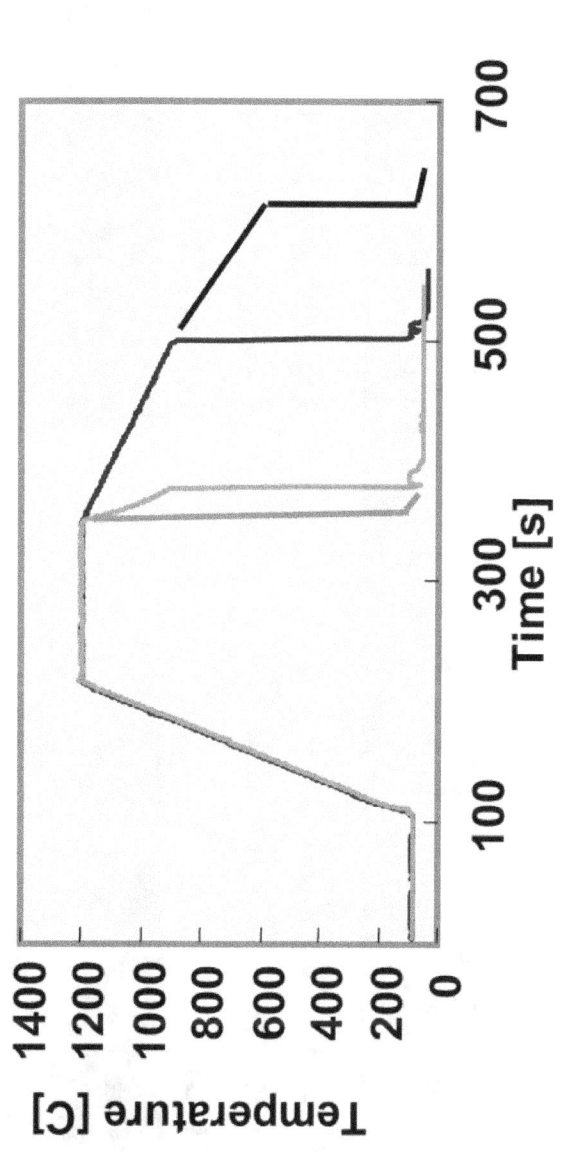

215

# Effect of Cooling Rate on Microstructure

Oxidized at 1200 C
Cooled at 2K/s to 900 C
Quenched

Oxidized at 1200 C
Cooled at 7 K/s to 900 C
Quenched

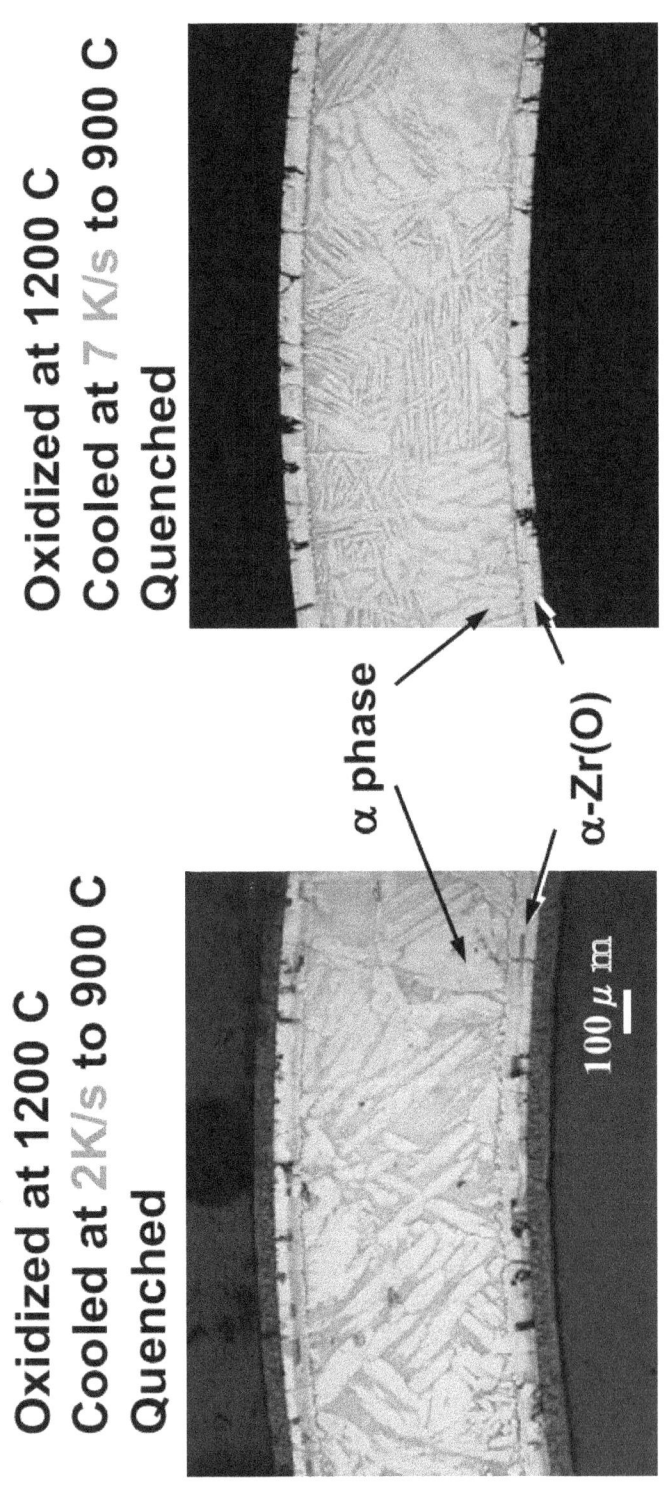

α phase

α-Zr(O)

100 μ m

Unit volume of α phase may be slightly greater in the test with a lower cooling rate.

# α Phase in Metallic Phase is Really Brittle?

## Oxidized at 1200 C for 72 s

- The α phase precipitated in metallic phase is harder than matrix prior-β phase.

- But, not so hard as α-Zr(O) outer layer.

- It is harder in specimen cooled with lower rate.

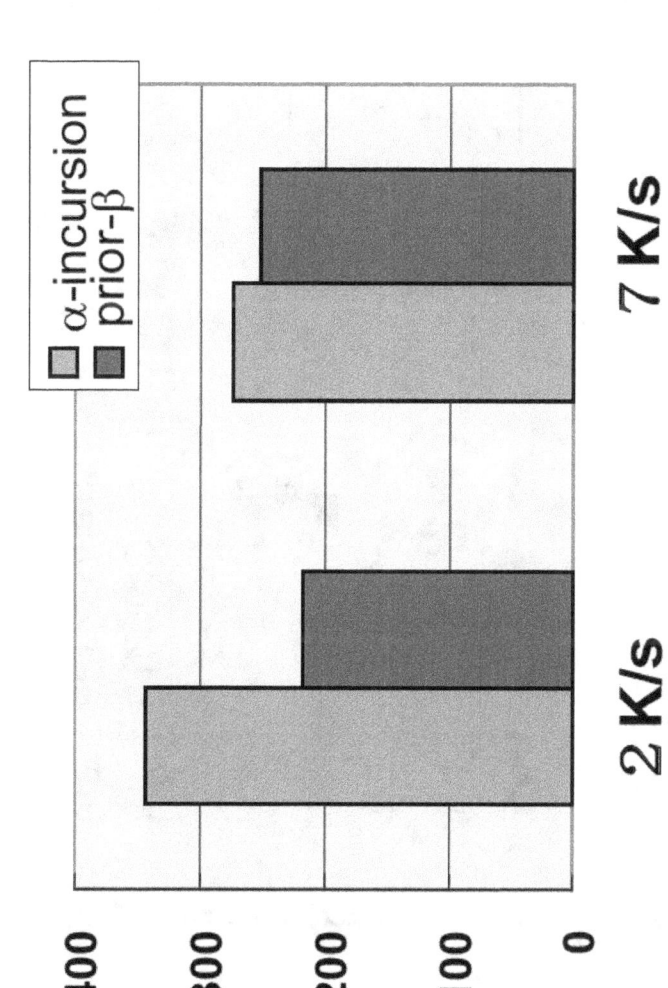

217

# Effect of Cooling Rate on Embrittlement

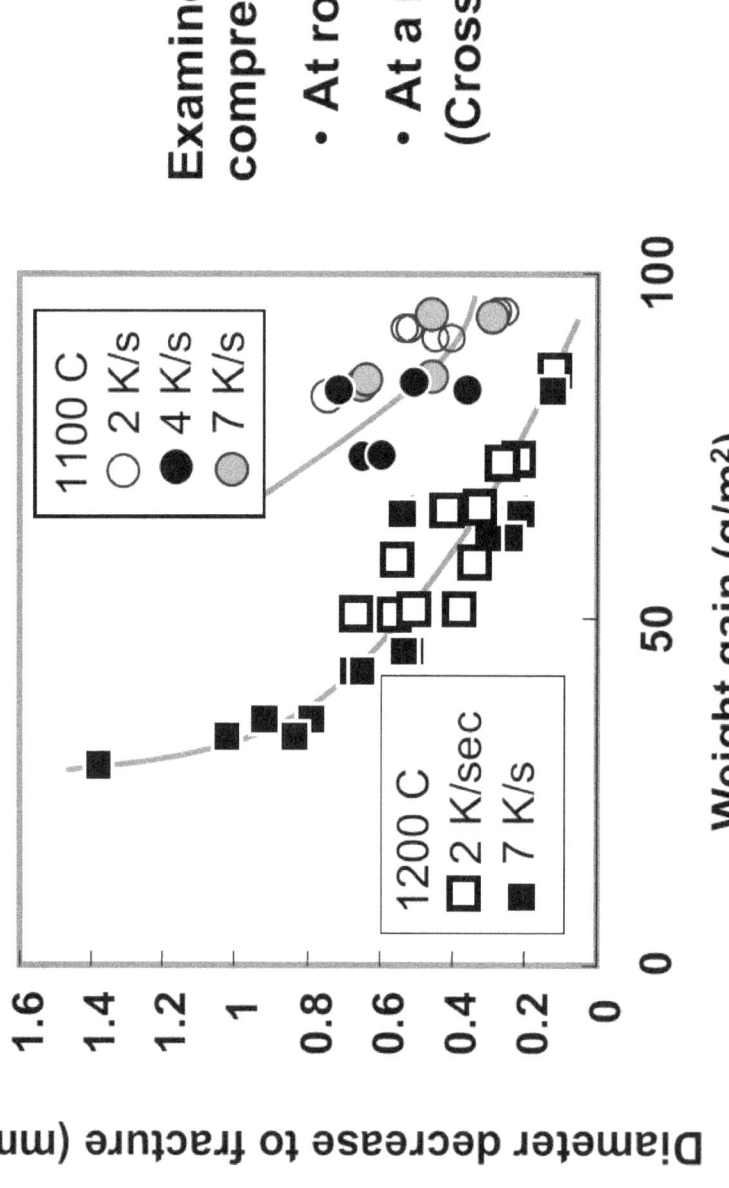

**Examined by ring compression test**

- At room temperature

- At a rate of 2mm/s (Cross head speed)

**Cooling rate has negligible effect on cladding embrittlement.**

# Effect of Quench Temperature on Microstructure

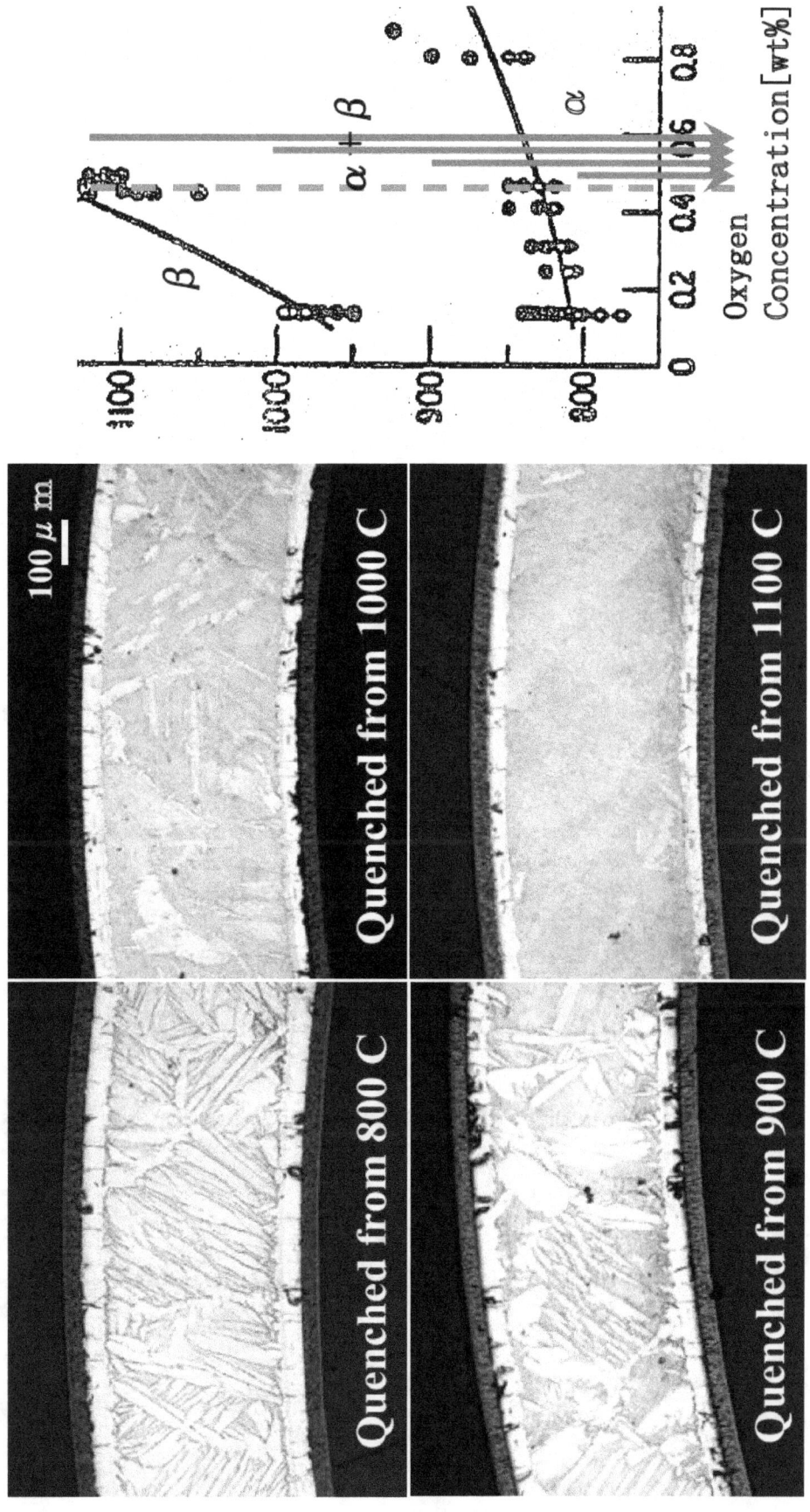

Cross sectional area of α phase decreases with increase in quench temperature.

219

# Effect of Quench Temperature on Embrittlement

**Oxidation temperature: 1100 C**

**Weight gain: 82 to 89 g/m²**

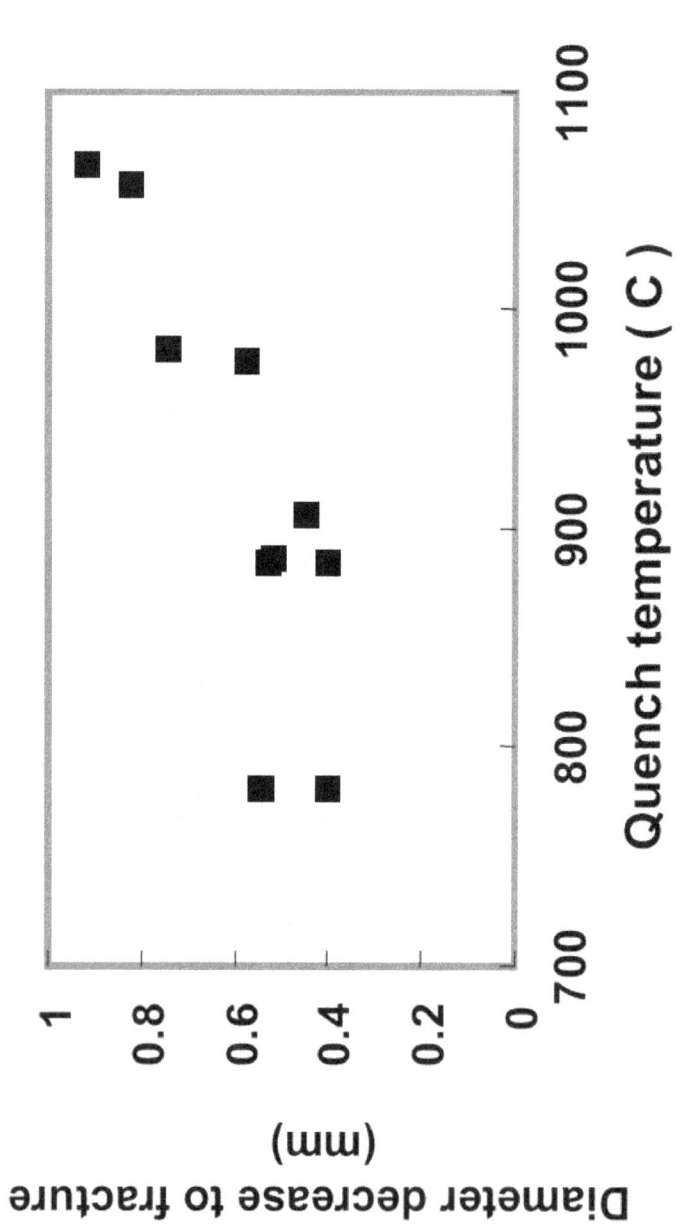

Quench temperature has obvious effect on cladding embrittlement.

*JAERI*

## Future plan

Integral tests with

- 55 GWd/t PWR cladding (ZIRLO™, MDA, NDA)

- 63 to 75 GWd/t PWR cladding (ZIRLO™, M5, MDA, NDA)

- 63 GWd/t BWR cladding (Zircaloy-2)

Burnup: Rod average

MDA: Zr-0.8Sn-0.2Fe-0.1Cr-0.5Nb
developed by Mitsubishi Heavy Industries, Ltd

NDA: Zr-1.0Sn-0.27Fe-0.16Cr-0.1Nb-0.01Ni
developed by Nuclear Fuel Industries, Ltd.
and Sumitomo Metal Industries, Ltd.

Separate tests

- Secondary hydriding (Effects of rupture opening size, gap size, internal corrosion, etc.)
- Axial restraint (including measurement at high temperature)

## Summary (1/2)

✓ Integral thermal shock tests simulating the whole LOCA sequence have been conducted with irradiated PWR fuel specimens as a part of a research program on high burnup fuel behaviors under LOCA conditions.

✓ In the burnup level of the present study, differences were not significant between irradiated and unirradiated specimens in terms of rupture and high-temperature oxidation.

✓ One of irradiated specimens, oxidized at 1453 K up to ~30% ECR, fractured during quenching. The conditions are consistent with the fracture threshold derived from unirradiated specimens with a similar hydrogen concentration.

JAERI

222

## Summary (2/2)

✓ The transient secondary hydriding is localized more towards the burst location in the irradiated specimen than in unirradiated samples.

✓ Specimens, oxidized to ~22% ECR, survived the quench. This indicates that failure boundary is not reduced significantly by PWR irradiation in the examined burnup level.

✓ Cooling rate in the slow cooling stage has negligible effect on cladding embrittlement, while quench temperature has obvious effect.

✓ The influence of further burnup extension and application of new cladding materials on fuel behavior under LOCA conditions will be investigated.

# Results from Studies on High Burn-up Fuel Behavior under LOCA Conditions

Fumihisa NAGASE and Toyoshi FUKETA

Department of Reactor Safety Research
Japan Atomic Energy Research Institute
Tokai-mura, Ibaraki-ken, 319-1195, Japan

## Abstract

The Japanese regulatory criterion for a loss-of-coolant-accident (LOCA) is based on a threshold of fuel rod fracture during quenching, which was experimentally determined under simulated LOCA conditions. In order to evaluate the fracture threshold of high burn-up fuel rods, JAERI performs integral thermal shock tests simulating LOCA conditions. The tests have been performed with pre-hydrided, unirradiated claddings and high burn-up fuel claddings irradiated to 39 and 44 GWd/t at a PWR. It was shown that fracture/no-fracture threshold primarily depends on the oxidation amount and that the threshold decreases with increases in hydrogen concentration and axial restraint during the quench. It was also shown that fracture conditions of the high burn-up fuel claddings are consistent with the fracture threshold derived from unirradiated claddings with similar hydrogen concentrations.

## 1. Introduction

Milestones for the Japanese LOCA criteria and key studies are shown in **Fig. 1**. The LOCA criterion on fuel safety, 15% cladding oxidation (ECR*), was established in 1975 and based on the concept of zero ductility of oxidized cladding, which was determined by ring compression tests, as in the U.S. After their establishment, it was found that oxidation of the cladding after rod-burst is accompanied by significant hydrogen absorption [1]. Since the absorbed hydrogen is generated by oxidation of cladding inner surface with steam that invaded from the burst opening, this phenomenon is called "Inner surface oxidation" including the hydrogen absorption. Ring compression tests were performed on specimens cut from the cladding that experienced rod-burst and subsequent oxidation to examine cladding embrittlement due to oxidation and hydrogen absorption. As a result, ductility of the ring specimens fell down to the zero-ductility range when the cladding was oxidized to several percent ECR, indicating that the significant hydrogen absorption caused by the inner surface oxidation enhances cladding embrittlement [2]. Accordingly, JAERI conducted "integral thermal shock tests" to evaluate thermal shock resistance of oxidized cladding under simulated LOCA conditions [3]. In the test, a short test rod was heated up, burst, oxidized in steam and quenched with flooding water. Obvious effect of the hydrogen absorption was seen on fracture of the cladding during the quench. Therefore, it is necessary to simulate LOCA conditions in order to evaluate fuel safety. At the same time, the results confirmed that the criterion of 15% ECR still

---

* ECR: Equivalent Cladding Reacted (Proportion of oxide layer thickness assuming that all of absorbed oxygen forms stoichiometric $ZrO_2$)

had safety margin, and the LOCA criteria were revised in 1981 referring the results of the integral thermal shock tests. Namely, the Japanese LOCA criterion for cladding oxidation is based on a threshold of fuel rod fracture during the quench, which was experimentally determined under simulated LOCA conditions. The basic concept is that coolabe geometry of the reactor core is ensured if fuel rods survive the quench after a high-temperature oxidation phase.

The LOCA criteria are based mainly on experiments conducted with fresh Zircaloy fuel claddings. Although burn-up effect was generally taken into account on the establishment, the level of fuel burn-up was rather low at that time. The fuel burn-up would be extended further. Corrosion and hydrogen absorption during the reactor operation would become more significant in the fuel cladding, resulting in degradation of the cladding mechanical property. The high burn-up cladding would be subject to fracture due to thermal shock during reflooding. Therefore, it is one of the most important issues to evaluate fracture conditions of the high burn-up fuel rod, though peak clad temperature becomes lower with the burn-up extension.

Under this circumstance, a systematic research program is being conducted at the Japan Atomic Energy Research Institute (JAERI) for evaluating high burn-up fuel behavior under LOCA conditions. The program consists of integral tests and separate effect tests such as oxidation tests and mechanical tests. Both non-irradiated and irradiated claddings are used in the study. Non-irradiated claddings are artificially oxidized or hydrided to simulate corrosion or hydrogen absorption during the reactor operation. Hydrogen effect has been especially examined in detail, because hydrogen absorption has generally a great impact on cladding mechanical property. The present paper summarizes results from the integral thermal shock tests with pre-hydrided, unirradiated claddings and high burn-up fuel claddings irradiated to 39 and 44 GWd/t at a PWR.

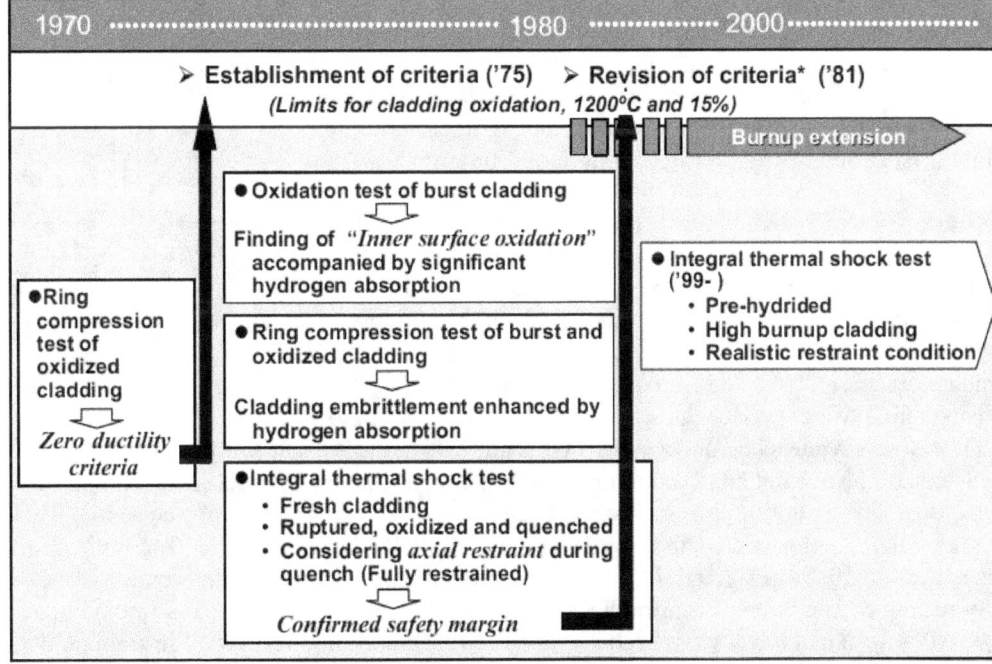

* Consequently, limits for cladding oxidation remained unchanged.

**Fig. 1 Milestones for the Japanese LOCA criteria and key studies**

225

## 2. Thermal shock resistance of unirradiated and irradiated claddings

### 2.1 Experimental procedure

**Figure 1** shows a schematic of a test rod. Length of the cladding is 600 mm for unirradiated cladding and 190 mm for irradiated cladding. Alumina pellets are inserted into the cladding to simulate heat capacity of $UO_2$ pellets. After welding the Zircaloy end caps to the claddings, the rods were pressurized to about 5 MPa with Ar gas at room temperature. **Figure 2** shows a schematic of the test apparatus. The main components are an Instron-type tensile testing machine, a quartz reaction tube, an infrared image furnace with four tungsten-halogen lamps, a steam generator, and a water supply system for flooding. The tensile testing machine is the main frame of the test apparatus. The other components are equipped on the testing machine. The test rod was mounted vertically in the center of the reaction tube for the test and the bottom end is fixed to the testing machine. By fixing the top end of the test rod, it can be axially restrained and the load change is measured.

**Figure 3** shows an example of temperature history during the test. The rod is heated up at a rate of 10 K/s. Steam introduction is started prior to the heat-up, and the steam flow is maintained at a supply rate of about 36mg/s during the oxidation. The steam supply rate is sufficiently high to oxidize cladding tubes without steam starvation. With both an increase in rod internal pressure and a decrease in cladding strength, the cladding tube balloons and ruptures at temperatures ranging 1050 to 1100 K during the heat up. The rod is isothermally oxidized after the rupture. Isothermal oxidation temperature and time ranges from 1430 to 1470 K and from 120 to 500s. Four Pt-Pt/13%Rh thermocouples are spot-welded on the outer surface of the cladding, as shown in Fig. 2, to control and measure the cladding temperature. The rod is cooled in the steam flow to about 970 K and is finally quenched with water flooding from the bottom. The average cooling rate is about 20 K/s from the oxidation temperature to 1170 K and about 5 K/s from 1170 to 970 K. Raising rate of water surface during quenching is 30 to 40 mm/s.

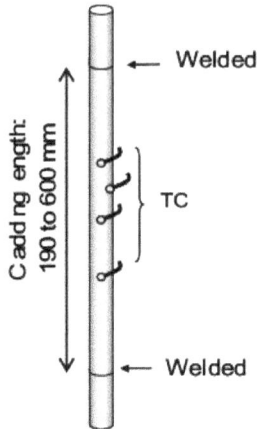

**Fig. 1 Test rod**

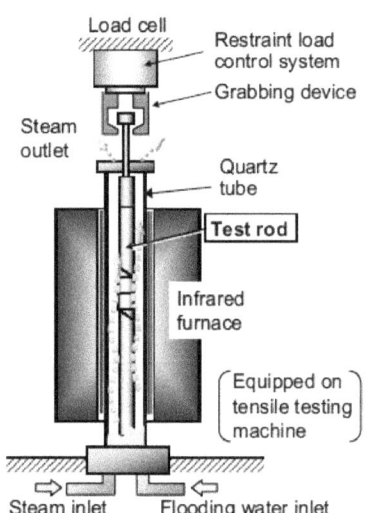

**Fig. 2 Test apparatus**

The test rod is quenched under restrained conditions in the present study. Under the restrained condition, both ends of the test rod are fixed to the tensile testing machine just before the cooling stage initiates. The tensile load history during the cooling and quenching phase of a test is indicated in **Fig. 4**. The tensile load increases as the rod is cooled and quenched. The tensile load was controlled and limited to three different levels of 390±50, 540±50, and 735±50N to realize intermediate constraint conditions, in addition to the fully restrained condition, for unirradiated claddings. The maximum load was limited to about 540 N (30 to 35 MPa for initial metal cross section) for irradiated claddings. The actual restraining conditions can be altered by fuel bundle design, accident sequence, peak clad temperature, the extent of cladding deformation and oxidation, etc. Then, maximum restraint load for irradiated claddings was conservatively determined referring to previous reports that estimated restrained conditions in bundle geometry [4-6].

Equivalent Cladding Reacted (ECR) is used as an indication of oxidation amount. The Baker-Just equation for the oxidation rate [7] was used to estimate the amount of oxygen absorbed during oxidation from oxidation temperature and time. Decrease of cladding thickness due to ballooning is taken into account in calculating ECR. Namely, ECR is calculated for reduced cladding thickness after ballooning.

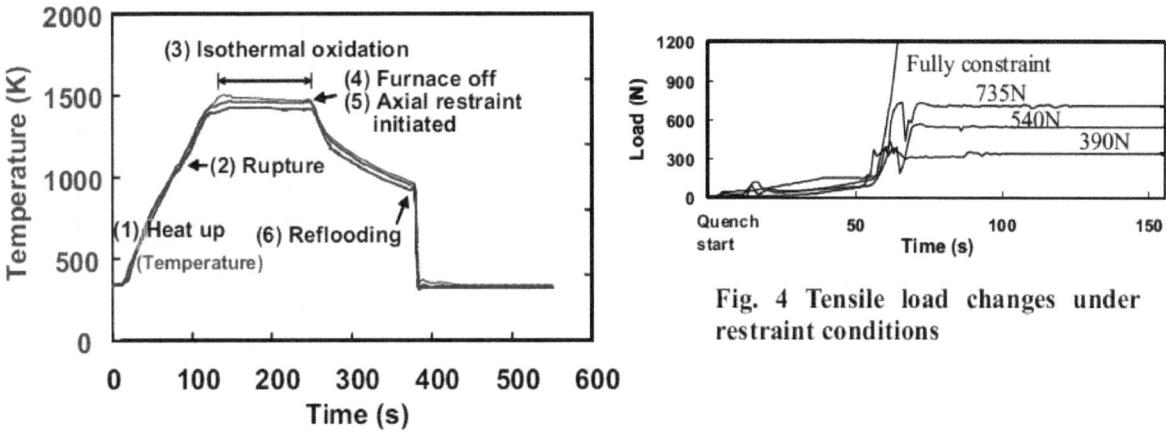

Fig. 3 Temperature history during test

Fig. 4 Tensile load changes under restraint conditions

## 2.2 Hydride effect on fracture condition

**Figures 5** shows fracture maps for two restraint load conditions (390 and 540 N) relevant to the ECR value and the initial hydrogen concentration. These figures show that the fracture threshold decreases with an increase in hydrogen concentration for the lower concentration range, while the fracture boundary is nearly independent of hydrogen concentration for the higher concentration range, which suggests embrittlement saturation in oxidized cladding despite an increase in the hydrogen concentration. **Figure 6** shows fracture maps relevant to ECR values and axial restrained loads for the hydrogen concentration ranging from 350 to 800 ppm. "Restraint load" used in the abscissas are those at fracture for the fractured cladding. These are the maximum restrained load under intermediate restraint conditions and the maximum generated load under the full restraint condition for the no-fracture cladding. The figure clearly shows that the fracture threshold decreases with an increase in the restrained load. Consequently, the

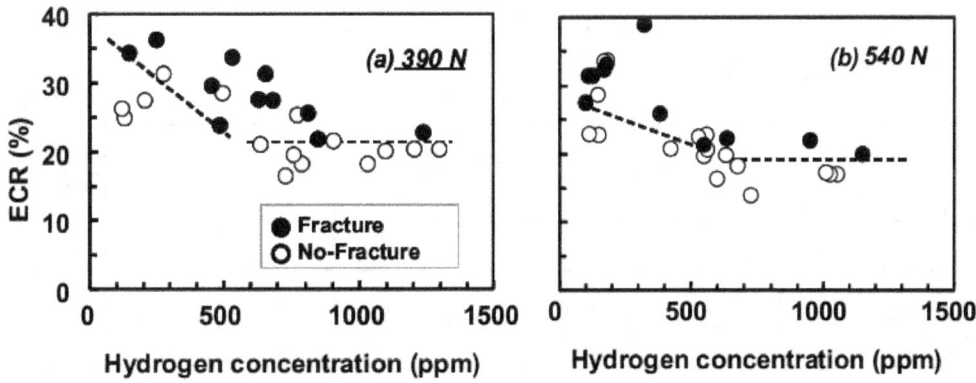

Fig. 5 Fracture maps for two restraint load conditions (390 and 540 N) relevant to the ECR value and the initial hydrogen concentration

fracture threshold is higher than 20% ECR for the whole hydrogen concentration range under the restrained conditions of 390 and 535 N. The ECR value of 20% is sufficiently higher than the limit in the Japanese ECCS acceptance criterion (15% ECR).

### 2.3 Fracture condition of high burnup fuel cladding

Two PWR fuel rods, irradiated to 39 and 44GWd/t (rod average) at Takahama unit-3 reactor, Kansai Electric Power Co., Inc. are currently subjected to the thermal shock tests. The cladding material is Zircaloy-4 containing 1.3 wt% of Sn. The initial outer diameter and thickness were 9.50

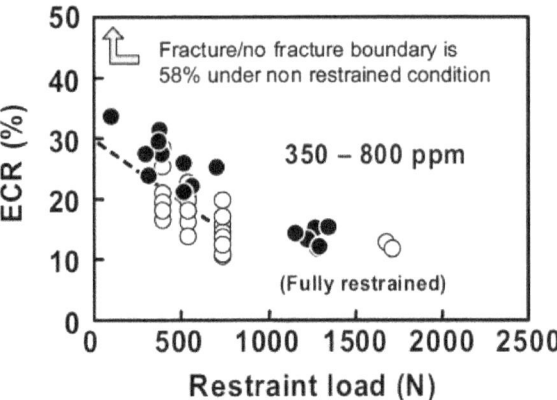

**Fig. 6 Fracture maps relevant to ECR values and axial restraint load**

and 0.57/0.64 mm (Sample No. Ax-x/Bx-x), respectively. Segments of 190 mm-long were cut from the rods and fuel pellets were mechanically removed. Accordingly, only the defueled claddings were subjected to the tests. Six tests have been conducted and information of the cladding tubes used is summarized in **Table 1**. Thickness of oxide layer formed during the reactor operation ranged 15 to 25 μm. Hydrogen concentration is estimated to range from 120 to 210 ppm assuming that 15% of hydrogen generated by corrosion was absorbed.

**Table 1   Summary of integral thermal shock tests with irradiated PWR fuel claddings**

| Test No. | 1 | 2 | 3 | 4 | 5 | 6 |
|---|---|---|---|---|---|---|
| Sample No. | A 3-1 | A1-2 | B L-3 | B I-3 | B I-5 | B L-7 |
| Burn-up (GWd/t) | 43.9 | 43.9 | 39.1 | 40.9 | 40.9 | 39.1 |
| Estimated initial Oxide (μm) [*1] | 20 | 25 | 18 | 18 | 15 | 15 |
| Estimated initial hydrogen (ppm) | 170 | 210 | 140 | 140 | 120 | 120 |
| Rupture temperature (K) | 1073 | 1024 | 1093 | 1058 | 1028 | 1077 |
| Rupture strain (%) | 14.1 | 27.7 | 24.3 | t.m. [*2] | t.m. [*2] | t.m. [*2] |
| Oxidation temperature (K) | 1449 | 1451 | 1427 | 1445 | 1303 | 1450 |
| Oxidation time (s) | 486 | 120 | 200 | 363 | 2195 | 543 |
| ECR (%) | 29.3 | 16.6 | 16.0 | 21.0 [*3] | 22.0 [*3] | 26.4 [*3] |
| Fractured / Survived | F | S | S | S | S | F |
| Load at fracture (N) | 498 | --- | --- | --- | --- | 385 |
| Maximum restraint load (N) | --- | 529 | 529 | 529 | 529 | --- |

*1. Estimated from data of sibling rod [8], *2. t.m.: to be measured, *3.Assumed 20% reduction of cladding thickness by ballooning

Top                                                    Bottom

| 3 cm |

| 1 cm |

Fig. 7 Post-test appearance of high burn-up
PWR cladding oxidized to about 29% ECR
and fractured during quench

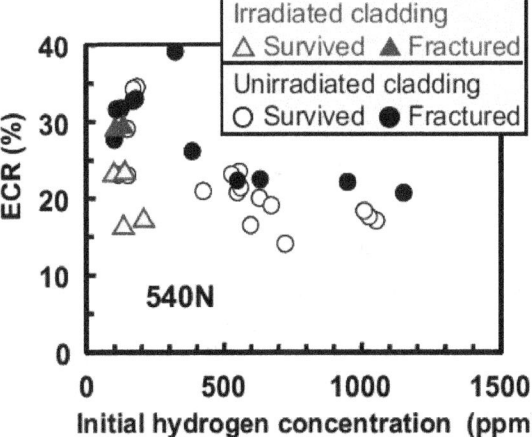

Fig. 8 Fracture maps for irradiated
and unirradiated claddings, relevant to
ECR values and initial hydrogen
concentration

Two cladding, oxidized to about 26 to 29% ECR, fractured during the quench. The post-test appearance is shown in **Fig. 7**. It is considered that cracking initiated at the rupture opening and propagated circumferentially. There is no difference in fracture position and direction of crack propagation between the irradiated and non-irradiated claddings. Fracture/no-fracture conditions of irradiated claddings are compared with those of unirradiated claddings in **Fig. 8**, relevant to the ECR value and the initial hydrogen concentration. Since the failure boundary of the non-irradiated claddings lies at about 28% ECR at about 200 ppm, the fracture of the irradiated claddings is consistent with the fracture criteria for non-irradiated claddings with a similar hydrogen concentration. Therefore, fracture boundary appears not to be reduced so significantly by irradiation to the examined burnup level. Four claddings oxidized to about 16 to 22% ECR, respectively, survived the quench. The fracture boundary is between 22 and 26% ECR for these high burn-up fuel claddings, and it is higher than the limit in the Japanese ECCS acceptance criterion (15% ECR).

Axial profiles of hydrogen concentration in a fractured irradiation cladding (A3-1) and pre-hydrided,

unirradiated claddings are shown in **Fig. 9**. It is known that significant amount of hydrogen is absorbed locally apart from the rupture position [2]. Steam invaded from the rupture opening oxides the inner surface of the cladding, and generated and stagnated hydrogen is absorbed positions where hydrogen partial pressure is very high. The figure indicates that the peak position of the secondary hydriding may be closer to the rupture position in the irradiated cladding. The cladding always fractures at the rupture position under the restrained condition. If the peak position is generally close to the rupture position, hydrogen concentration at the fracture position is higher and it may affect the fracture condition. Therefore, the secondary hydriding in high burnup fuel claddings should be carefully investigated.

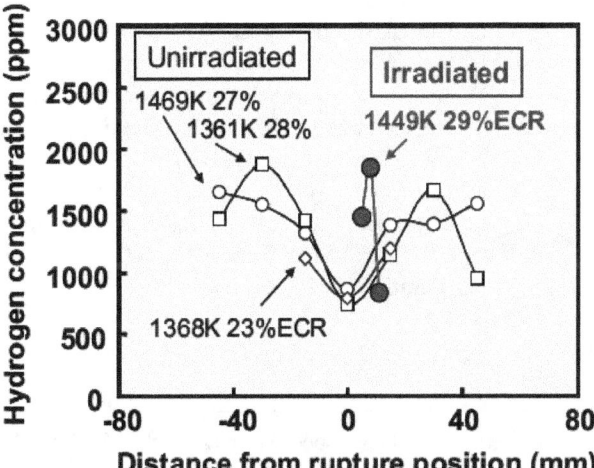

Fig. 9 Axial profiles of hydrogen
concentration in unirradiated and irradiated
claddings after integral thermal shock tests

## 2.4 Future plan

The JAERI plans to perform LOCA-related studies with PWR and BWR claddings (MDA[1], NDA[2], ZIRLO™, M5, and Zircaloy-2) highly irradiated to about 70GWd/t. The influence of further burnup extension and new alloys will be investigated in detail.

## 3. Summary

In order to evaluate the fracture threshold of high burn-up fuel, JAERI performs integral thermal shock tests, simulating LOCA conditions. The tests were performed with pre-hydrided, unirradiated claddings and high burn-up fuel claddings irradiated to 39 and 44 GWd/t at a PWR. Two of irradiated claddings, oxidized at about 1450 K up to 26 and 29% ECR, fractured during quenching. The conditions are consistent with the fracture threshold derived from unirradiated claddings with a similar hydrogen concentration. Claddings, oxidized to about 22% ECR, survived the quench. This indicates that failure boundary is not reduced significantly by PWR irradiation in the examined burnup level. The transient secondary hydriding is localized more towards the rupture opening in the irradiated cladding than in unirradiated claddings. The influence of further burnup extension and application of new cladding materials on fuel behavior under LOCA conditions will be investigated.

## Acknowledgments

The integral thermal shock tests were carried out as the collaboration program between JAERI and Japanese PWR utilities.

## REFERENCE

1.  H. Uetsuka et al., J. Nucl.Sci.& Tech. Vol.20, No.9, pp.705, 1981.
2.  H. Uetsuka et al., report JAERI-M 9445, April, 1981 (Text in Japanese).
3.  H. Uetsuka et al., J. Nucl.Sci.& Tech., Vol.20, No.11, pp 941, 1983.
4.  K. Honma et al., ANS Annual Meeting, Milwaukee, Wisconsin, June 17-21, 2001.
5.  T. Murata et al., ANS Annual Meeting, Milwaukee, Wisconsin, June 17-21, 2001.
6.  N. Waeckel et al, OECD/NEA SEG/FSM LOCA Meeting March 22-23, 2001, Aix-en-Province, March, 2001.
7.  L. Baker and L.C. Just, ANL-6548, Argonne National Laboratory, (1962).
8.  "1998 Fiscal Year Report for NUPEC High Burnup Irradiation Test," Nuclear Power Engineering Corporation, 1999.

---

[1] Mitsubishi Developed Alloy (Zr-0.8Sn-0.2Fe-0.1Cr-0.5Nb) was developed by Mitsubishi Heavy Industries, Ltd.

[2] New Developed corrosion resistance Alloy (Zr-1.0Sn-0.27Fe-0.16Cr-0.1Nb-0.01Ni) was developed by Nuclear Fuel Industries, Ltd. and Sumitomo Metal Industries, Ltd.

# Realistic Assessment of Fuel Rod Behavior Under Large-Break LOCA Conditions

M. E. Nissley, C. Frepoli, K. Ohkawa
Westinghouse Electric Company

Fuel clad swelling and rupture can occur during a loss of coolant accident (LOCA), depending on the core heatup transient and the pressure differential across the cladding. Clad rupture will lead to release of fission products from the fuel, and double-sided metal-water reaction (oxidation) within the ballooned region. In order to simplify the radiological dose calculations, it is typically assumed that 100% of the rods in the core fail. However, it is instructive to consider what a realistic failure fraction might be under more representative conditions. The objective of this study will be to assess the extent of failure and the consequences for the cladding oxidation for the large break LOCA scenario, with and without detailed treatment of uncertainties.

The first assessment will use a deterministic calculation of a large break LOCA under normal operating (baseload) conditions, using the realistic computer program WCOBRA/TRAC. In order to ensure some cladding rupture, a full train of ECCS will be assumed lost (worst single failure) and bounding rod power conditions in the lead fuel assembly will be used. Estimates of the extent of rupture throughout the core will be made by considering peak cladding temperature dependence on rod power, rupture temperature as a function of cladding pressure differential, burnup effects on rod internal pressure, and a core-wide census of rod power and burnup. A comparison of the maximum local oxidation within and away from the ballooned region will also be made.

The second assessment will use the results from a best-estimate plus uncertainties analysis of a large break LOCA, performed using methods consistent with US design basis LOCA regulatory requirements. Uncertainties in thermal-hydraulic models, plant operating conditions, and fuel rod models are accounted for in this method by simultaneously sampling from the uncertainty distributions of each parameter for each transient case. The plant operating conditions considered in the uncertainty analysis include transient power distributions, such that more severe axial shapes and higher linear heat rates are considered than in the first assessment. The extent of rupture within the uncertainty cases will be examined, and conclusions drawn relative to the threshold for rupture. Maximum local oxidation within and away from the ballooned region will be reviewed for the most limiting cases, and those results will be assessed for their dependence on the related fuel rod uncertainty parameters (burst strain, degree of fuel relocation, etc.).

The information presented in these assessments should be interpreted as illustrative and representative. Extent of rupture and degree of oxidation are highly dependent on the transient conditions, which are highly dependent on plant-specific parameters such as core power, nuclear peaking factors, ECCS capacity and other factors.

# Realistic Assessment of Fuel Rod Behavior Under Large-Break LOCA Conditions

M. E. Nissley, C. Frepoli and K. Ohkawa

Nuclear Safety Research Conference
Washington, DC
October 25-27, 2004

# Presentation Objectives

**Part I:** Review Factors Affecting Extent of Rupture and Oxidation

**Part II:** Deterministic Analysis for Normal Operation

- Compare baseload and licensing basis results
- Estimation of extent of rupture for baseload operation

**Part III:** Review of Results from a Statistical Analysis

BNFL

 Westinghouse

# Part I: Factors Affecting Extent of Rupture

Number of Rods at High Temperature

- Assembly loading pattern drives rod power census
- Plant ECCS design limits achievable temperatures

Cladding Pressure Differential

- Initial backfill pressure
- Fission gas generation, release with burnup
- Effects of integral poisons
- Break size (small break vs. large)

# Factors Affecting Extent of Oxidation

Time at Elevated Temperature

- ECCS Design
- Core Power Level
- Steam Generator Tube Plugging
- Containment Design

Double-Sided Reaction at Burst Elevation

- May or May Not be Same as PCT Elevation

# Part II: Deterministic Assessment

## Objectives

- Estimate extent of cladding rupture if LOCA initiated from normal operating conditions

- Compare realistic and licensing basis results for PCT and oxidation

BNFL

 Westinghouse

# Method

1. Select Transient Case from Existing Licensing Basis Analysis that Approximates Normal Operating Conditions for US Plants

2. Estimate Burst Temperature Threshold as a Function of Burnup

3. Use Rod Power Census from Core Depletion Calculations to Synthesize Estimate of Extent of Rupture

BNFL

 Westinghouse

237

# Plant Selection

4-Loop Plant, 3600 MWt, 17x17 VANTAGE-5

- Peak linear heat rate limit = 15.3 kw/ft (FQ = 2.6)

- Hot channel average heat rate limit = 10.0 kw/ft (FdH = 1.70)

- Licensing basis results:
  - PCT$^{95\%}$ = 1140°C
  - ECR = 12% (burst elevation)

BNFL

 Westinghouse

# Case Selection

| Peaking Factor | Tech Spec Limit | Actual @ Limiting Burnup | This Study |
|---|---|---|---|
| FQ | 2.60 | 1.80 | 2.10 |
| FdH | 1.70 | 1.50 | 1.73 |
| $P_{HA}$ | N. A. | 1.42 | 1.66 |

BNFL  Westinghouse

# Axial Power Shape (Baseload)

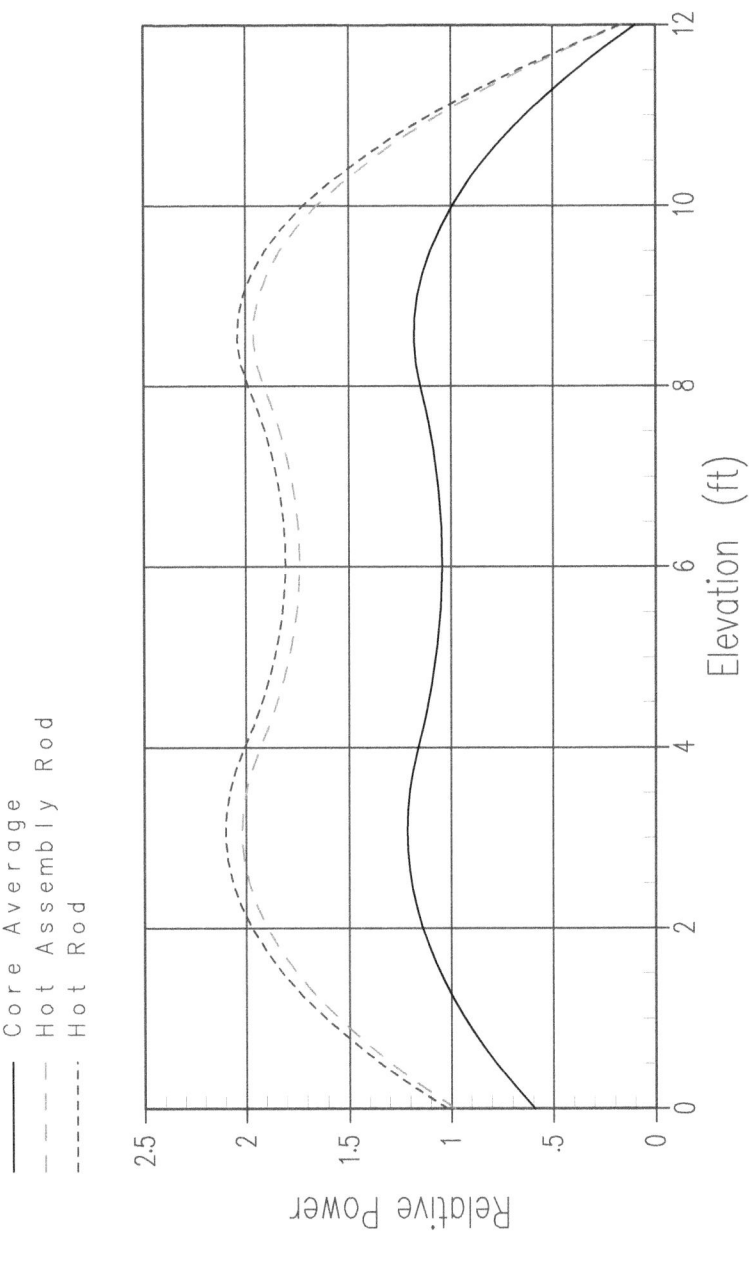

BNFL

Westinghouse

# PCT Response for Baseload Shape

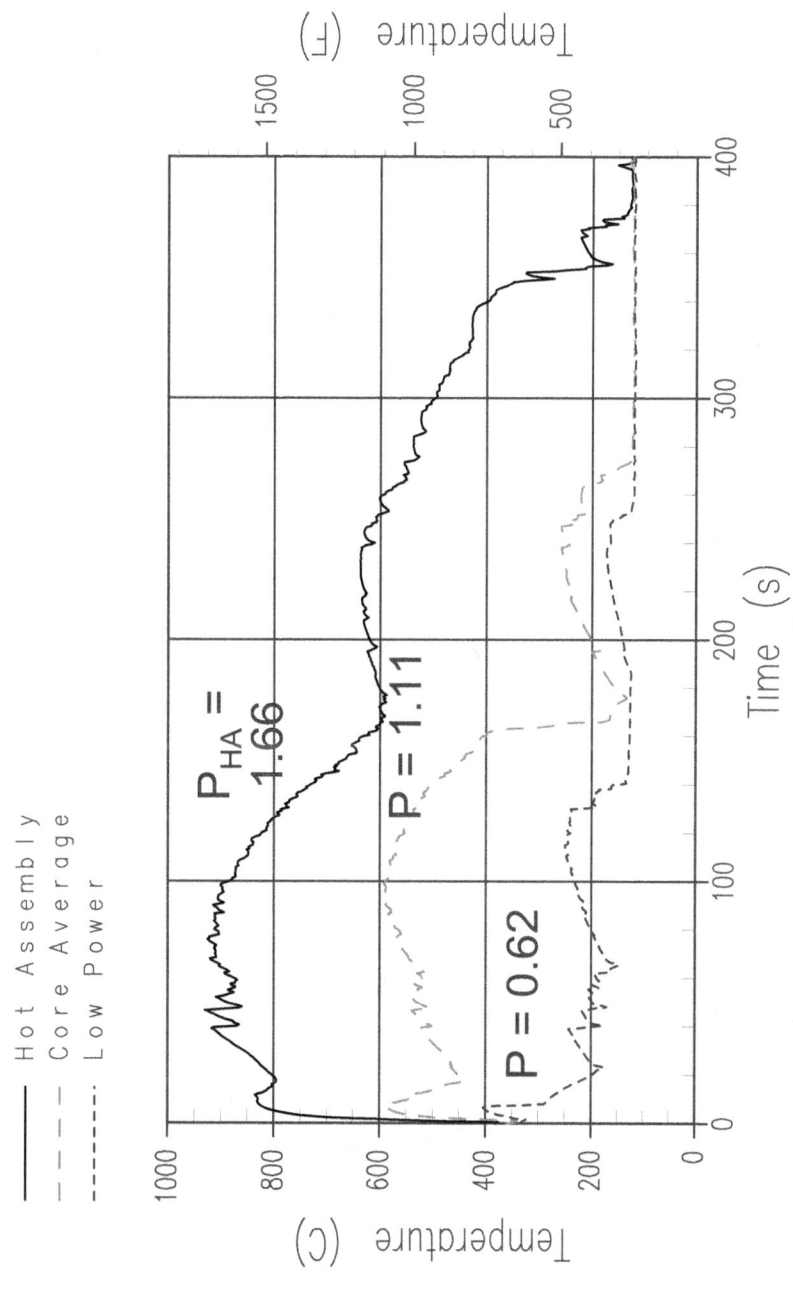

BNFL

Westinghouse

# Hot Assembly PCT and Burst

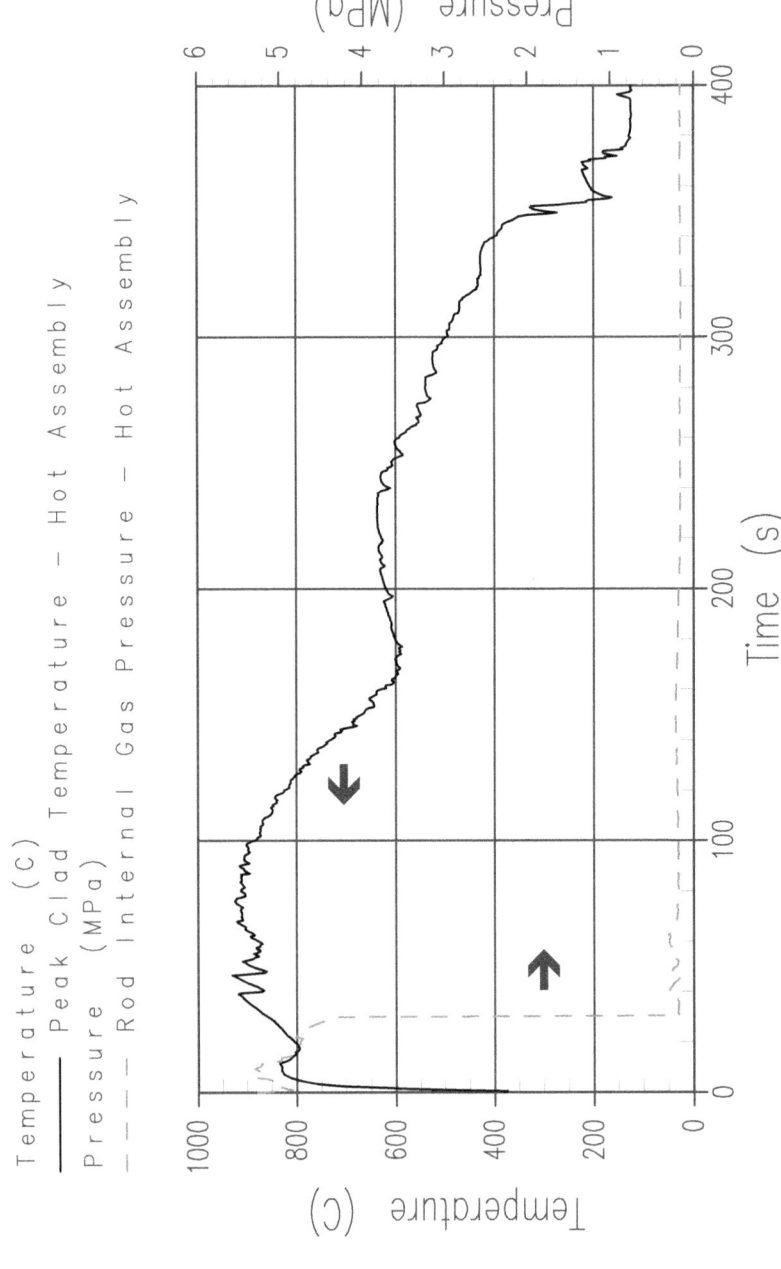

Temperature (C)
—— Peak Clad Temperature – Hot Assembly
Pressure (MPa)
–––– Rod Internal Gas Pressure – Hot Assembly

BNFL

Westinghouse

# Rupture Temperature vs. Burnup

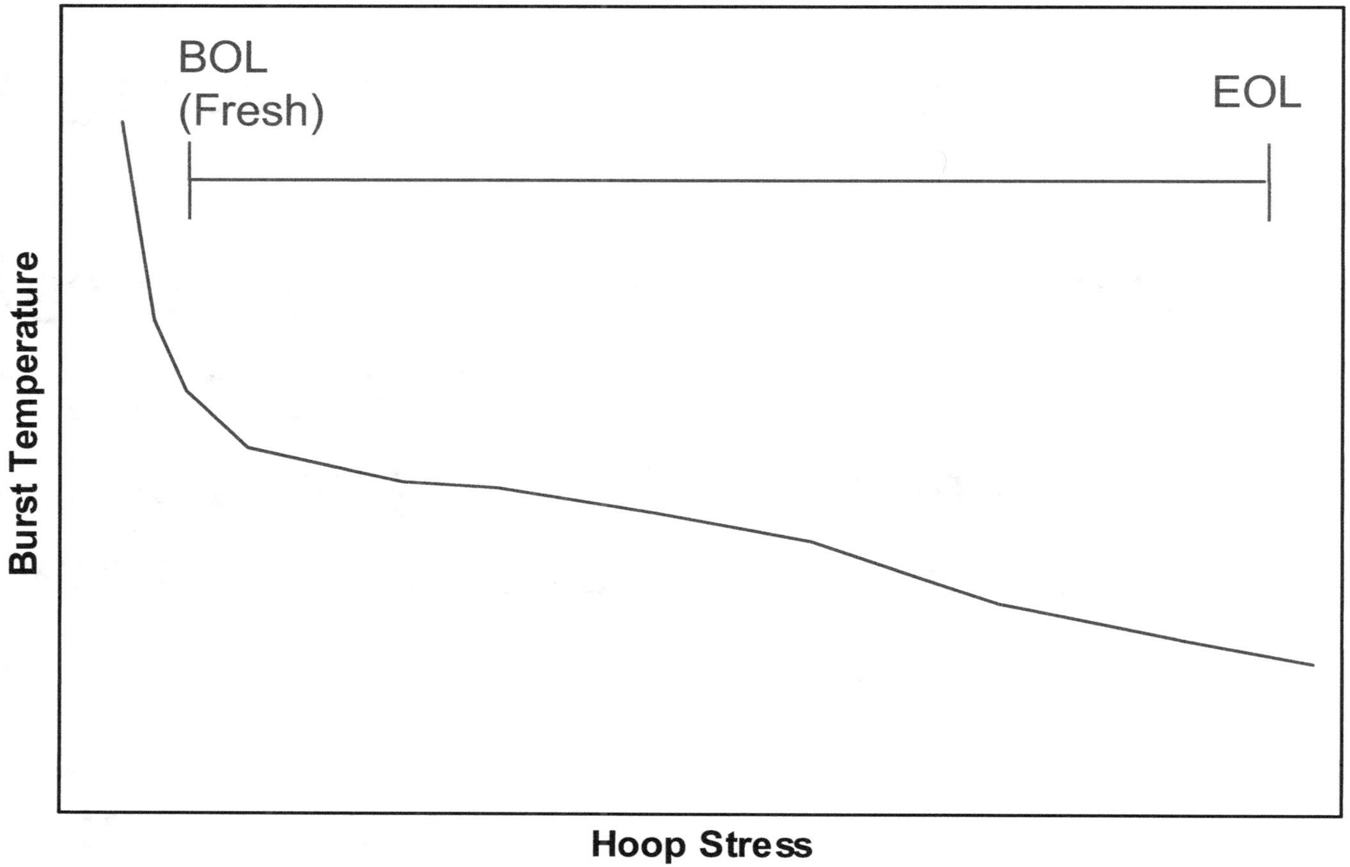

BNFL       Westinghouse

243

# Rupture Temperature Threshold Conclusions

Conservative Threshold Established for Temperatures Resulting in Rupture

- $\geq$ 816°C (1500°F) for fresh fuel
- $\geq$ 760°C (1400°F) for once- or twice-burned fuel

Next, Review Core Power Census

244

BNFL

 Westinghouse

# Fuel Loading Pattern (Quarter Core Symmetry)

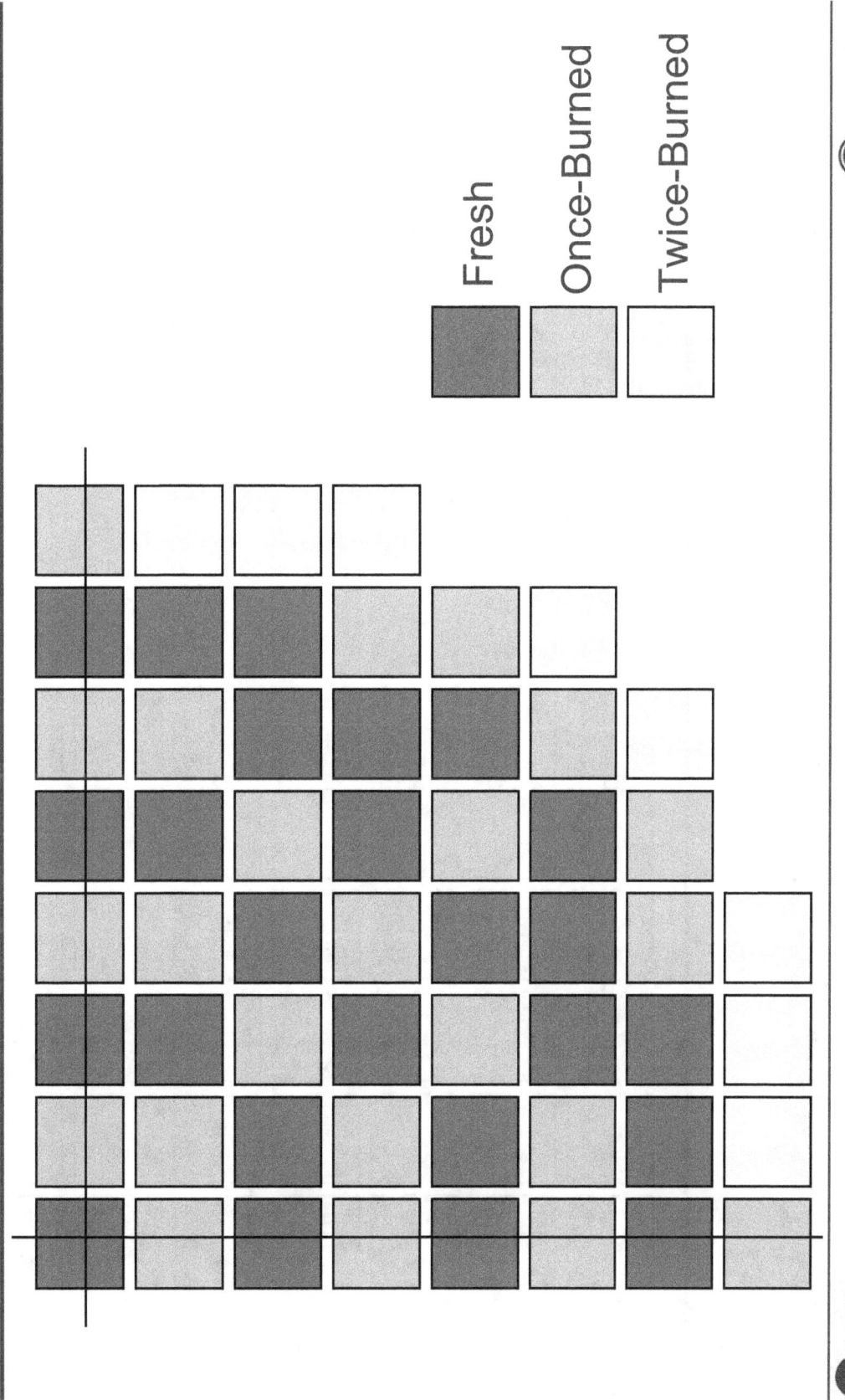

Fresh

Once-Burned

Twice-Burned

# Assembly Power Distribution (Limiting BU)

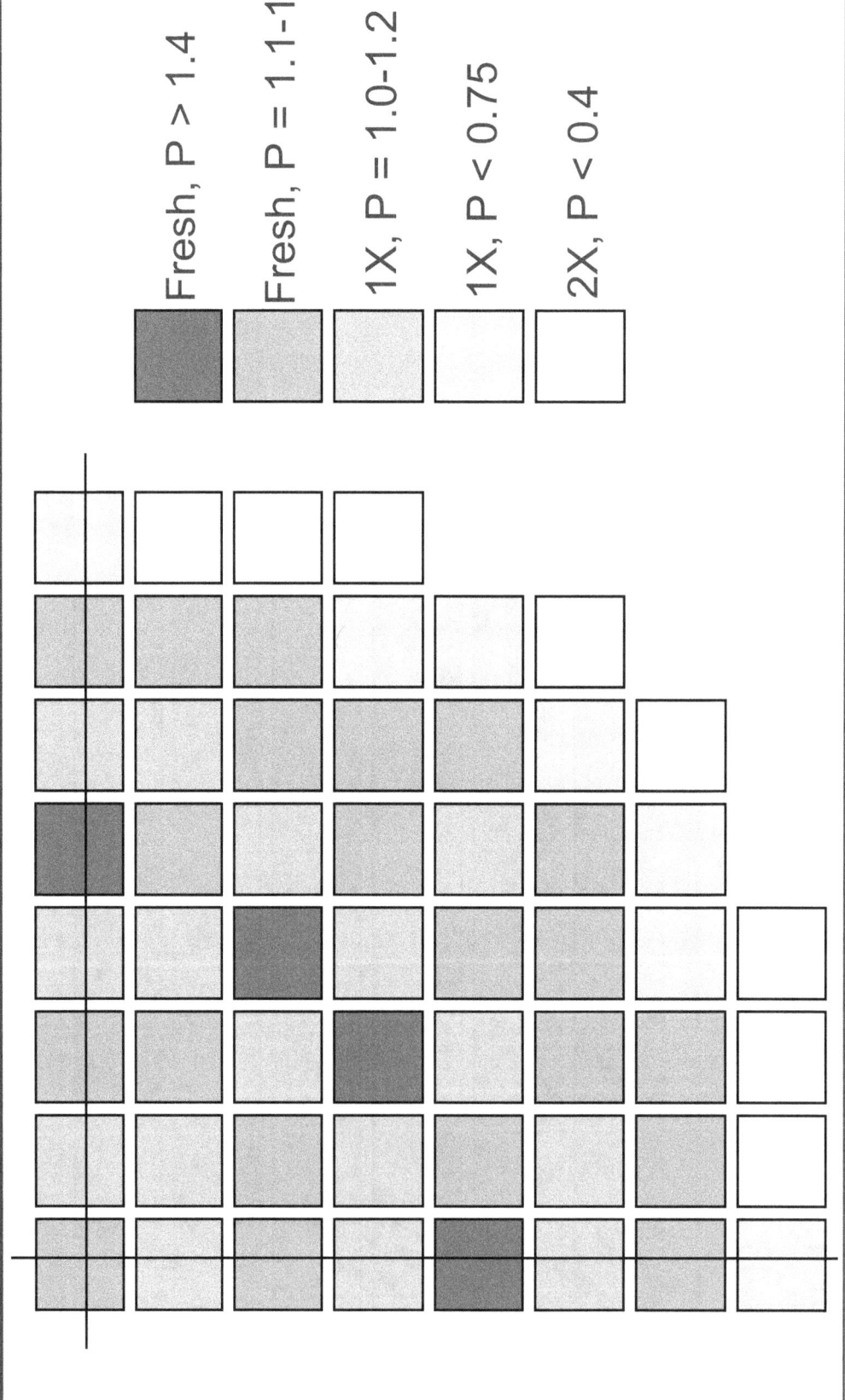

Fresh, P > 1.4

Fresh, P = 1.1-1.4

1X, P = 1.0-1.2

1X, P < 0.75

2X, P < 0.4

BNFL

Westinghouse

# Extent of Rupture for Baseload Operation

Fresh Fuel Assessment

- Rupture will not occur unless P > ~ 1.45
- 12 fresh assemblies exceed P = 1.4 at limiting burnup

Burned Fuel Assessment

- Rupture will not occur unless P > ~ 1.35
- No burned assemblies exceed P = 1.3 at any burnup

Conclusion: < 10% of Rods in Core will have Rupture

BNFL

Westinghouse

# Realistic vs. Licensing Basis Results

|  | PCT | ECR (Burst) | ECR (Non-Burst) |
|---|---|---|---|
| Licensing | 1140°C | 12% | 6% |
| Realistic | 944°C | 1.4% | 0.8% |

BNFL

 Westinghouse

# Part III: Results of Statistical Analysis

Most Recent Westinghouse BEPU Methodology Uses Non-Parametric Sampling Method Based on Order Statistics

Demonstration Analysis for Licensing Used 59 Cases to Estimate PCT at 95/95 Level

- Each case sampled:
  - system-wide physical models (break flow, etc.)
  - local fuel rod models (heat transfer, relocation, etc.)
  - plant system state (power shape, etc.)

# Plant Selection

4-Loop Plant, 3216 MWt, 15x15 VANTAGE+

- Peak linear heat rate limit = 17 kw/ft (FQ = 2.5)

- Hot channel average heat rate limit = 11.7 kw/ft (FdH = 1.72)

- Demonstration case results:
  - PCT = 1037°C
  - ECR = 2.1% (burst elevation)

250

BNFL

Westinghouse

# Limiting Case: PCT and Burst

Temperature (C)
—— Peak Clad Temperature – Hot Assembly
Pressure (MPa)
– – – Rod Internal Gas Pressure – Hot Assembly

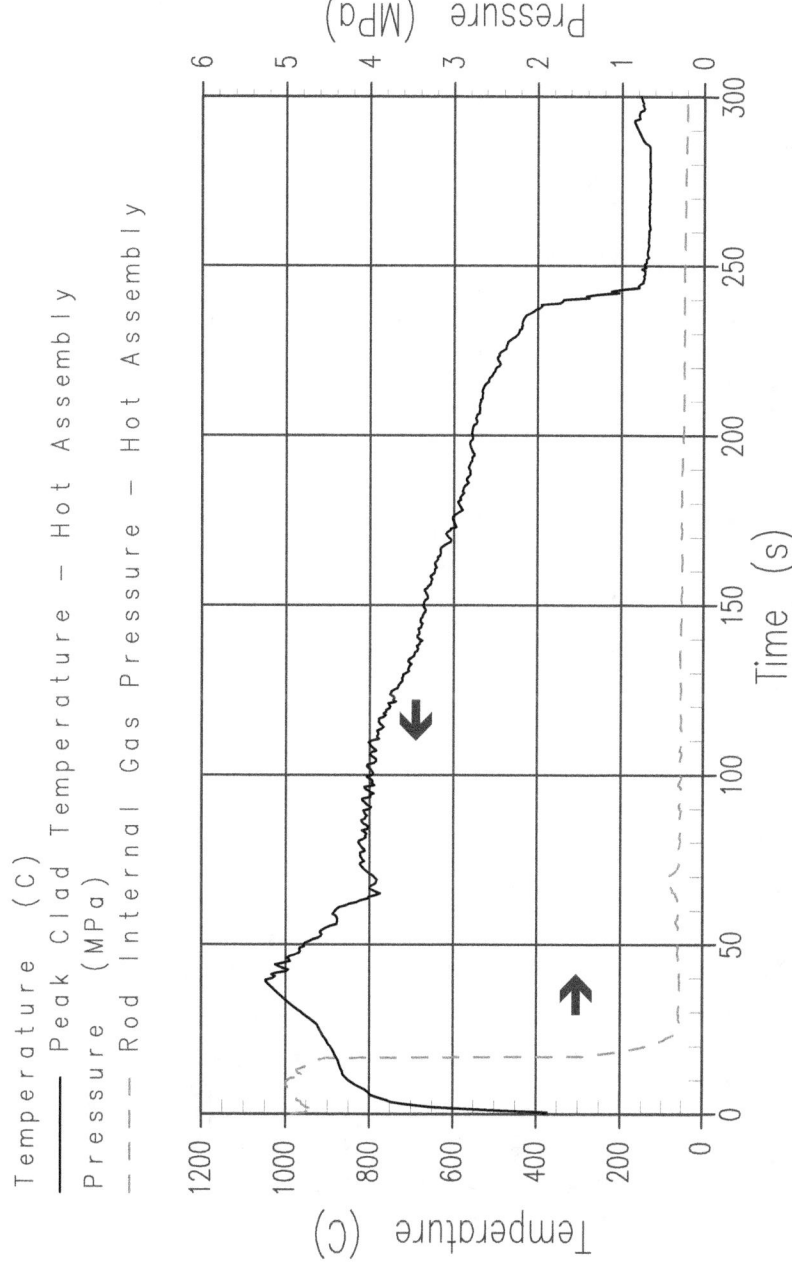

# Limiting Case:
# Axial Distribution of PCT and Oxidation

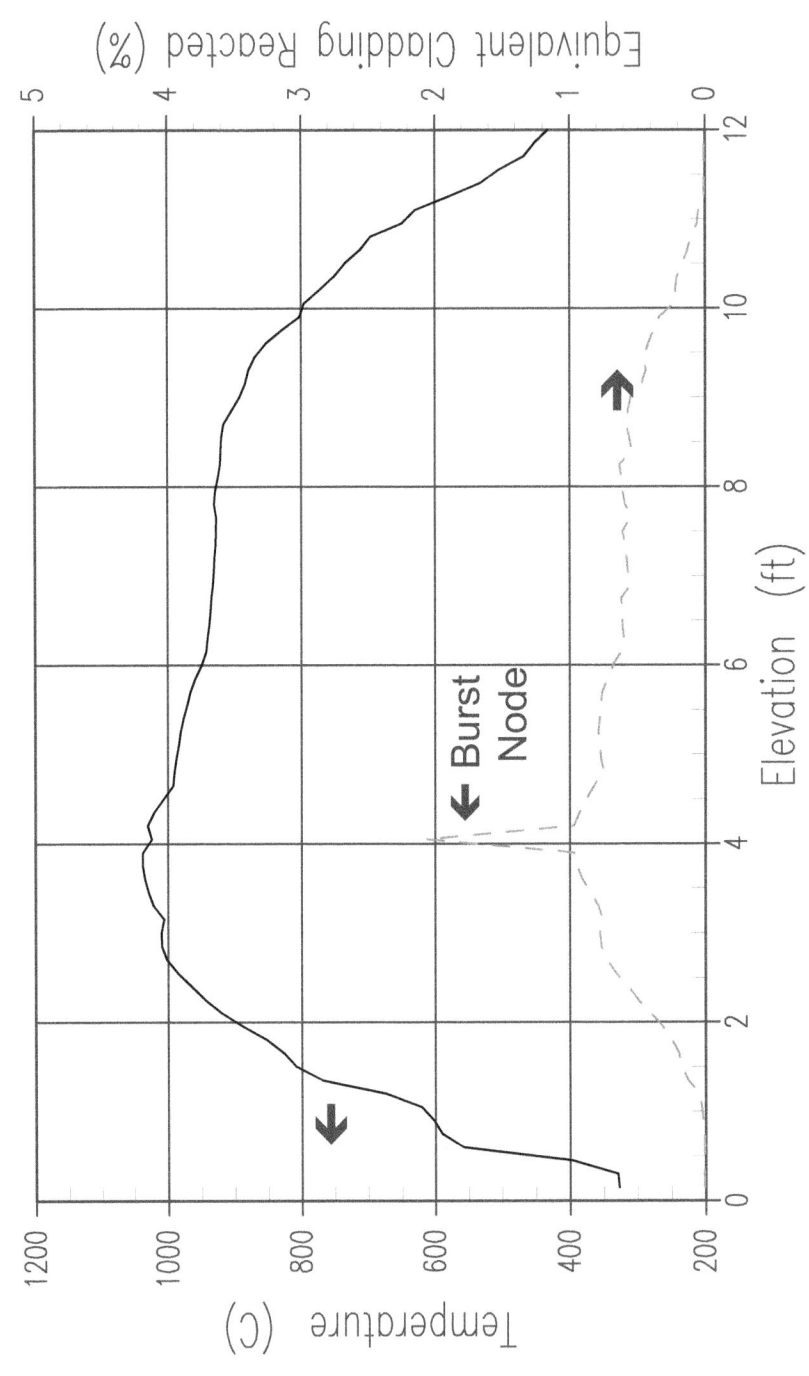

# Case 3: PCT and Burst

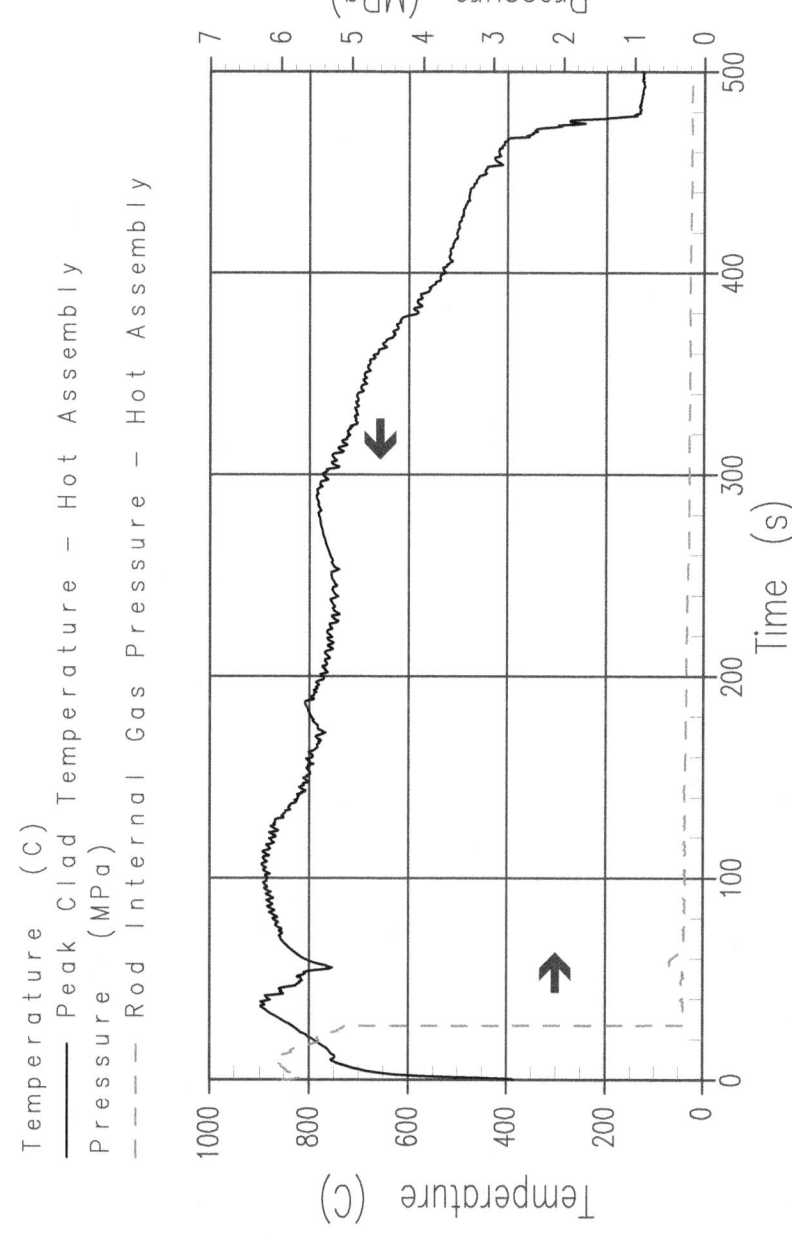

Temperature  (C)
—— Peak Clad Temperature – Hot Assembly
Pressure  (MPa)
– – – Rod Internal Gas Pressure – Hot Assembly

BNFL

Westinghouse

# Case 3:
## Axial Distribution of PCT and Oxidation

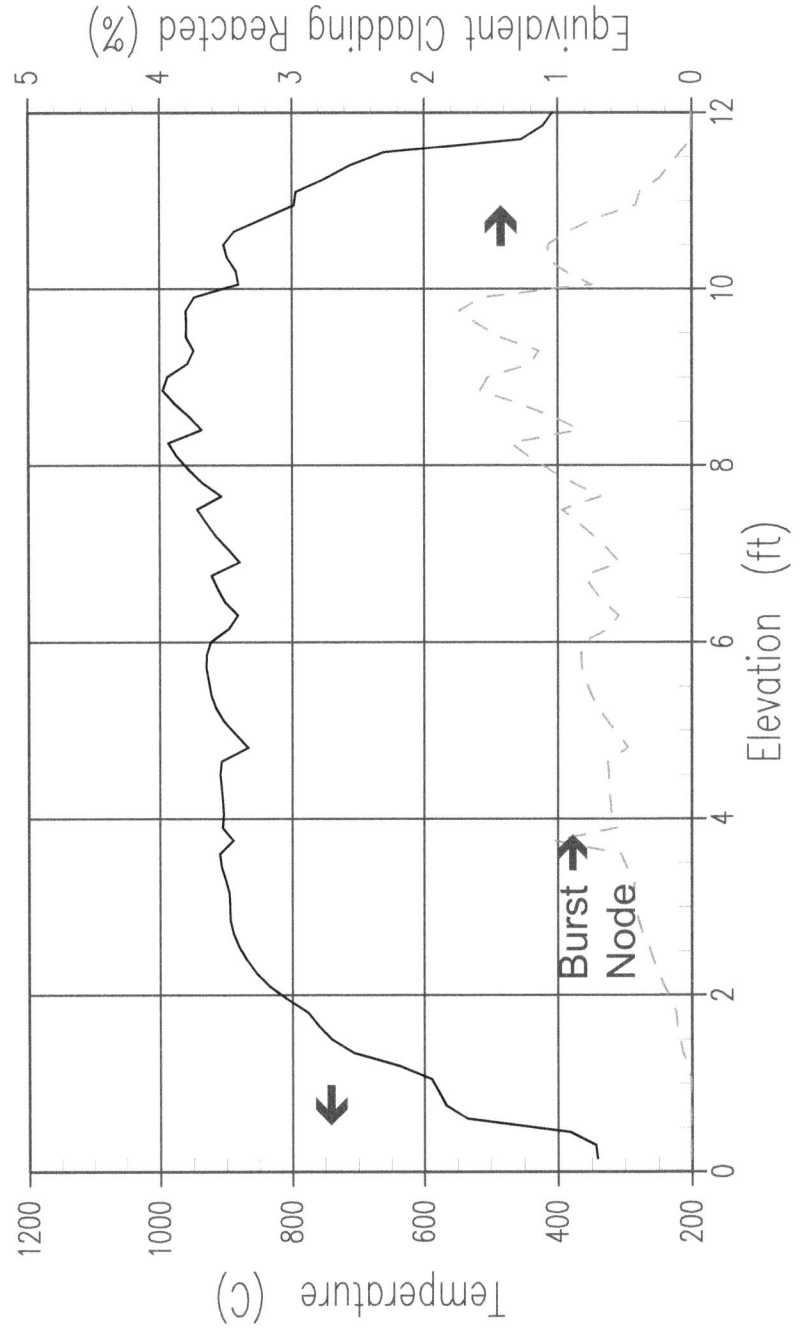

BNFL  Westinghouse

# Summary of Most Limiting Cases

| PCT (°C) | 1037 | 1035 | **995** | 973 | 964 | 959 | 933 | 928 | 925 |
|---|---|---|---|---|---|---|---|---|---|
| ECR (Burst), % | **2.1** | 1.6 | **1.0** | 1.4 | 1.2 | 1.3 | 1.4 | 1.0 | 1.0 |
| ECR (Non-Burst) | **1.0** | 1.0 | **1.7** | 0.7 | 0.6 | 1.2 | 1.3 | 0.5 | 0.6 |
| PCT @ Burst? | No | No | **No** | No | No | No | No | No | No |
| Burst Strain (%) | **40** | 44 | **47** | 44 | 38 | 48 | 60 | 39 | 24 |
| Packing Fraction | **0.66** | 0.78 | **0.65** | 0.75 | 0.62 | 0.67 | 0.71 | 0.69 | 0.67 |

BNFL

Westinghouse

# Conclusions

## Deterministic Assessment of Baseload Operation

- Significant margin to licensing basis results
  - PCT ~ 200°C lower
  - Oxidation greatly reduced (< 2% vs. 12%)

- Extent of rupture relatively low (< 10%)

BNFL

 Westinghouse

# Conclusions (cont'd)

## Statistical Analysis – Burst Node Conclusions

- PCT elevation is not at the burst for limiting cases

- Maximum ECR generally is

- Would be instructive to repeat at higher temperatures
  - Note that this study bounds temperatures expected for high burnup $UO_2$ fuel

BNFL

Westinghouse

# Realistic Assessment of Fuel Rod Behavior Under Large-Break LOCA Conditions

M. E. Nissley, C. Frepoli, K. Ohkawa
Westinghouse Electric Company

## Abstract

Ballooning and rupture of nuclear fuel rod cladding during a postulated loss of coolant accident is explicitly accounted for in most Evaluation Models used to perform design basis analyses. Of particular interest are the combined effects of the resulting thinning of the cladding and the double-sided metal-water reaction on the peak cladding temperature and maximum local oxidation. Cladding rupture and oxidation predictions for a large break LOCA scenario are examined in order to gain insights into typical PWR transient and design basis analysis results. The first assessment investigates the extent of core-wide fuel cladding rupture in a deterministic analysis of a transient initiated from normal operating conditions. The peak cladding temperature, and the maximum local oxidation within and away from the rupture location, are then compared with the results from a design basis analysis performed for the same plant. Dramatic reductions are seen when more realistic, yet conservative, assumptions are used. The second assessment examines the extent of local oxidation within and away from the rupture location for the most limiting cases from a design basis analysis performed using a non-parametric order statistics method. The results are also reviewed to examine whether the limiting PCT elevation and the rupture elevation were coincidental. For this analysis the maximum local oxidation is generally at the rupture, while the PCT occurs away from the rupture.

## Introduction

Ballooning and rupture of nuclear fuel rod cladding can occur during a loss of coolant accident (LOCA), depending on the core heatup transient and the pressure differential across the cladding. Cladding rupture will lead to release of fission products from the fuel, and double-sided metal-water reaction (oxidation) within the ballooned region. In order to simplify the radiological dose calculations, it is typically assumed that 100% of the rods in the core fail. However, it is instructive to consider what a realistic failure fraction might be under more representative conditions. The objective of this study will be to assess the extent of failure and the effect of rupture on the peak cladding temperature (PCT) and maximum local oxidation for the large break LOCA scenario, with and without detailed treatment of uncertainties.

## Extent of Fuel Cladding Rupture

The first assessment used a deterministic calculation of a large break LOCA under normal operating (baseload) conditions, with the goal of estimating the extent of fuel cladding rupture throughout the core. The system response was analyzed using the realistic system thermal-hydraulic computer program WCOBRA/TRAC [1]. WCOBRA/TRAC is an improved version of the COBRA/TRAC code, originally developed at Pacific Northwest Laboratory [2] by combining the COBRA-TF code [3] and the TRAC-PD2 code [4].

The extent of fuel cladding rupture during a large break LOCA is dependent on a number of factors. Foremost is the number of rods in the core that achieve high cladding temperatures during the transient.

This is a function of the rod power census and the ECCS design capability. Another important factor is the cladding pressure differential for the various rods, which is affected by the initial backfill pressure, the fission gas generation with burnup, the effects of integral poisons, and the break size. A larger number of fuel rods experiencing rupture would typically be expected for large breaks than small breaks, due to the lower system pressure at the time of elevated cladding temperatures.

This study considered a double-ended cold leg break in a 4-loop Westinghouse plant rated at 3600 MWt, operating with 17x17 VANTAGE-5+ fuel. In order to ensure some cladding rupture, a full train of ECCS was assumed to be lost (worst single failure) and bounding baseload rod power conditions in the lead fuel assembly were used. Table 1 compares the peaking factors assumed in this study with the Technical Specification limits, and the actual maximum values predicted to occur throughout the cycle. The assumed peaking factors are on the order of 15% higher than the maximum expected values. Figure 1 shows the axial power shape assumed for this study.

Estimates of the extent of rupture throughout the core were made by considering peak cladding temperature dependence on rod power, rupture temperature as a function of cladding pressure differential, burnup effects on rod internal pressure, and a core-wide census of rod power and burnup. Figure 2 shows the peak cladding temperature response predicted by WCOBRA/TRAC for the hot assembly average rod (relative power of 1.66), a core balance rod (relative power of 1.11), and a low power rod (relative power of 0.62). The cladding temperature excursion during reflood (~ 100 seconds) is seen to increase approximately linearly with rod power. Figure 3 shows the rod internal pressure response for the hot assembly average rod, and indicates that cladding rupture occurs for high powered rods during the refill period of this transient scenario. Figure 4 illustrates the general trend for the cladding rupture temperature to decrease as the differential pressure across the cladding increases. The data supporting Figure 4 are proprietary for the Westinghouse ZIRLO$^{TM}$ cladding, but the trend is similar to those previously published for Zircaloy cladding [5]. The cladding rupture data supporting this trend, combined with fuel performance code predictions of rod internal pressure throughout life, indicate that a threshold of about 820°C (~ 1500°F) can be established as the minimum expected rupture temperature for fresh fuel in its first cycle of irradiation. Similarly, a threshold of about 760°C (~ 1400°F) can be established as the minimum expected rupture temperature for once- or twice-burned fuel.

A review of Figure 2 indicates that with these threshold cladding rupture temperatures, fresh fuel will not rupture in this scenario unless the rod relative power exceeds about 1.45 (corresponding to 820°C). Fuel in its second or third cycle of irradiation will not rupture unless the rod relative power exceeds about 1.35 (corresponding to 760°C). Figure 5 shows the predicted assembly-wise power distribution (quarter-core symmetry) at the burnup at which this core has the maximum number of high power assemblies. This loading pattern is representative of reload designs in use today. Only 12 fresh assemblies exceed a relative power of 1.40 at this limiting burnup. None of the previously irradiated assemblies exceeds a relative power of 1.30. With this information it can be estimated that less than 10% of the core (12 of 193 assemblies) would achieve sufficient cladding temperatures to have cladding rupture.

It is also instructive to compare the results from this deterministic assessment of a conservative baseload power distribution (peaking factors ~ 15% above actual cycle maximums) with the design basis results. The design basis results in this study used an NRC-approved uncertainty methodology [1] that is closely patterned after the Code Scaling, Applicability and Uncertainty (CSAU) methodology developed under the guidance of the NRC [6]. Uncertainties in thermal-hydraulic models, plant operating conditions, and fuel rod models are accounted for in this method using a combination of response surface equations and Monte Carlo sampling techniques [7]. Table 2 compares the peak cladding temperature, and the equivalent cladding reacted at, and away from, the cladding rupture location. (Equivalent cladding reacted, or ECR, is the same as maximum local oxidation.) The design basis results include the effect of uncertainties in the physical models and the plant operating conditions. They support the Technical Specification peaking factors shown in Table 1, which are much higher than the actual predicted maximums. They also support highly skewed power distributions which would not occur under normal baseload operation, but could occur under extreme load following situations. Not only is the peak cladding temperature reduced substantially with the conservative baseload power distribution, but the equivalent cladding reacted is reduced to negligible amounts.

Review of Statistical Analysis Results

The second assessment used the results from a best-estimate plus uncertainties analysis of a large break LOCA, performed using a non-parametric order statistics method that was recently approved by the NRC [8, 9]. Uncertainties in thermal-hydraulic models, plant operating conditions, and fuel rod models are accounted for in this method by simultaneously sampling from the uncertainty distributions of each parameter for each transient case. The plant operating conditions considered in the uncertainty analysis include transient power distributions, such that more severe axial shapes and higher linear heat rates are considered than in the deterministic case used for the first assessment. The plant used in this study was a 4-loop Westinghouse plant rated at 3216 MWt, operating with 15x15 VANTAGE+ fuel. This study used a sampling of 59 separate large break LOCA transients, each with its own combination of randomly sampled uncertainty parameters. According to the statistical theory, the most limiting of the 59 cases will bound at least 95 percent of the actual PCT distribution, with 95 percent confidence.

The goal of this assessment was to examine the extent of local oxidation within and away from the ballooned region for the most limiting cases, and assess to what degree the limiting PCT elevation and the cladding rupture elevation were coincidental. Table 3 shows the results for all of the cases above 925°C (1700°F). Below this threshold, oxidation levels are very low. The most limiting PCT case (1037°C) is seen to also correspond to the maximum local oxidation case (2.1%). The maximum local oxidation occurred at the rupture elevation in this case, but the PCT did not.

Figure 6 shows the PCT and rod internal pressure response for the hot assembly average rod for the limiting PCT case (Case 1). These results were calculated by WCOBRA/TRAC, and do not include the effect of hot rod model uncertainties. Figure 7 shows the PCT and local oxidation throughout the transient as a function of elevation along the hot rod. These results were calculated by HOTSPOT, which is the fuel rod conduction code used by Westinghouse to account for hot rod model uncertainties. The cladding rupture elevation differs only slightly from the limiting PCT elevation. Their close proximity is due to the relatively short time duration at high temperatures after rupture. The oxidation "spike" at the rupture elevation is due to the thinning of the cladding (40% burst strain), and the double-sided metal-water reaction.

Case 3 in Table 3 is the only one that has the limiting ECR away from the rupture elevation. Figure 8 shows the PCT and rod internal pressure response for the hot assembly average rod for this case. Figure 9 shows the PCT and local oxidation throughout the transient as a function of elevation along the hot rod. The approximate 100°C increase in PCT between Figures 8 and 9 is due primarily to the extremely low reflood heat transfer multiplier sampled for this case. In contrast to Figure 7, the rupture elevation is seen to be much lower than the limiting PCT elevation. This is due to the relatively longer time duration at high temperatures after cladding rupture. The dips in the oxidation profile high in the core are due to grid heat transfer enhancement reducing the local cladding temperature.

Summary and Conclusions

These assessments of cladding fuel rupture and oxidation predictions lead to the following observations:

- The extent of core-wide fuel cladding rupture which would actually be expected in a large break LOCA is far less than the 100% assumed by many US licensees in their radiological dose calculations. Even assuming the worst single failure and a conservative normal operating power shape with linear heat rates 15% higher than predicted, less than 10% of the rods in the core were estimated to have cladding failures.

- Significant margins exist between realistic estimates of PCT and ECR, and those resulting from design basis analyses. Even assuming the worst single failure and a conservative normal operating power shape with linear heat rates 15% higher than predicted, the PCT was reduced by ~ 200°C, and the ECR was reduced to negligible amounts compared to the design basis analysis results.

- The rupture location tends to have the maximum ECR, due to thinning of the cladding and double-sided oxidation.

- PCT frequently occurs away from the rupture location, for plants that have a LOCA transient response similar to Figures 6 and 8 (e.g., 4-loop plants with large dry containment designs).

The information presented in these large break LOCA assessments should be interpreted as illustrative and representative. Extent of rupture and degree of oxidation are highly dependent on the transient conditions, which are highly dependent on plant-specific parameters such as core power, nuclear peaking factors, ECCS capacity and other factors.

References

[1] Bajorek, S. M., et al., 1998, "Code Qualification Document for Best Estimate LOCA Analysis," Technical Reports WCAP-12945-P-A (Proprietary) and WCAP-14747 (Non-Proprietary), Westinghouse Electric Company, LLC, Pittsburgh PA.

[2] Thurgood, M. J., et al., 1983, "COBRA/TRAC – A Thermal-Hydraulics Code for Transient Analysis of Nuclear Reactor Vessels and Primary Coolant Systems," NUREG/CR-3046, Pacific Northwest Laboratory, Richland WA.

[3] Thurgood, M. J., et al., 1980, "COBRA-TF Development," 8th Water Reactor Safety Information Meeting.

[4] Liles, D. R., et al., 1981, "TRAC-PD2, An Advanced Best Estimate Computer Program for Pressurized Water Reactor Loss-of-Coolant Accident," NUREG/CR-2054, Los Alamos National Laboratory, Los Alamos NM.

[5] Powers, D. A., and Meyer, R. O., 1980, "Cladding Swelling and Rupture Models for LOCA Analysis," NUREG-0630, U. S. Nuclear Regulatory Commission, Washington DC.

[6] Boyack, B. E., et al., 1989, "Quantifying Reactor Safety Margins: Application of Code Scaling, Applicability, and Uncertainty Evaluation Methodology to a Large-Break Loss-of-Coolant Accident," NUREG/CR-5249, Idaho National Engineering Laboratory, Idaho Falls ID.

[7] Young, M. Y., et al., 1998, "Best Estimate Analysis of the Large Break Loss of Coolant Accident," ICONE-6252, Proceedings of the 6th International Conference on Nuclear Engineering, Westinghouse Electric Company, LLC, Pittsburgh PA.

[8] Nissley, M. E., et al., 2003, "Realistic Large-Break LOCA Evaluation Methodology Using the Automated Statistical Treatment of Uncertainty Method," Technical Reports WCAP-16009-P (Proprietary) and WCAP-16009-NP (Non-Proprietary), Westinghouse Electric Company, LLC, Pittsburgh PA.

[9] Letter, H. N. Berkow (NRC) to J. A. Gresham, "Final Safety Evaluation for WCAP-16009-P, Revision 0, Realistic Large-Break LOCA Evaluation Methodology Using the Automated Statistical Treatment of Uncertainty Method," November 5, 2004.

Table 1. Nuclear Peaking Factors Considered in Deterministic Assessment of Extent of Rupture

| Nuclear Peaking Factor[1] | Technical Specification Limit | Actual Maximum At Limiting Burnup | This Study |
|---|---|---|---|
| FQ | 2.60 | 1.80 | 2.10 |
| FdH | 1.70 | 1.50 | 1.73 |
| $P_{HA}$ | N.A. | 1.42 | 1.66 |

FQ = Total Peaking Factor = (Maximum Linear Heat Rate) / (Core Average Linear Heat Rate)

FdH = Enthalpy Rise Peaking Factor = (Maximum Rod Power) / (Average Rod Power)

$P_{HA}$ = Hot Assembly Relative Power = (Hot Assembly Power) / (Average Assembly Power)

Table 2. Conservative Baseload Operation Results Compared with Design Basis Results

| | Peak Cladding Temperature | Equivalent Cladding Reacted (Burst) | Equivalent Cladding Reacted (Non-Burst) |
|---|---|---|---|
| Design Basis | 1140°C | 12% | 6% |
| Conservative Baseload Conditions | 944°C | 1.4% | 0.8% |

Table 3. Results Summary for Most Limiting Cases

| Case | 1 | 2 | 3 | 4 | 5 | 6 | 7 | 8 | 9 |
|------|-----|------|-----|-----|-----|-----|-----|-----|-----|
| PCT (°C) | **1037** | 1035 | **995** | 973 | 964 | 959 | 933 | 928 | 925 |
| ECR (Burst), % | **2.1** | 1.6 | **1.0** | 1.4 | 1.2 | 1.3 | 1.4 | 1.0 | 1.0 |
| ECR (Non-Burst) | **1.0** | 1.0 | **1.7** | 0.7 | 0.6 | 1.2 | 1.3 | 0.5 | 0.6 |
| PCT @ Burst? | **No** | No | **No** | No | No | No | No | No | No |
| Burst Strain (%) | **40** | 44 | **47** | 44 | 38 | 48 | 60 | 39 | 24 |
| Packing Fraction* | **0.66** | 0.78 | **0.65** | 0.75 | 0.62 | 0.67 | 0.71 | 0.69 | 0.67 |

\* Fraction of the available volume at the rupture elevation that contains pellet fragments following relocation

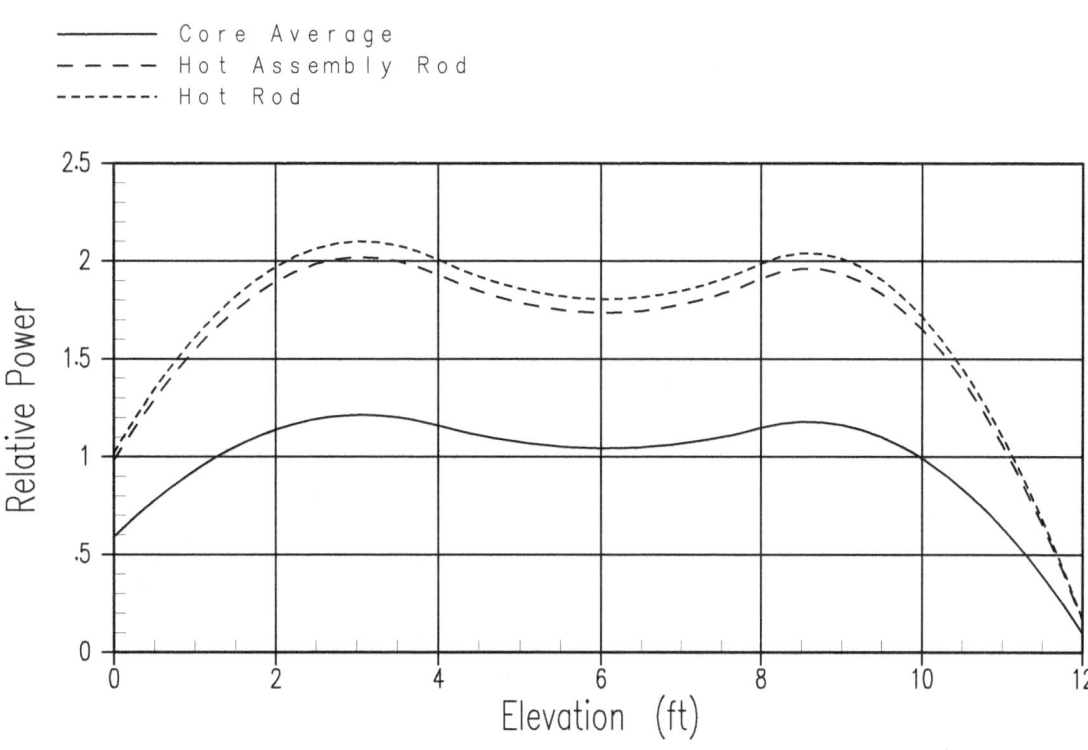

Figure 1. Axial Power Shape Used in Deterministic Assessment of Extent of Rupture

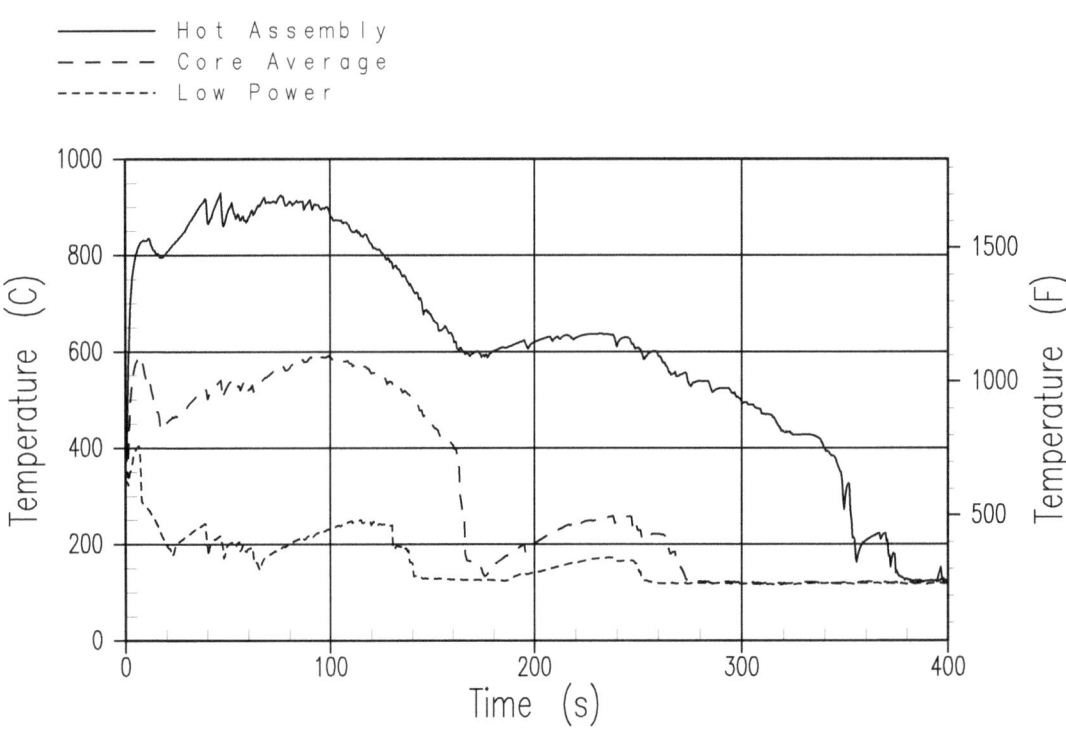

Figure 2.  Peak Cladding Temperature Response for Deterministic Assessment of Extent of Rupture

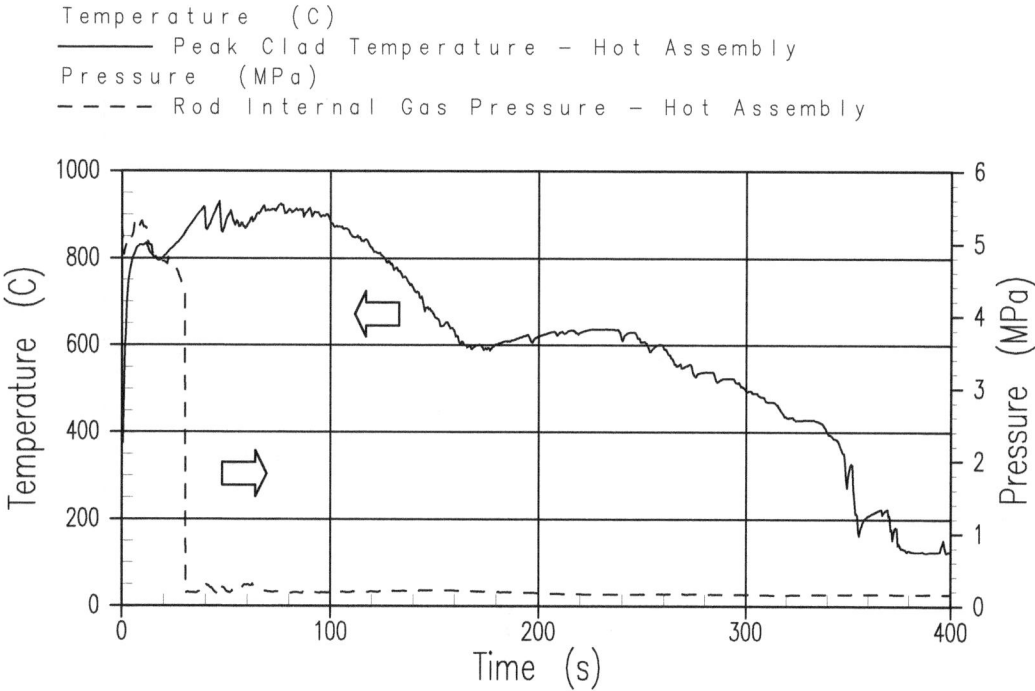

Figure 3. Hot Assembly Rod Burst for Deterministic Assessment of Extent of Rupture

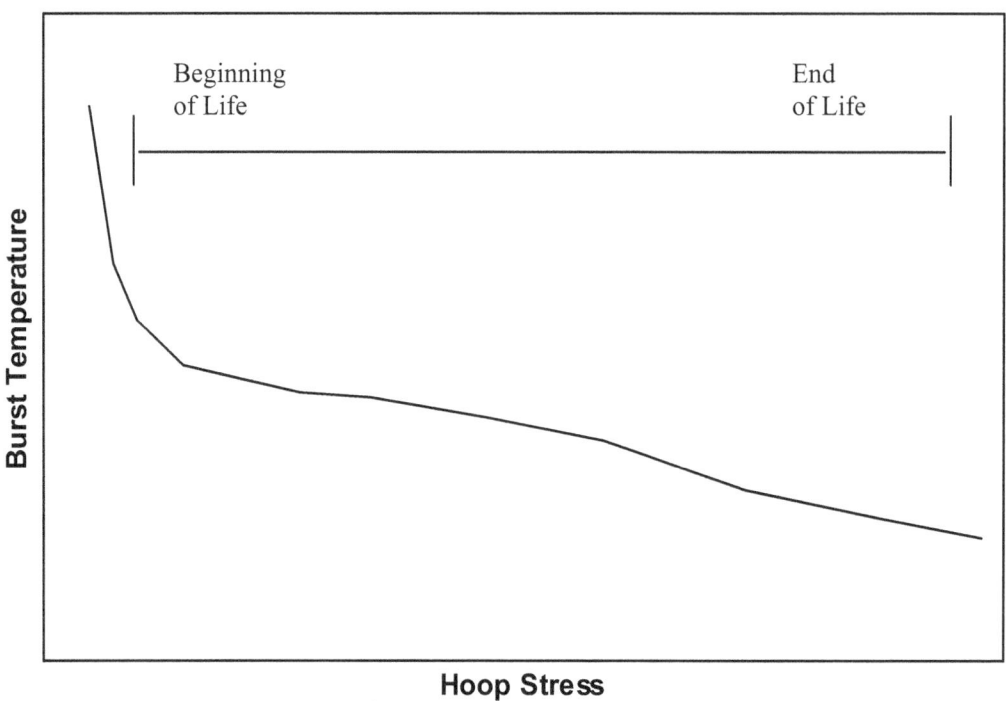

Figure 4. Cladding Rupture Temperature Dependence on Differential Pressure (Hoop Stress) and Burnup

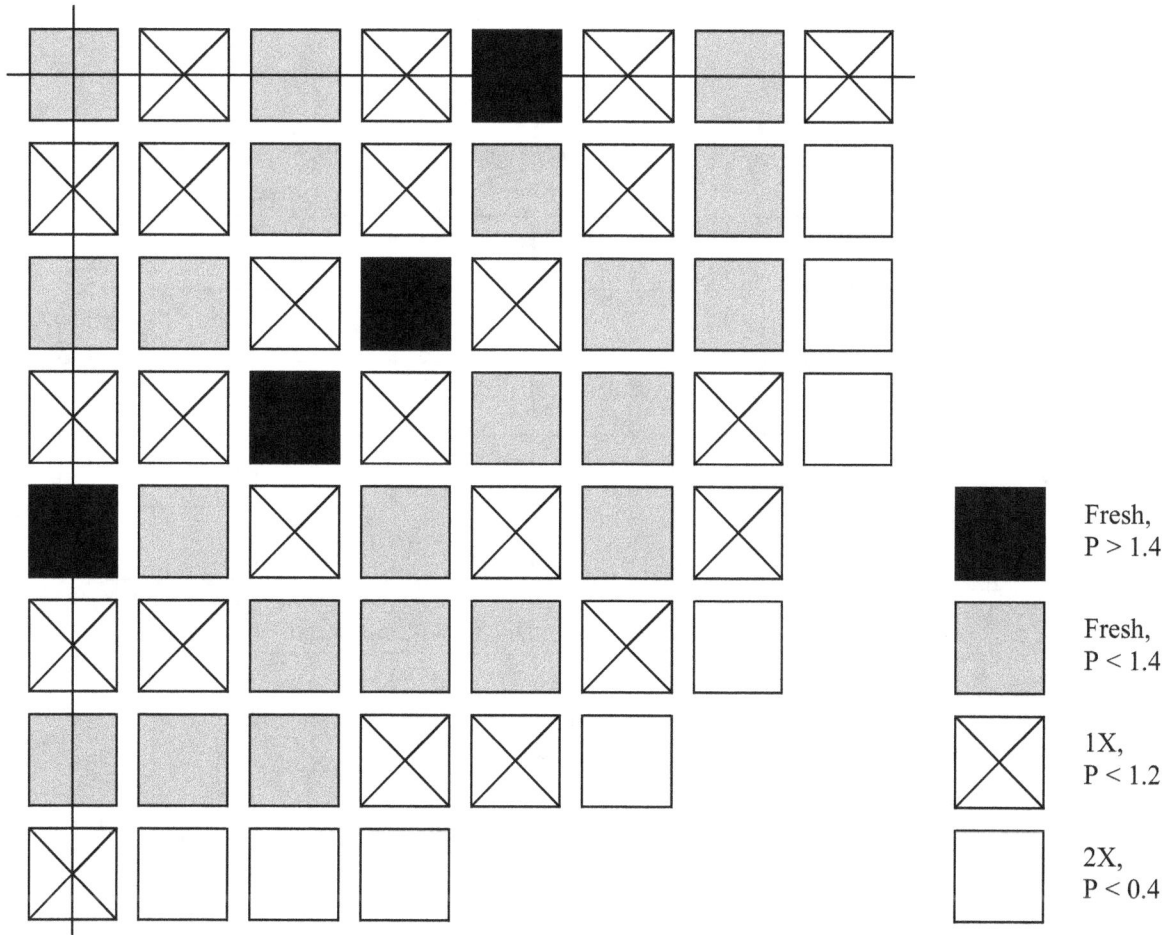

Figure 5. Assembly Power Distribution at Limiting Burnup

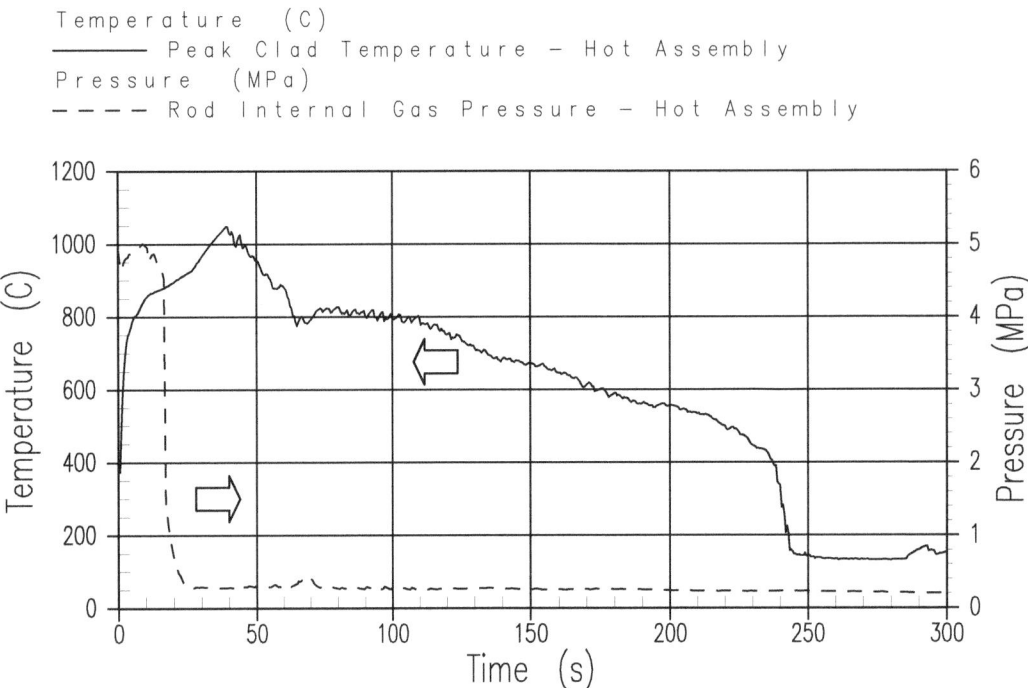

Figure 6.  PCT and Rod Internal Pressure for Limiting PCT Case

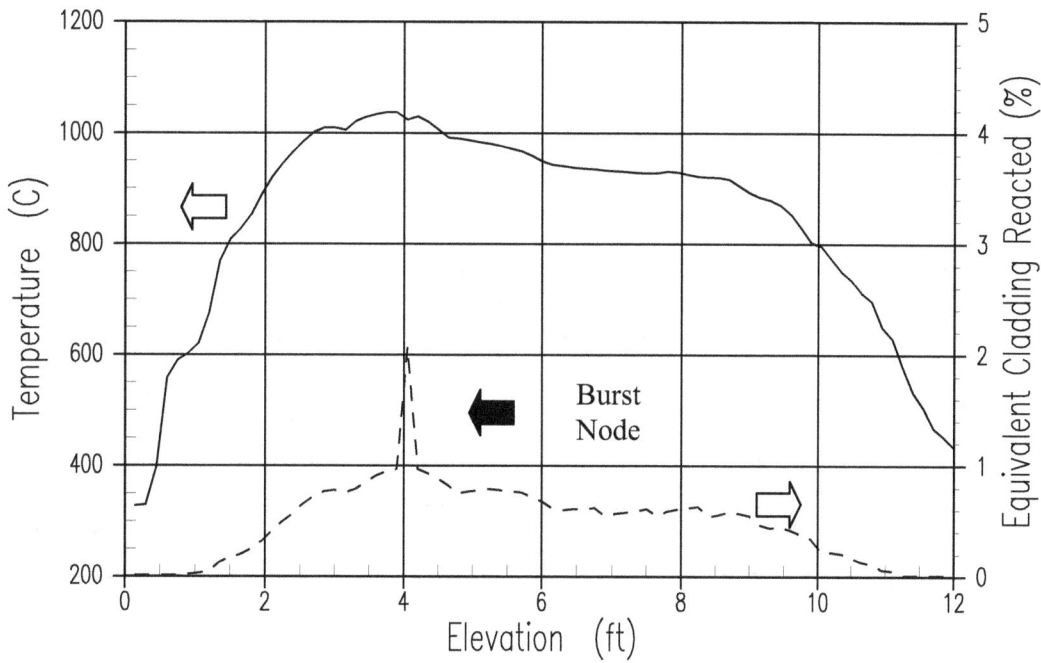

Figure 7.  Axial Distribution of PCT and Oxidation for Limiting PCT Case

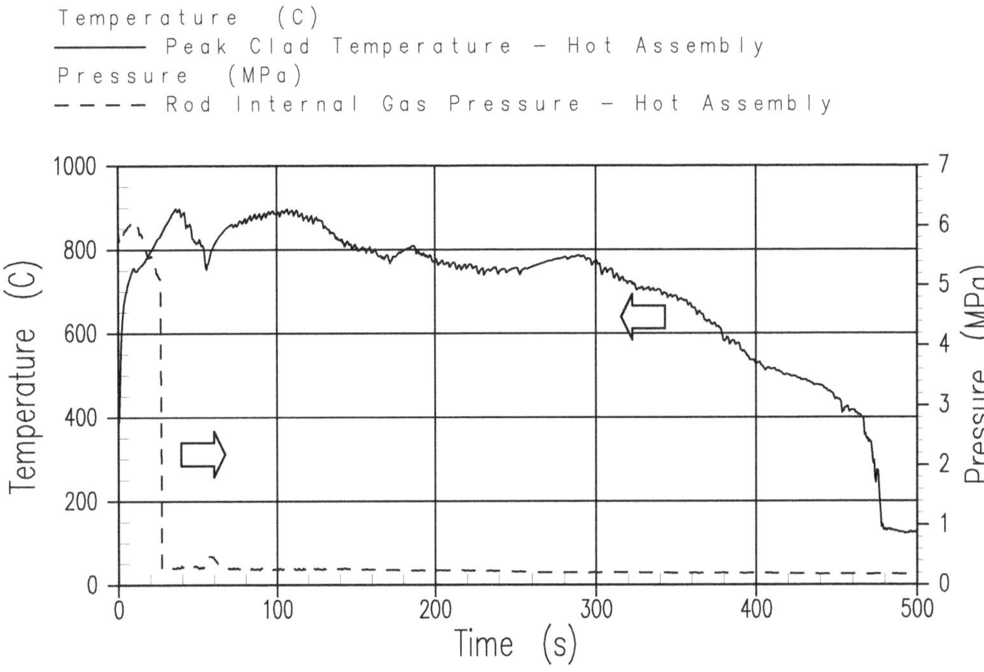

Figure 8. PCT and Rod Internal Pressure for Limiting Non-Burst Oxidation Case

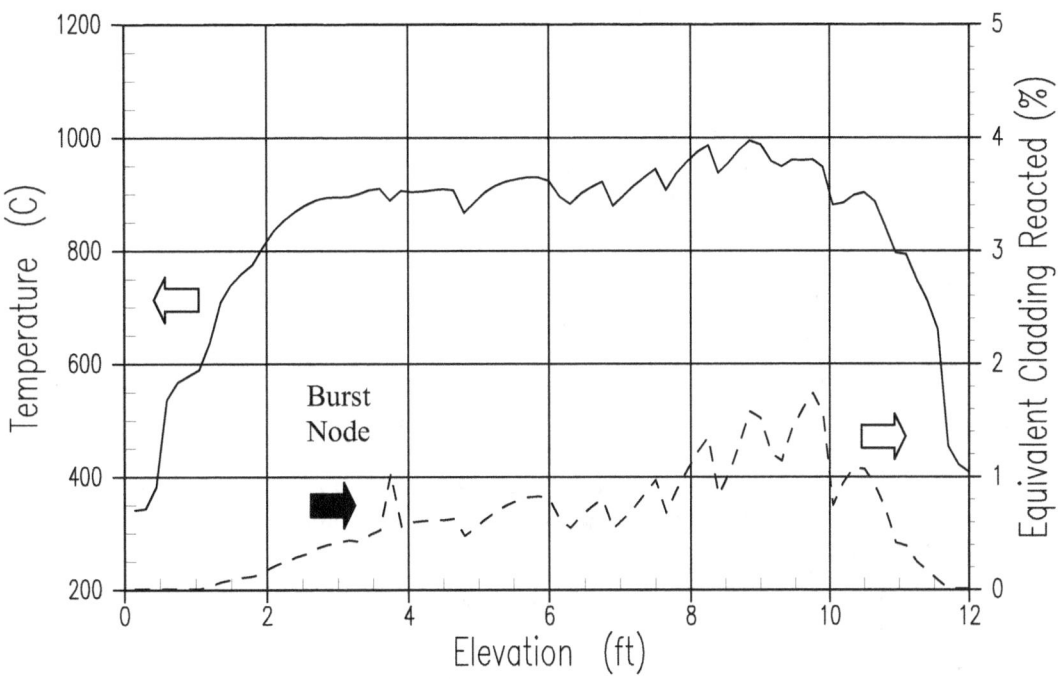

Figure 9. Axial Distribution of PCT and Oxidation for Limiting Non-Burst Oxidation Case

# Post-Quench Ductility of Advance Alloy Cladding

**Michael C. Billone, Yong Yan, and Tatiana A. Burtseva**

Argonne National Laboratory (ANL), Argonne, IL 60439

Diametral (ring)-compression screening tests have been conducted to assess the ductility of 17×17 Zry-4, ZIRLO and M5 samples oxidized to 0-20% ECR at 1000 C, 1100 C and 1200 C. The 25-mm-long samples were exposed individually to two-sided steam oxidation in the same test apparatus for the same test times, slow cooled to 800 C and water-quenched. Test times were calculated using the Cathcart-Pawel weight-gain correlation and a reference wall thickness of 0.57-mm (Zry-4 and ZIRLO). Based on sample weight increase (normalized to the surface area), weight gain was determined and compared to Cathcart-Pawel (CP) predictions. As expected, good agreement was achieved among Zry-4, ZIRLO and M5 and the CP predictions for the 1100 C- and 1200 C-oxidized samples, while differences in weight gain vs. time were observed for the alloys oxidized at 1000 C. After ≈3400 s at 1000 C, the weight gains of M5 and ZIRLO were ≈36% and ≈20% less than Zry-4, respectively. For lower test times, the M5 weight gain was consistently lower than Zry-4, while the ZIRLO and Zry-4 weight gains were about the same. The experimental weight gains, along with the sample thickness (0.61 mm for M5), were used to determine experimental ECR values.

Similar tests were performed with E110 tubing (0.71-mm) at 1000 C and 1100 C to characterize the onset of breakaway oxidation, subsequent hydrogen pickup, and decrease of post-oxidation ductility. Weight gain for as-received E110 could not be determined accurately because of the early (<300 s at 1000 C) breakaway oxidation resulting in oxide flaking and spalling. However, polished and/or machined-and-polished (0.58-0.69 mm wall) E110 exhibited stable oxide growth for oxidation times up to ≈300 s at 1000 C and >1000 s at 1100 C. For these samples, the E110 weight gains were similar to M5: lower than Zry-4 at 1000 C and about the same as Zry-4 at 1100 C.

Ring-compression samples (8-mm-long) were cut from the oxidized samples and tested initially at room temperature and 0.033 mm/s displacement rate (0.35%/s diametral strain rate). Load-displacement curves were analyzed by the traditional offset-displacement method to determine plastic ductility. It was found that this method over-predicts plastic displacement, determined directly from pre- and post-test diameter measurements along the loading direction, by ≤0.2 mm (2% strain). For rings with offset strains < 3%, direct measurement of post-test diameter after the first through-wall crack proved to be a more reliable measure of ductility. Rings with permanent strains < 1% were classified as brittle.

Zry-4, ZIRLO and M5 exhibited ductile behavior (offset strains ≥ 3% and/or permanent strains ≥ 1%) after oxidation at 1000 C and 1100 C for CP-calculated ECR ≥ 17%. The 1000 C results are interesting in that the all three alloys exhibit ≈3% offset-strain ductility after oxidation for the same test time (≈3400 s) at 1000 C, even though the measured ECR values were 22.4%, 18.0% and 13.3% for Zry-4, ZIRLO and M5, respectively. These results suggest that oxidation time at 1000 C and CP-calculated ECR correlate better with ductility than ECR based on actual weight gain, especially for M5. For as-received E110, embrittlement occurs after oxidation at 1000 C for ≈625 s, corresponding to an average hydrogen pickup of ≈300 wppm. The hydrogen concentration in this oxidized sample was highly non-uniform (25-560 wppm) in the circumferential and axial directions with the high hydrogen concentrations occurring under local areas of breakaway oxidation. The results suggest that hydrogen entering E110 through cracks in the oxide layer is essentially "frozen" in position during the course of the test. For polished and machined-and-polished E110 oxidized at 1100 C for ≤1011 s, the material was ductile up to 19% CP-ECR based on the machined-and-polished wall thickness of 0.58 mm. While surface polishing was found to stabilize oxide growth on E110 surfaces, pre-etching with solutions containing HF tended to de-stabilize oxide growth at earlier test times than observed for as-received E110 tubing.

For Zry-4, ZIRLO, M5 and E110 (RRC-KI/RIAR data) samples oxidized at 1200 C, the room-temperature offset strains decreased rather abruptly from 5 to 10% ECR. Based on interpolation of the permanent strain data, the embrittlement ECR values at room-temperature were: ≈10% for Zry-4 and ≈12% for ZIRLO and M5. Based on the RRC-KI/RIAR offset strain data, the embrittlement ECR for E110 oxidized at 1200 C was ≈8%.

Zry-4, ZIRLO and M5 alloys oxidized at 1200 C were retested at 135 C and 0.35%/s, as well as 3.5%/s for Zry-4. The enhancement in post-quench ductility with test temperature was remarkable. As shown in Figs. 1 and 2, ZIRLO and M5 (extrapolated) retained significant ductility for measured and CP-calculated ECR values >17%. Zry-4 also maintained post-quench ductility for ECR > 17% under these conditions. At the higher strain rate of 3.5%/s, Zry-4 was also ductile for ECR > 17%. The implication of these results is that as-fabricated Zry-4, ZIRLO and M5 satisfy the LOCA embrittlement criteria during and shortly after quench. Although testing of the 17x17 cladding alloys at 100 C has not yet been conducted, the results from testing of 15x15 Zry-4 at RT, 100 C and 135 C (Y. Yan, this meeting) suggest that these alloys will retain ductility at the longer-term post-quench temperatures of ≈100 C.

However, in-reactor corrosion results in hydrogen pickup. This hydrogen can enhance embrittlement directly, as well as indirectly by increasing the oxygen solubility and embrittlement in the prior-beta layer of the post-quench cladding alloy. Preliminary test results at 135 C with prehydrided 17x17 and 15x15 (Y. Yan, this meeting) Zry-4 samples oxidized at 1200 C indicate that ≈5%-ECR Zry-4 embrittles at ≈600 wppm H, that ≈8-9%-ECR Zry-4 embrittles at ≈350 wppm H and ≈11%-ECR Zry-4 embrittles at ≈300 wppm H. Future work will focus on the post-quench ductility at 100-135 C of prehydrided-nonirradiated ZIRLO and M5 and high-burnup Zry-4, ZIRLO and M5.

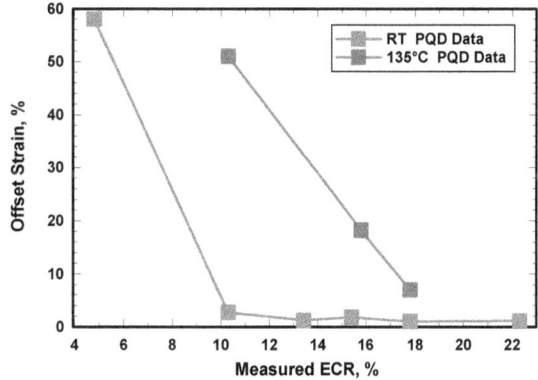

Fig. 1. Post-quench ductility of ZIRLO oxidized at 1200 C and ring-compressed at RT and 135 C.

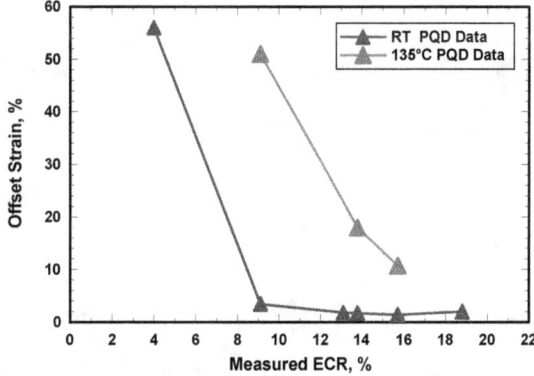

Fig. 2. Post-quench ductility of M5 oxidized at 1200 C and ring-compressed at RT and 135 C.

# Post-Quench Ductility of Advanced Alloy Cladding

M. C. Billone, Y. Yan and T. Burtseva

Energy Technology Division

Nuclear Safety Research Conference
Washington, DC
October 25-27, 2004

**Argonne National Laboratory**

A U.S. Department of Energy
Office of Science Laboratory
Operated by The University of Chicago

# Scope of Advanced-Alloy Program

- **US Licensing Issues Addressed**

  - 10 CFR 50.46 embrittlement criteria for maintaining residual ductility in Zircaloy (Zry) and ZIRLO cladding: PCT $\leq$ 1204°C, ECR $\leq$ 17%

  - Confirm for ZIRLO and compare to Zry-4 (both in Appendix K)

  - Confirm for M5 ("licensed by exception") and compare to Zry-4

- **Approach**

  - Oxidize in steam at 1000-1200°C, slow-cool to 800°C, quench

  - Determine ductile-to-brittle transition ECR from ductility vs. ECR at RT, 100°C (long-term core T) and 135°C (short-term core T)

  - Repeat tests for prehydrided and high-burnup Zry, ZIRLO, M5

- **Materials Research**

  - ZIRLO and M5 oxidation kinetics (vs. Zry-4) for 5-20% ECR

  - ZIRLO and M5 post-quench ductility (vs. Zry-4) for 5-20% ECR

  - Develop understanding of E110 oxide-growth instability/H-pickup

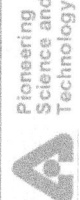

# *Perspective on ANL Advanced-Cladding-Alloy Data*

- **Consistency in Test Methods and Data Interpretation for All Alloys in Program**

- **Documented Results are Available (e.g., NRC ADAMS)**

  - Data available for independent assessment and interpretation

    Temperature control and monitoring

    Weight gain vs. time at temperature

    Ring-compression load vs. displacement; pre- and post-test diameter

    Metallography; microhardness; hydrogen pickup; limited SEM-TEM

  - Independent interpretation by NRC, industry and their partners

- **ANL Interpretation of Results will be Presented**

  - Ductility vs. "Measured ECR" (to assess mechanistic correlation)

  - Ductility vs. "Best-Estimate-Model ECR": Cathcart-Pawel (CP)

  - ANL Ductility criteria: offset strain $\geq 3\%$ or permanent strain $\geq 1\%$

Pioneering
Science and
Technology

Nuclear
Regulatory
Commission

# Nonirradiated Zr-Alloy Tubing and Cladding

| Material | Do, mm | h, mm | Oxygen wt.% | Outer Surface Finish & Roughness |
|---|---|---|---|---|
| Zry-4 | | | | |
| 15×15 (HBR) | 10.82 | 0.79 | 0.14 | Tubing, 0.36 μm |
| 15×15 (low-Sn) | 10.77 | 0.76 | 0.14 | Cladding, 0.31 μm |
| 17×17 (low-Sn) | 9.50 | 0.57 | 0.12 | Cladding, 0.14 μm |
| 17×17 ZIRLO | 9.50 | 0.57 | 0.12 | Cladding, 0.11 μm |
| 17 ×17 M5 | 9.50 | 0.61 | 0.145 | Cladding, 0.12 μm |
|  | 9.50 | 0.57 | 0.145 | Cladding, 0.12 μm |
| E110 | 9.17 | 0.71 | 0.05 | Tubing, 0.35 μm |
|  | 9.07 | 0.68 | 0.05 | Cladding: etch.-anodized |
| E110-ANL | ≈9.1 | ≈0.6 | 0.05 | Mach.-Pol., 0.13 μm. |
| 9×9 Zry-2 | 11.18 | 0.71 | 0.11 | Cladding, 0.14 μm |
| 10×10 Zry-2 | 10.3 | 0.66 | TBD | Cladding, 0.11 μm |

Nuclear
Regulatory
Commission

Pioneering
Science and
Technology

# *Advanced-Alloy Post-Quench Ductility Research*

- ## Steam Oxidation to ≤ 20% ECR (CP-Calculated)

  - ≤3400 s (1000°C), ≤1100 s (1100°C), and ≤400 s (1200°C)
  - 25-mm-long, undeformed samples, two-sided oxidation
  - Heat to target temperature ($T_s$) with no overshoot, hold at $T_s$ for $t_s$
  - furnace-cool ($\approx$10°C/s) to 800°C, quench (bottom-flooding) to 100°C
  - Use measured weight gain ($\Delta w$) and initial wall thickness ($h_o$) to calculate "measured" ECR

- ## Ring-Compression Tests

  - 8-mm-long ring, room temperature, 2 mm/min. (0.033 mm/s) disp. rate
  - Determine offset displacement ($\delta_p$) to first through-wall crack
  - If $\delta_p < 0.3$ mm, measure change in diameter ($d_p$) in loading direction
  - For screening, material is ductile for $\delta_p/D_o \geq 3\%$ or $d_p/D_o \geq 1\%$
  - If $d_p/D_o < 1\%$ for ECR <17%, repeat test at 135°C and 100°C

280

# Temperature History for 1000°C Steam Oxidation for 17×17 Cladding Samples (Quench not Shown)

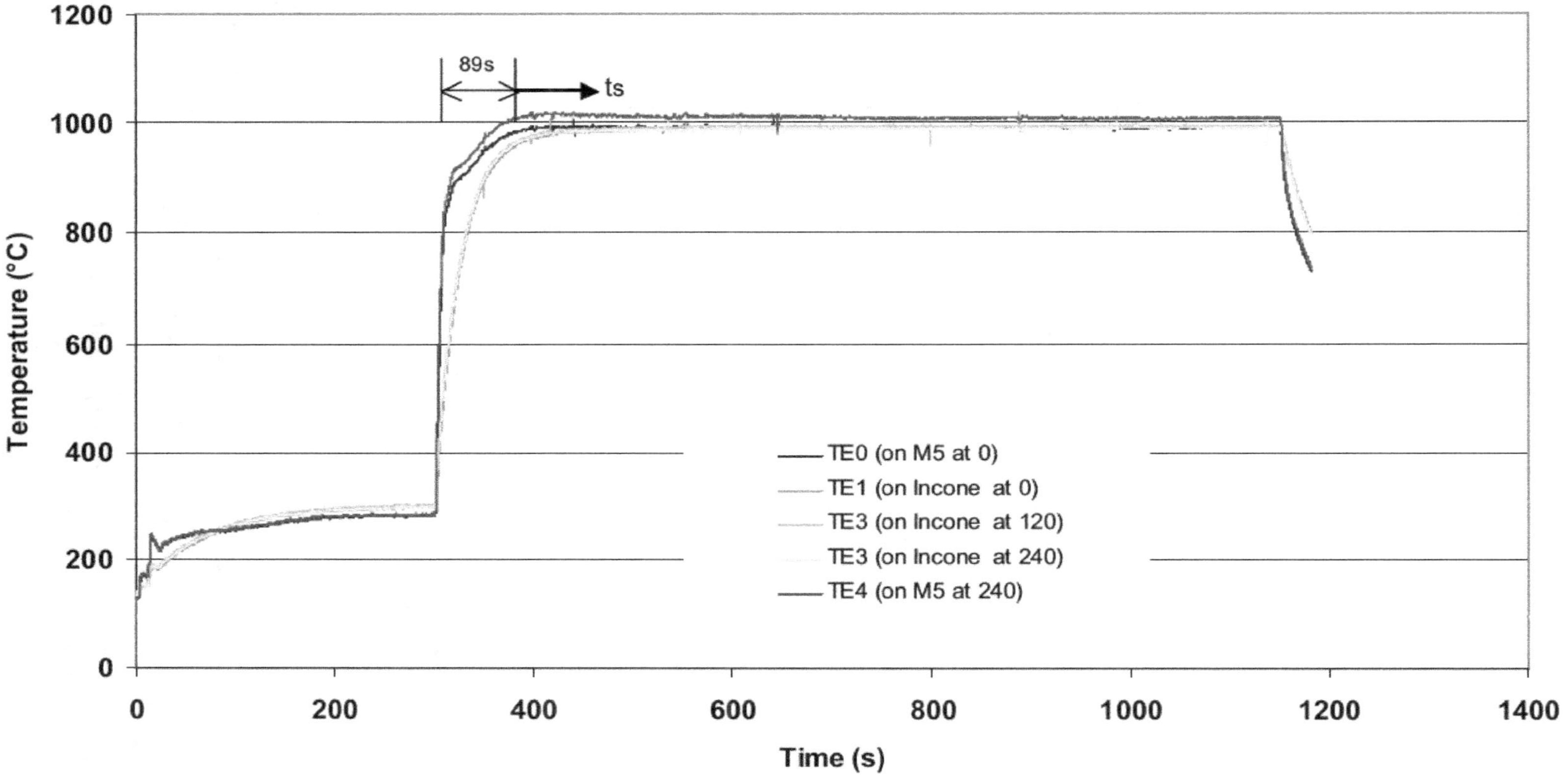

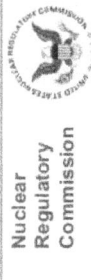

# Temperature History for 1200°C Steam Oxidation for 17×17 Cladding Samples (Quench not Shown)

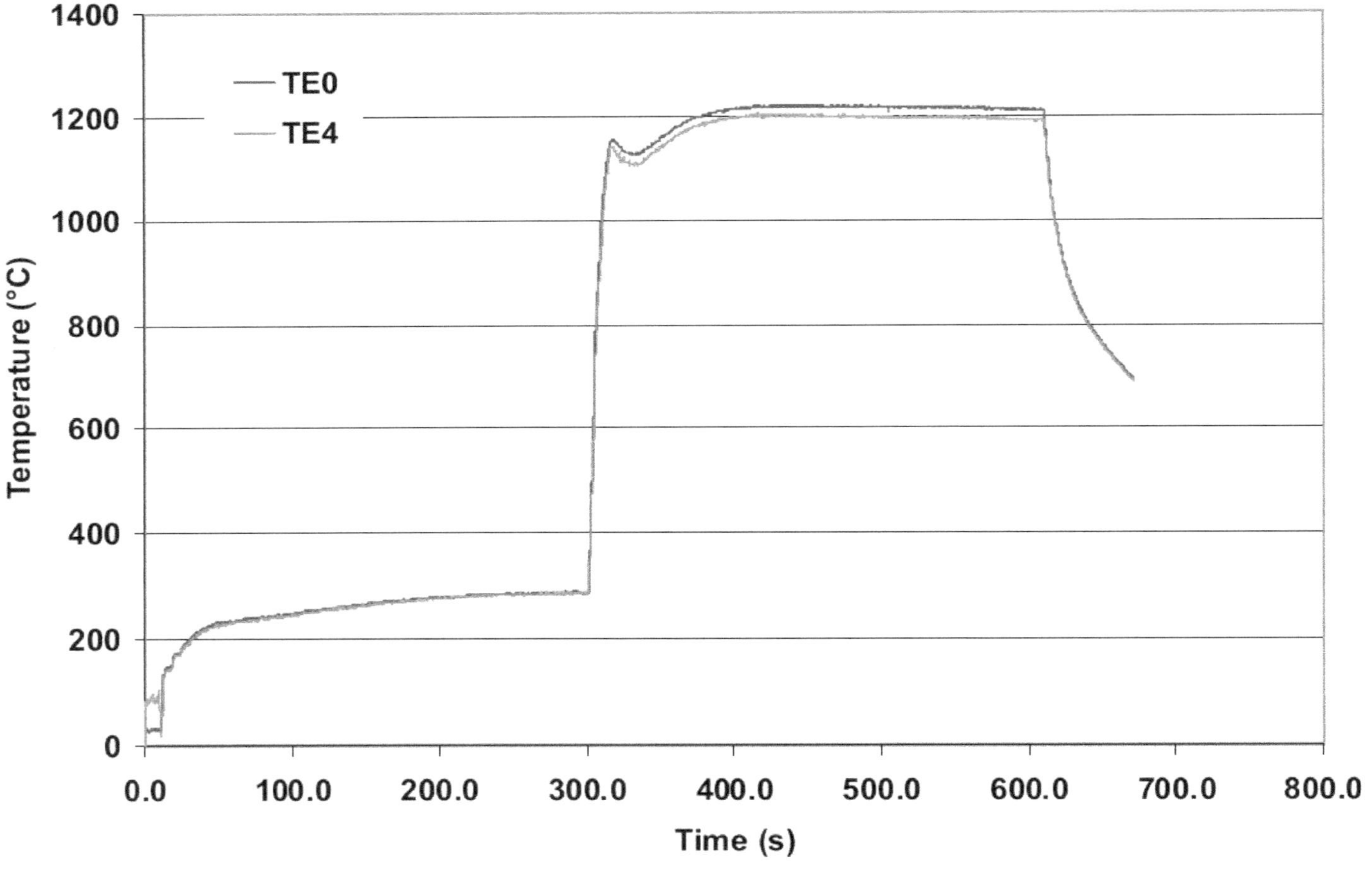

# Diametral (Ring) Compression Screening Test for Post-LOCA-Quench Ductility (PQD)

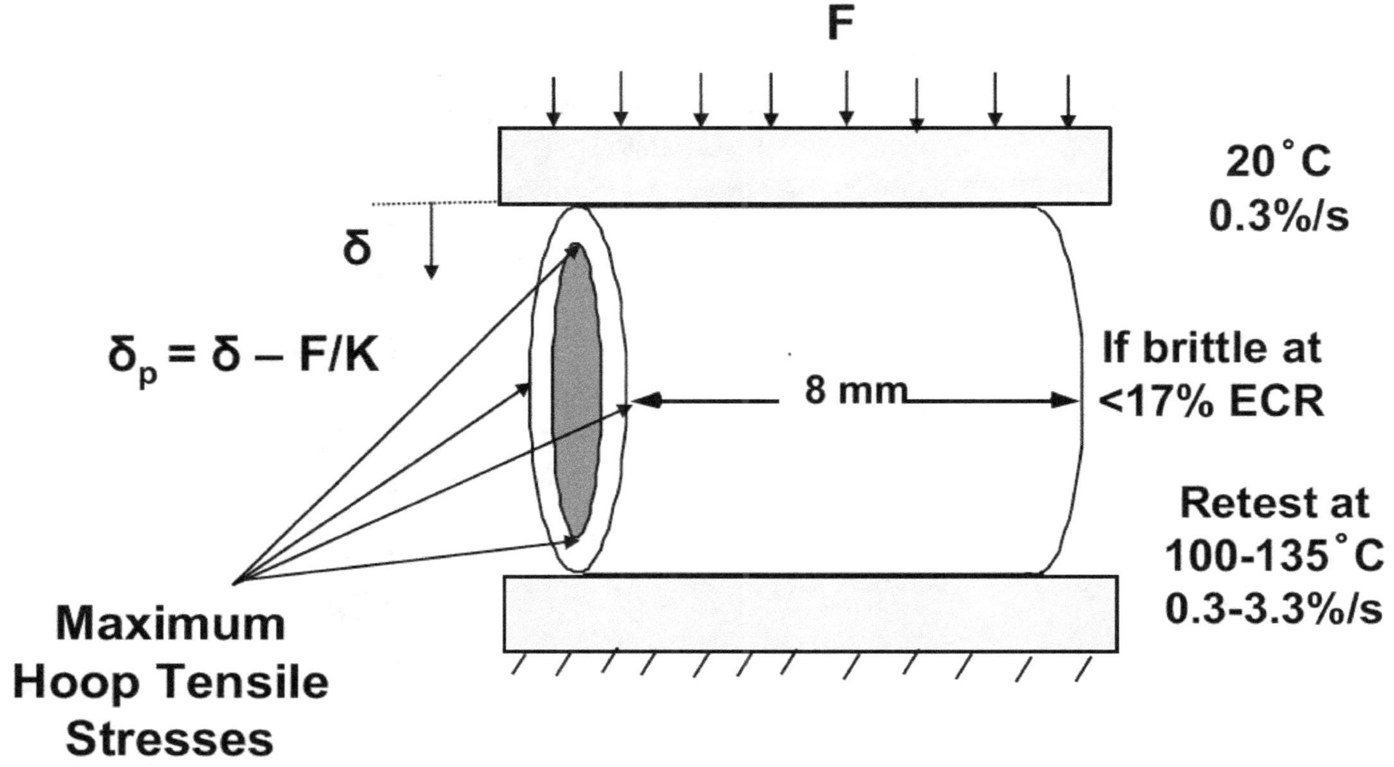

F

20°C
0.3%/s

If brittle at
<17% ECR

Retest at
100-135°C
0.3-3.3%/s

8 mm

$\delta_p = \delta - F/K$

δ

Maximum
Hoop Tensile
Stresses

Nuclear
Regulatory
Commission

Pioneering
Science and
Technology

# *Offset Displacement Methodology (for $\delta_p \geq 0.3$ mm)*

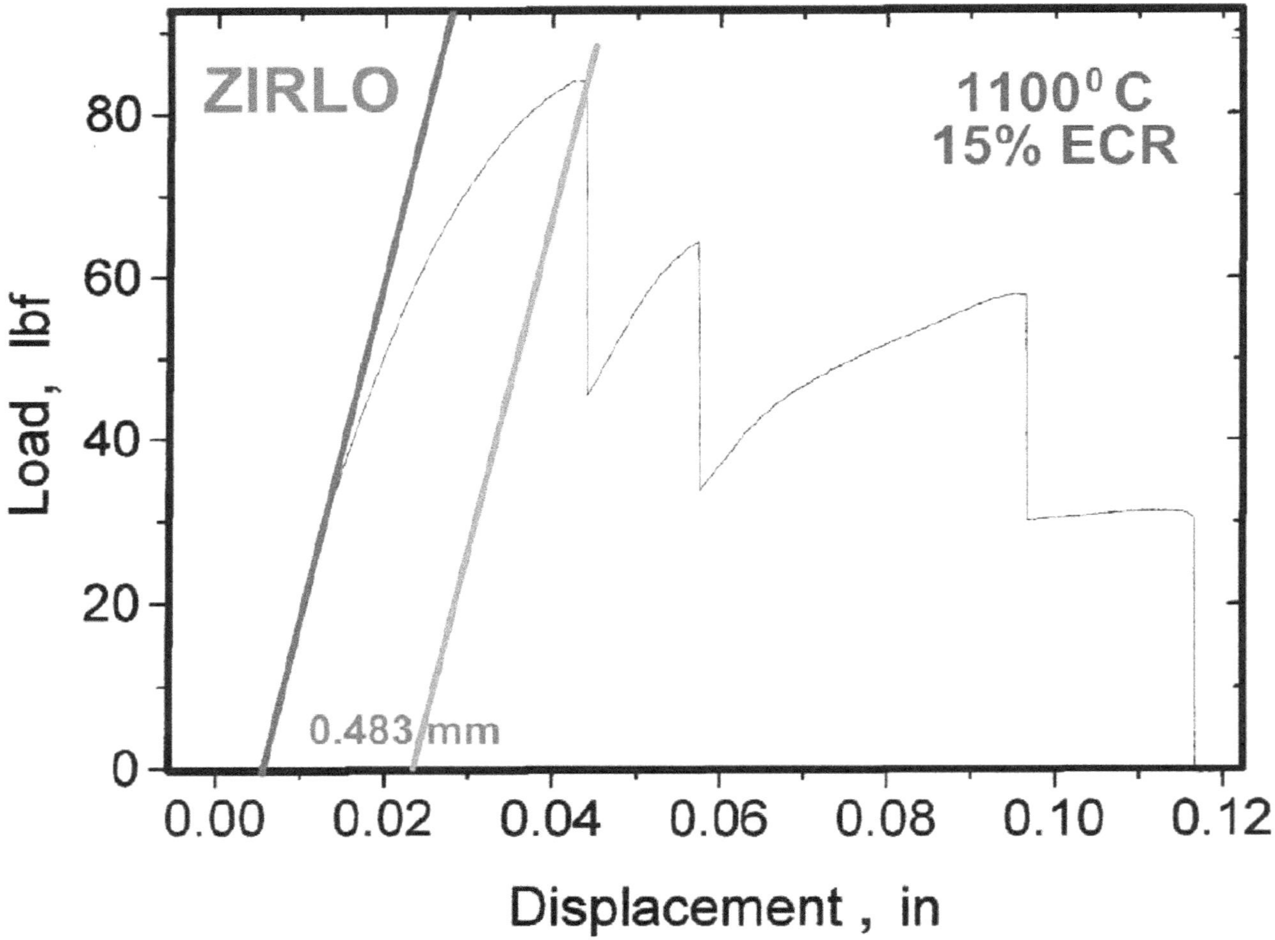

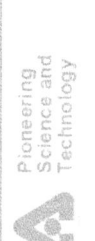

ZIRLO
1100° C
15% ECR
0.483 mm
Load, lbf
80
60
40
20
0
0.00
0.02
0.04
0.06
0.08
0.10
0.12
Displacement , in

# *Permanent Strain Methodology (for $\delta_p < 0.3$ mm)*

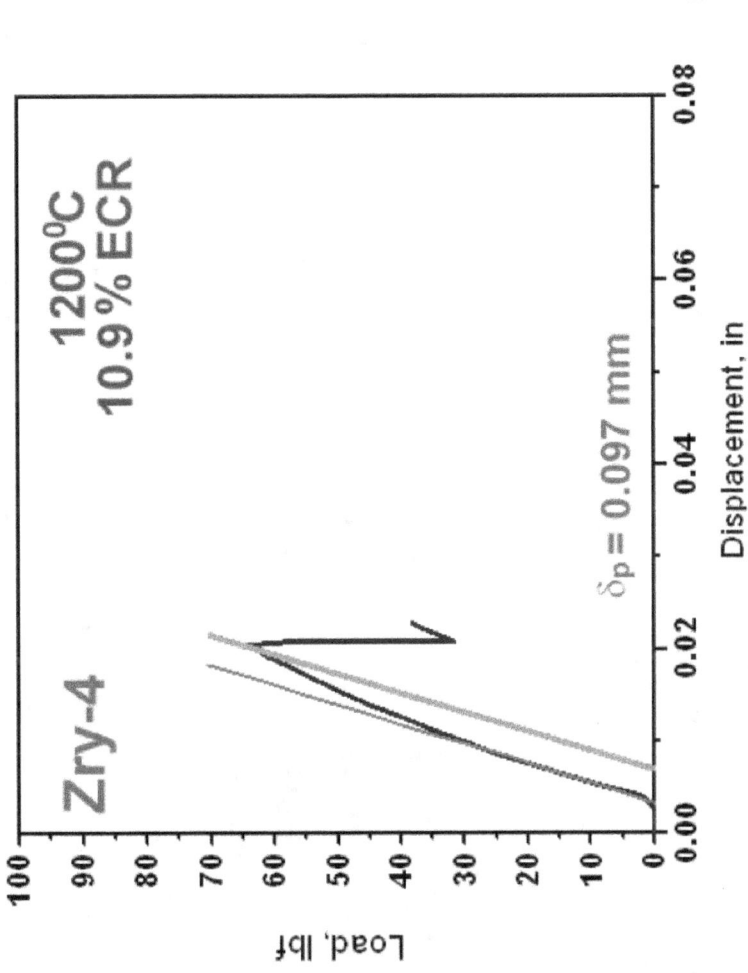

Zry-4

1200°C
10.9% ECR

$\delta_p = 0.097$ mm

Load, lbf

Displacement, in

Stop test after first crack
(tight through-wall)

Measure $d_p = D_{oi} - D_{of}$
In loading direction

$\mathbf{d_p} = 0.05$ mm

Offset Strain = 1.0%

Permanent Strain = 0.5%

Based on Screening Criteria, Material is Brittle

Nuclear
Regulatory
Commission

# Advanced-Alloy Results

- ## Weight Gain Kinetics

  - At 1100°C and 1200°C, Zry-4, M5, ZIRLO and polished-E110 data are in agreement with Cathcart-Pawel (CP) model predictions

  - At 1000°C, ZIRLO < Zry-4 for >8 mg/cm²; M5 << Zry-4

  Machined-and-polished E110 ($\approx$0.6-mm wall) $\approx$ M5 (0.6-mm wall)

- ## Room-Temperature Post-Quench Ductility Results

  - **Residual ductility $\geq$ 3% for Zry-4, ZIRLO and M5 alloys oxidized at 1000°C and 1100°C to 17% ECR (CP-calculated);** low transition ECR for E110 oxidized at 1000°C due to oxide instability and H-pickup

  - H-pickup is low (<40 wppm for $\leq$17% CP-ECR); except for E110

  - Metallography/microhardness data support ring-compression data

  - For 1000°C, ductility correlates better with CP-calculated ECR than with measured ECR for M5 and for ZIRLO at >13% ECR

  - For 1200°C, <1% ductility at <13% ECR for Zry-4, ZIRLO and M5

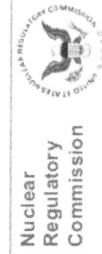

# Post-Quench Ductility at Room Temperature
## M5 & Zry-4 Oxidized at 1000°C vs. Measured ECR

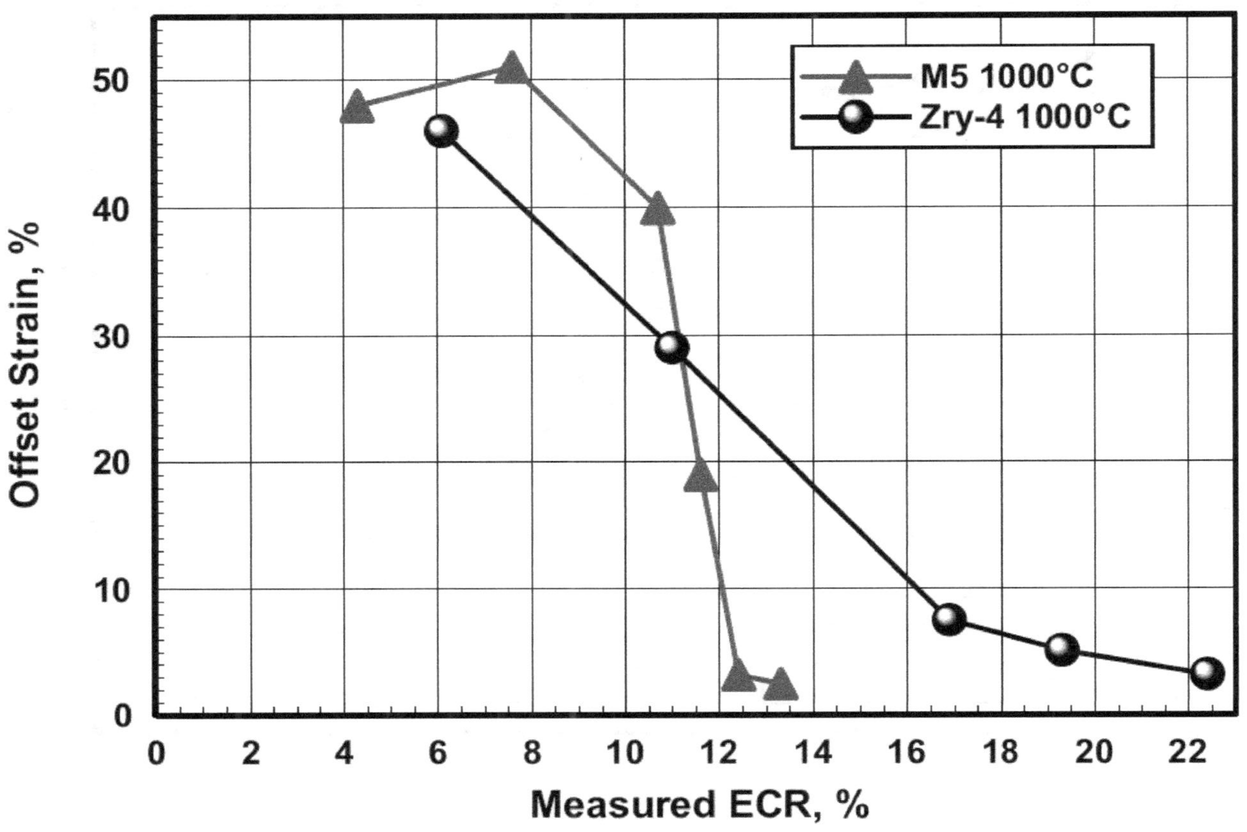

# Post-Quench Ductility at Room Temperature
## M5 & Zry-4 Oxidized at 1000°C vs. Calculated ECR

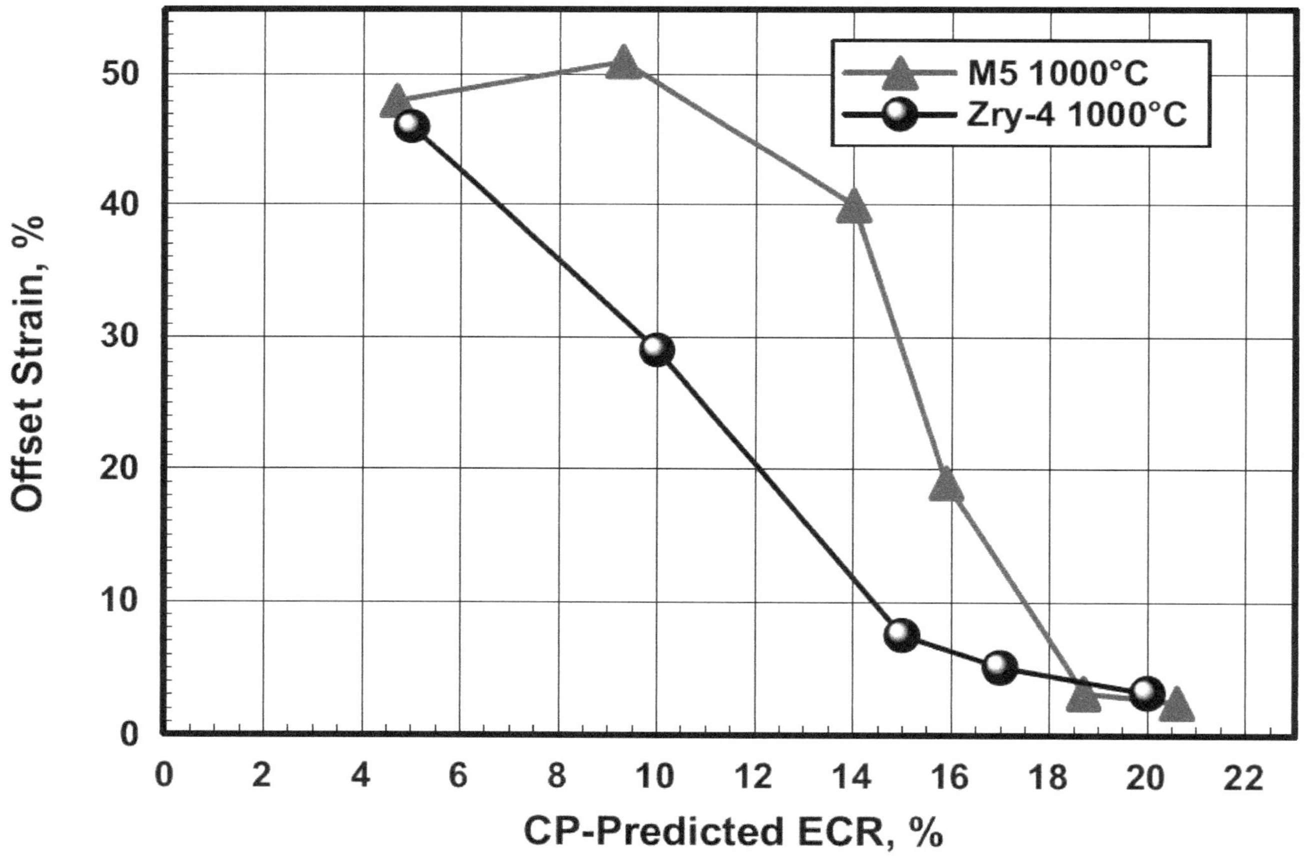

# Post-Quench Ductility at Room Temperature
## E110 Tubing Oxidized at 1000°C vs. Calculated ECR

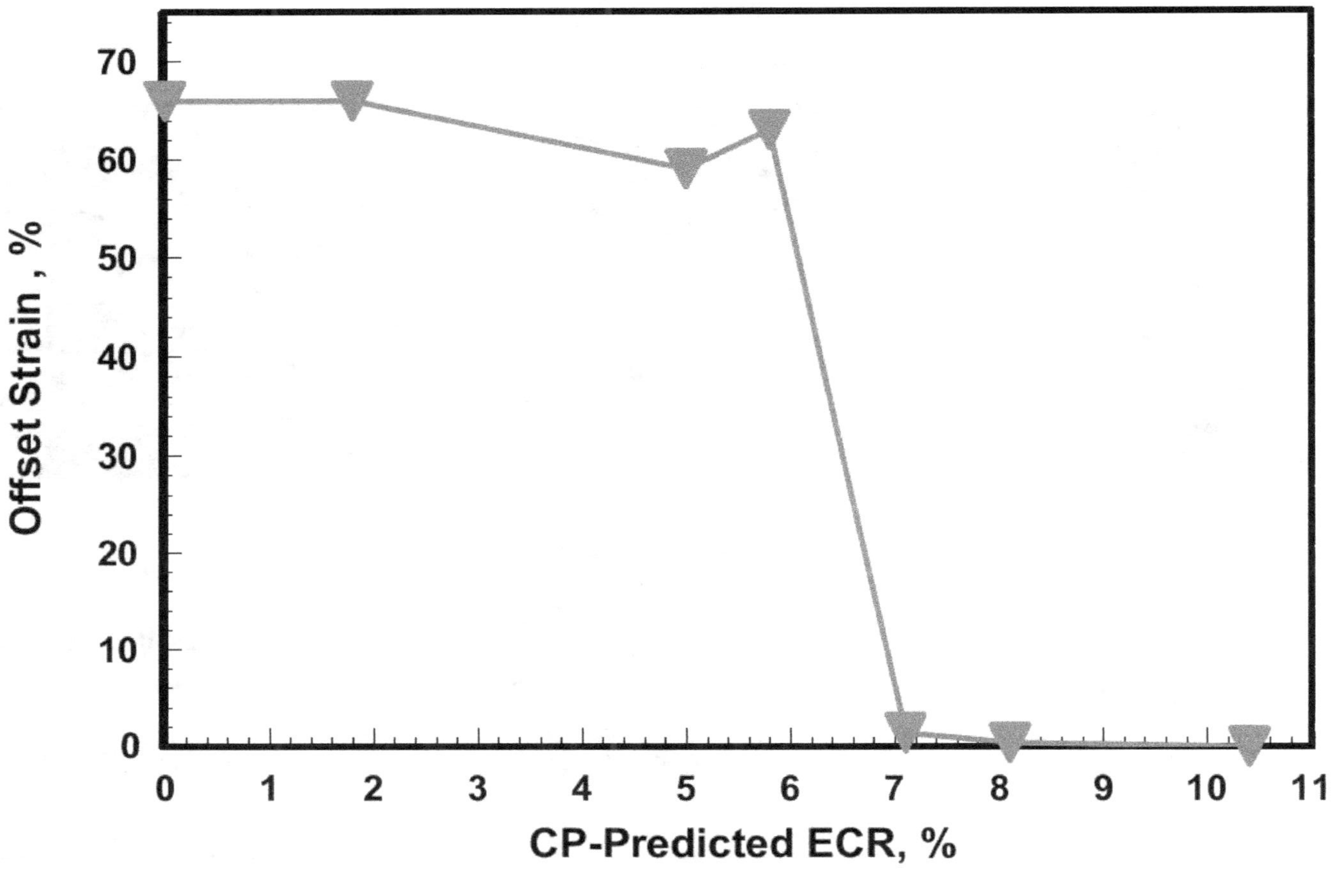

289

# E110 after 300 s and 1400 s at 1000°C

"

White" Spots
In
Black Matrix

75-s Ramp
5-s Hold Time

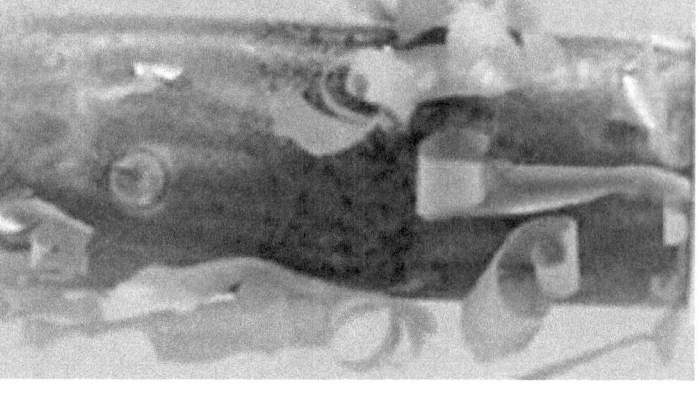

5% ECR
≈300 s
120±50 wppm H

10% ECR
≈1400 s
>4000 wppm H

# Post-Quench Ductility at Room Temperature ZIRLO & Zry-4 Oxidized at 1000°C vs. Calculated ECR

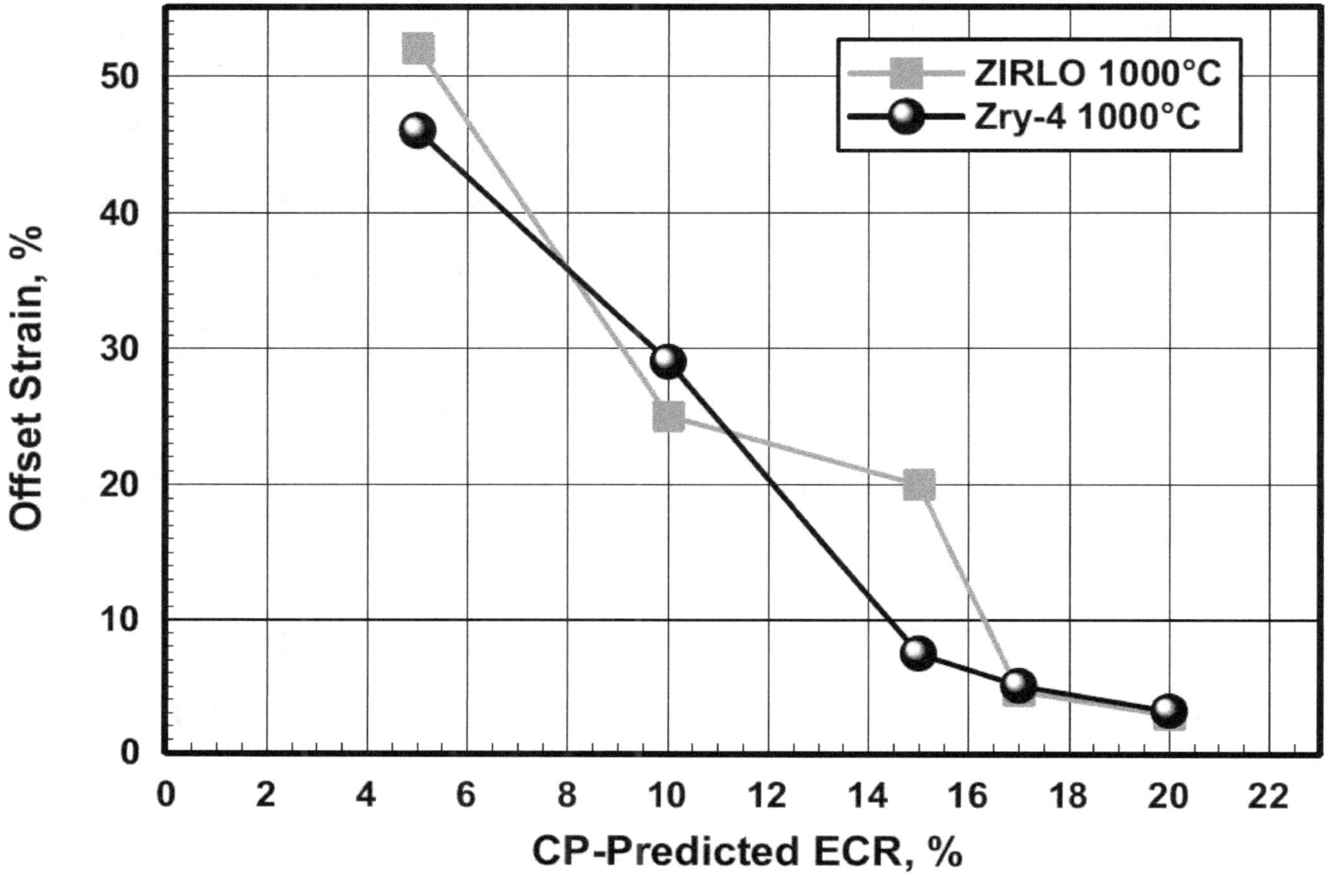

# Room-Temperature (RT) PQD Results for Alloys Oxidized at 1200°C

- **Zry-4**
  - 8% CP-ECR: 4.0% offset strain; 2.2% permanent strain (ductile)
  - 10% CP-ECR: 1.0% offset strain; 0.5% permanent strain (brittle)

- **M5**
  - 9% CP-ECR: 3.4% offset strain; 1.6% permanent strain (ductile)
  - 12% CP-ECR: 1.8% offset strain; 0.9% permanent strain (brittle)

- **ZIRLO**
  - 10% CP-ECR: 2.7% offset strain; 1.4% permanent strain (ductile)
  - 13% CP-ECR: 1.2% offset strain; 0.7% permanent strain (brittle)

- **Machined-and-Polished E110**
  - 10% CP-ECR: 2.1% offset strain; 1.1% permanent strain (transition)
  - 13% CP-ECR: 1.0% offset strain; 0.7% permanent strain (brittle)

292

Nuclear
Regulatory
Commission

Pioneering
Science and
Technology

# RT Offset Strain for 1200°C-Oxidatized ZIRLO vs. Zry-4

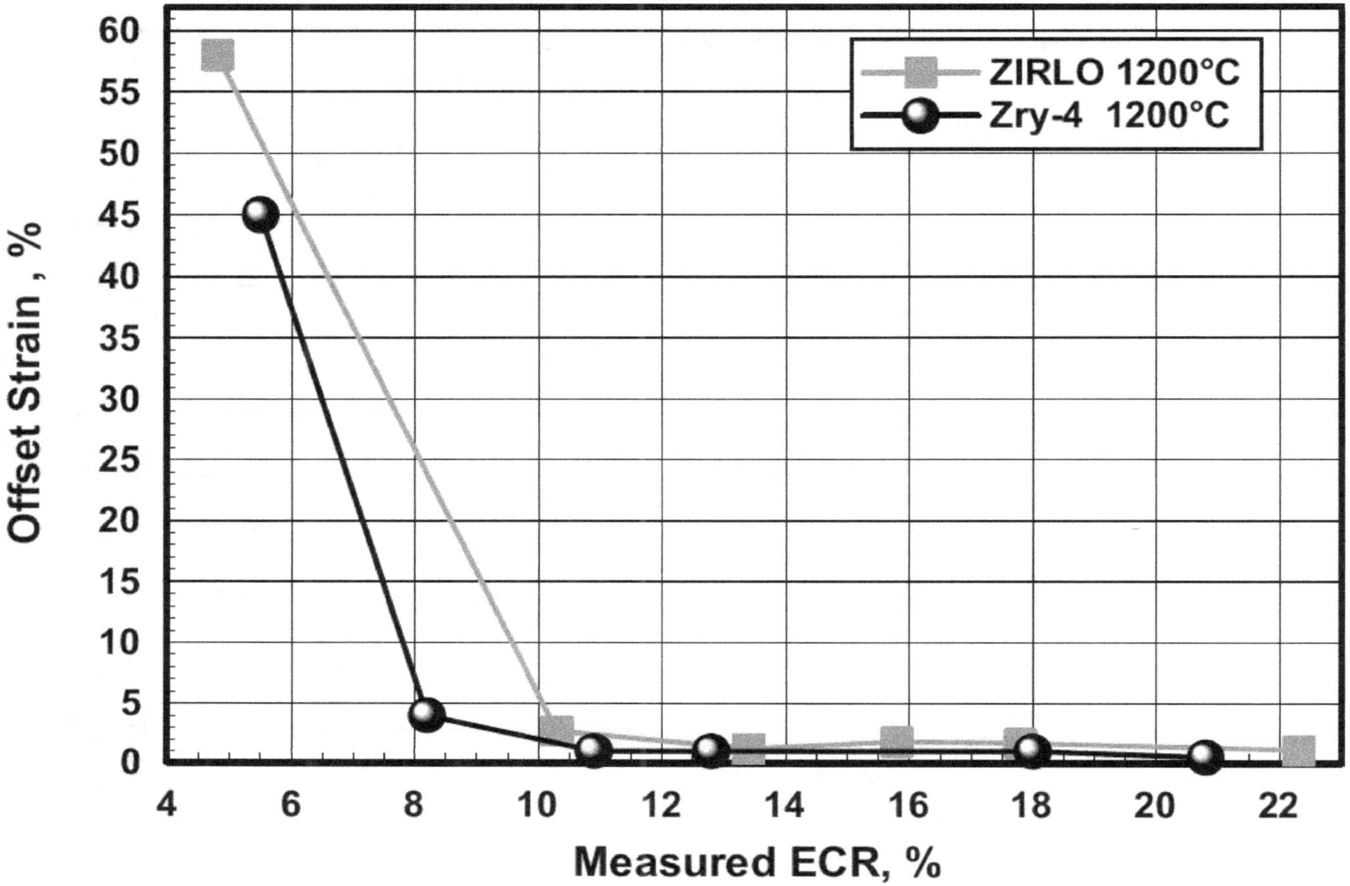

Measured ECR, %

Offset Strain , %

Legend:
- ZIRLO 1200°C
- Zry-4 1200°C

# RT Offset Strain for 1200°C-Oxidatized M5 vs. Zry-4

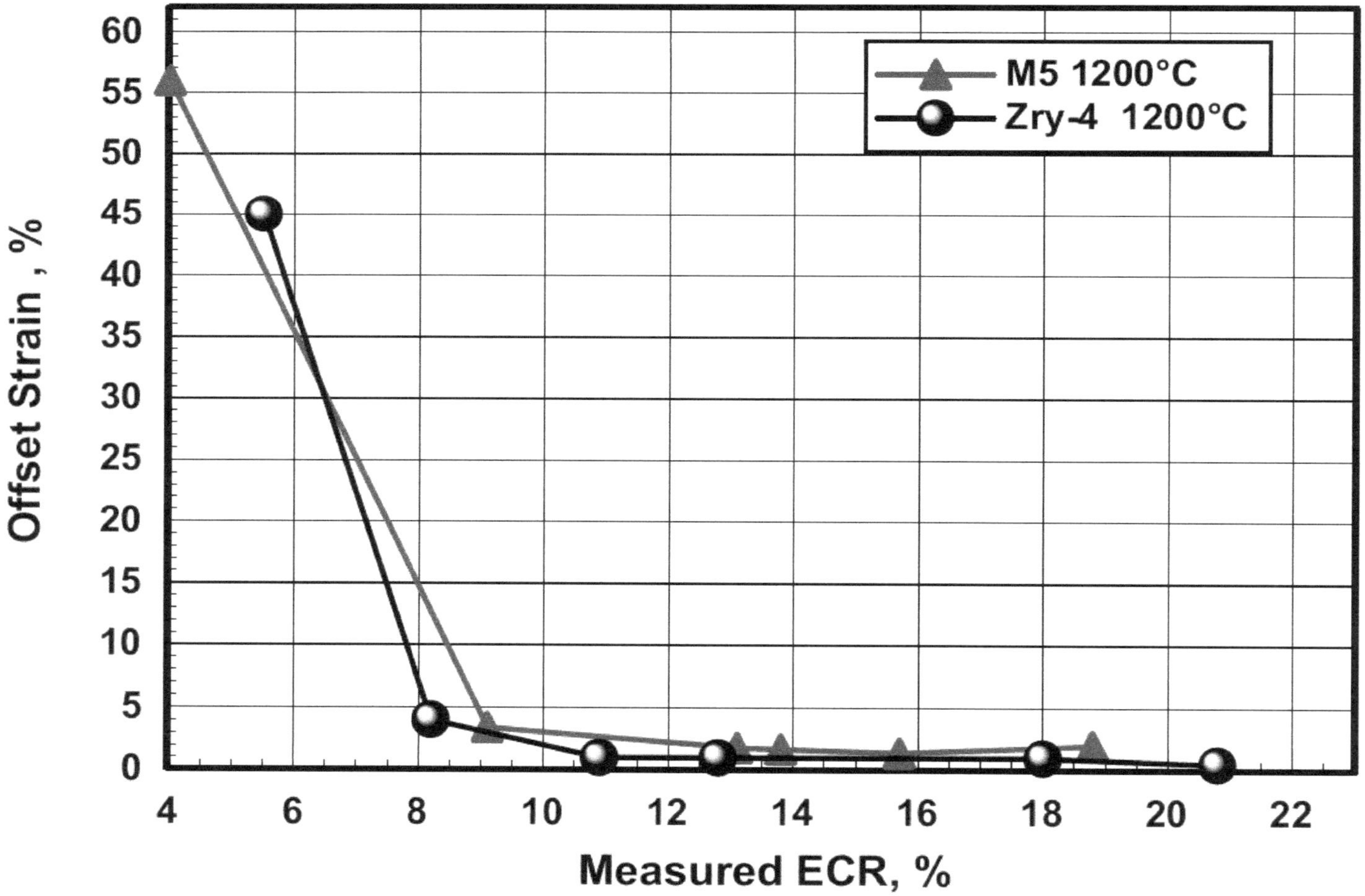

# Ring-Compression Testing at 135°C

- **All Alloys Exhibited Significant Enhancement in Ductility vs. ECR at 135°C**

- **As-Received (Non-Irradiated) Alloys Satisfy LOCA Embrittlement Criteria during Quench (T ≥ 135°C)**

- **Mechanism for Enhancement is not Clear**
  - Similar results observed by CEA/EdF/Framatome and others

- **However, Irradiated Alloys Contain Hydrogen**
  - Significant hydrogen embrittlement of E110 oxidized at 1000°C
  - CEA/EdF/Framatome results indicate significant embrittlement effects of hydrogen for Zry-4 (300, 600 wppm) and M5 (200 wppm)
  - Prehydrided, non-irradiated Zry-4 tested at ANL

    Results are preliminary: non-uniform prehydriding; old test train

    Tests will be repeated with better hydriding and temperature control

# 135°C Offset Strain for 17x17 Zry-4 Oxidized at 1200°C

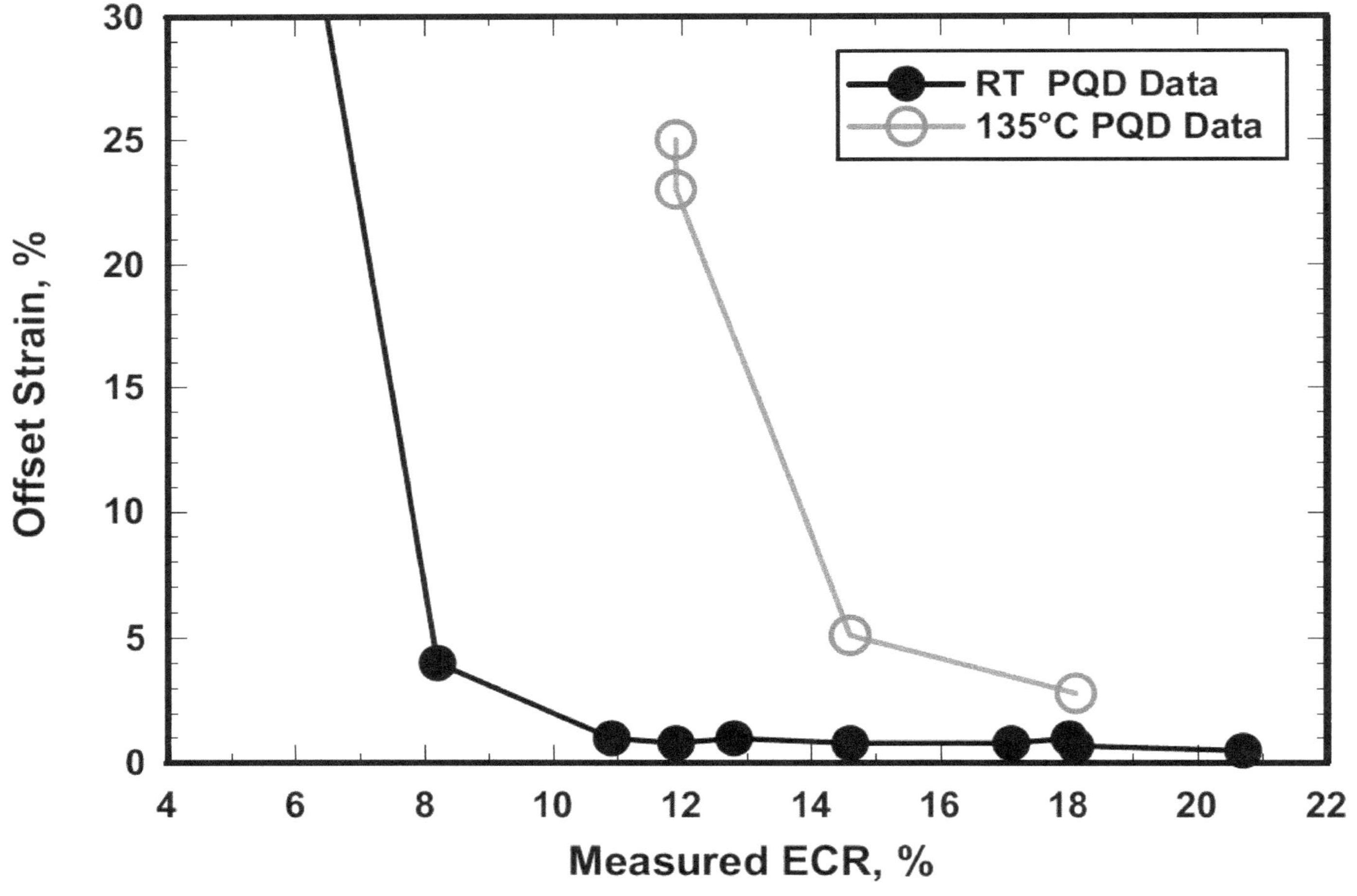

# *135°C Offset Strain for 17x17 ZIRLO Oxidized at 1200°C*

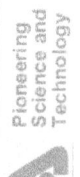

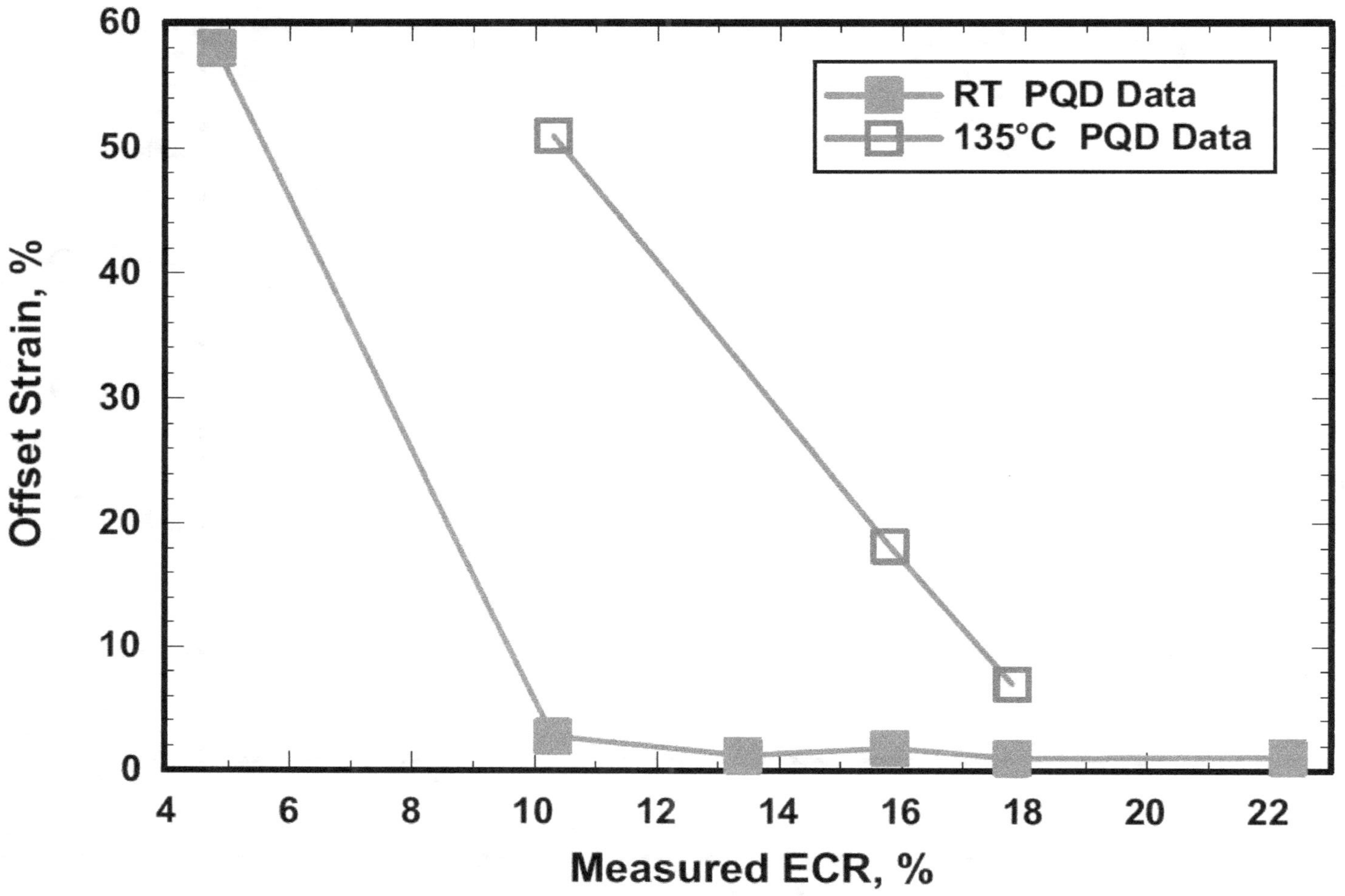
RT PQD Data
135°C PQD Data
Offset Strain, %
0
10
20
30
40
50
60
4
6
8
10
12
14
16
18
20
22
Measured ECR, %

# 135°C Offset Strain for 17x17 M5 Oxidized at 1200°C

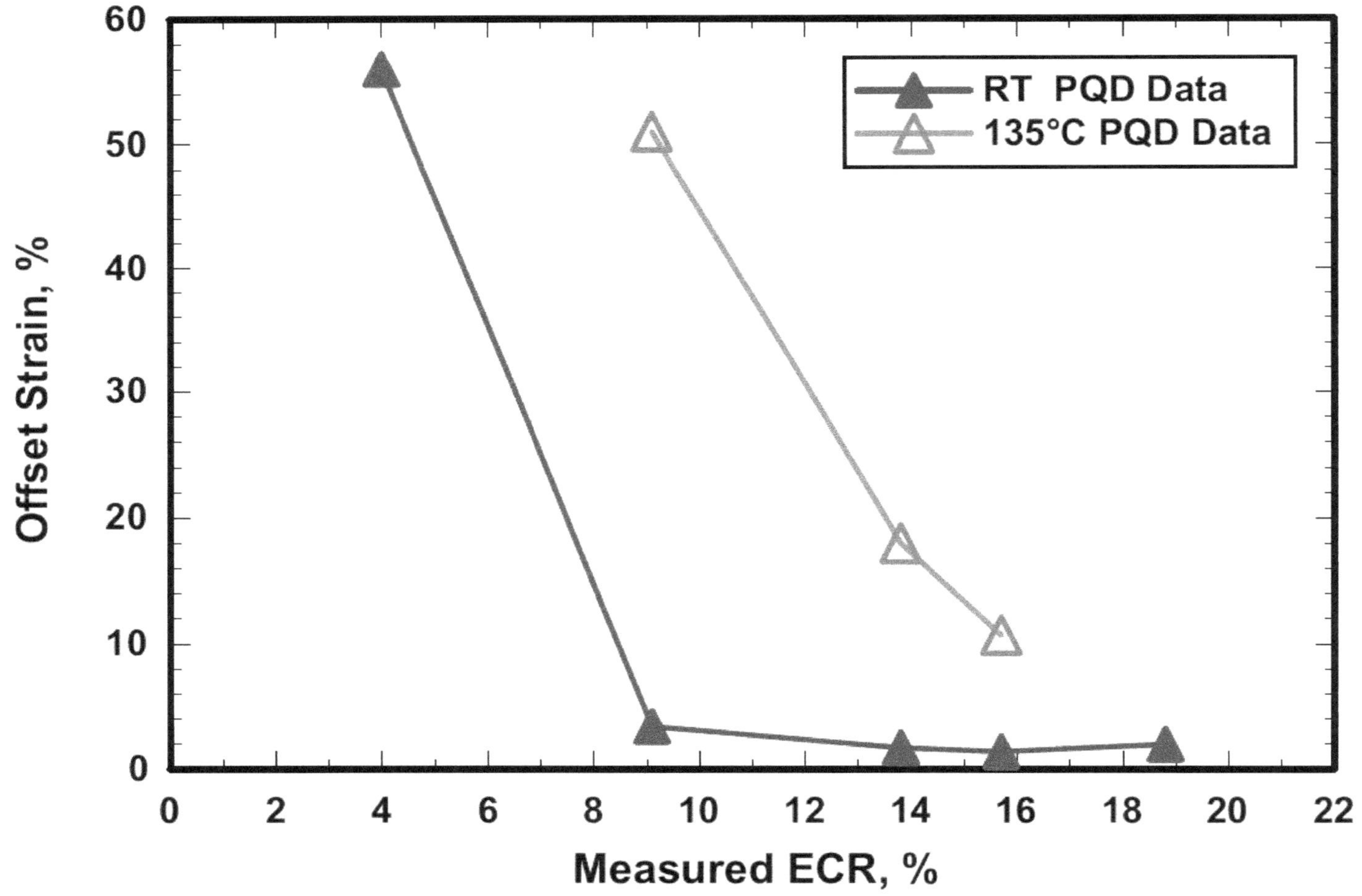

Pioneering
Science and
Technology

Nuclear
Regulatory
Commission

# 135°C Offset Strain for Machined-and-Polished E110 Oxidized at 1200°C

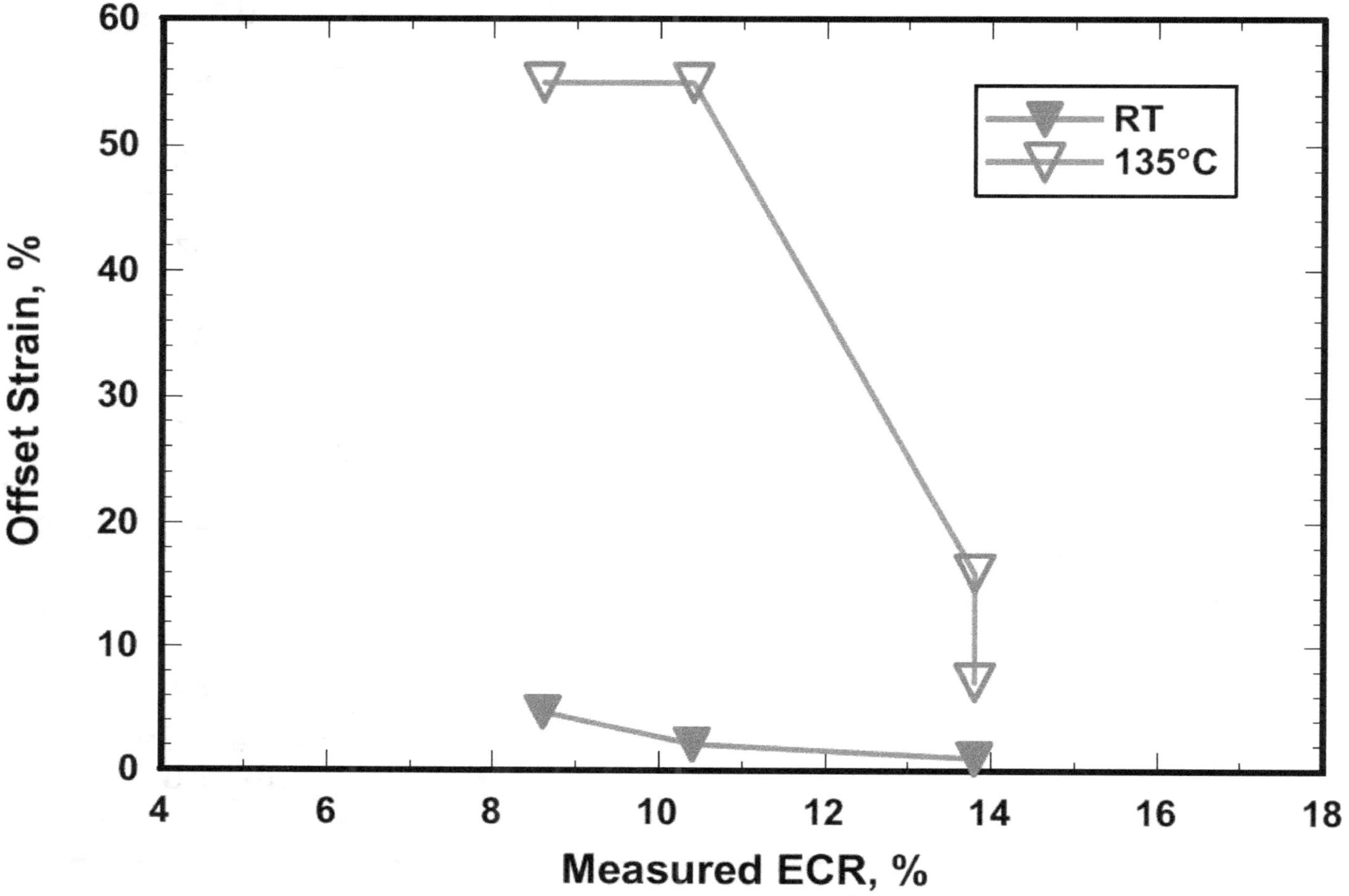

Nuclear
Regulatory
Commission

Pioneering
Science and
Technology

# CEA-EdF-FANP RT PQD Data for Zry-4 Oxidized at 1200°C: As-Fabricated & Pre-Hydrided

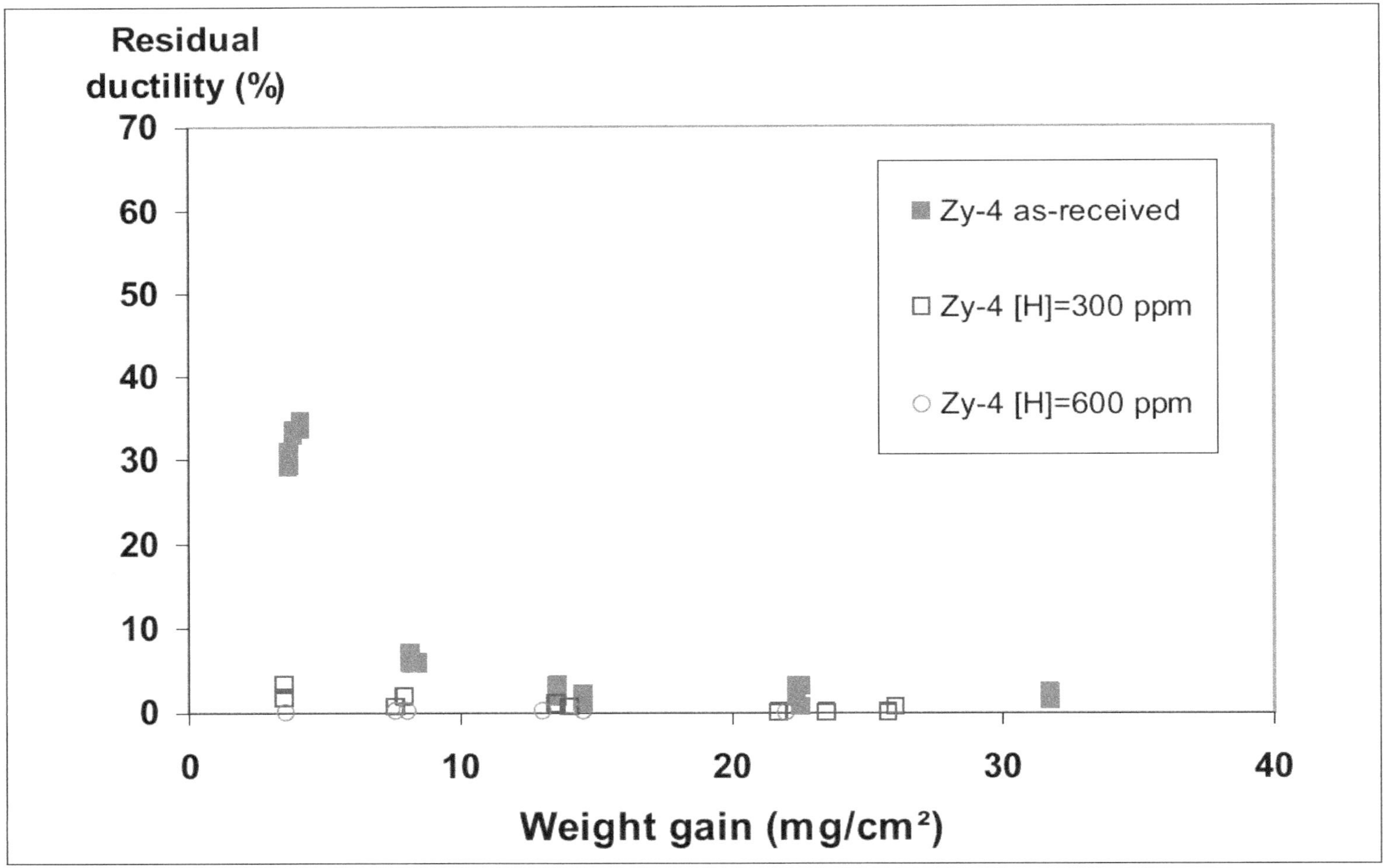

**Residual ductility (%)** vs **Weight gain (mg/cm²)**

Legend:
- Zy-4 as-received
- Zy-4 [H]=300 ppm
- Zy-4 [H]=600 ppm

Nuclear
Regulatory
Commission

Pioneering
Science and
Technology

# 135°C Offset Strain for Prehydrided 17x17 Zry-4 Oxidized at 1200°C

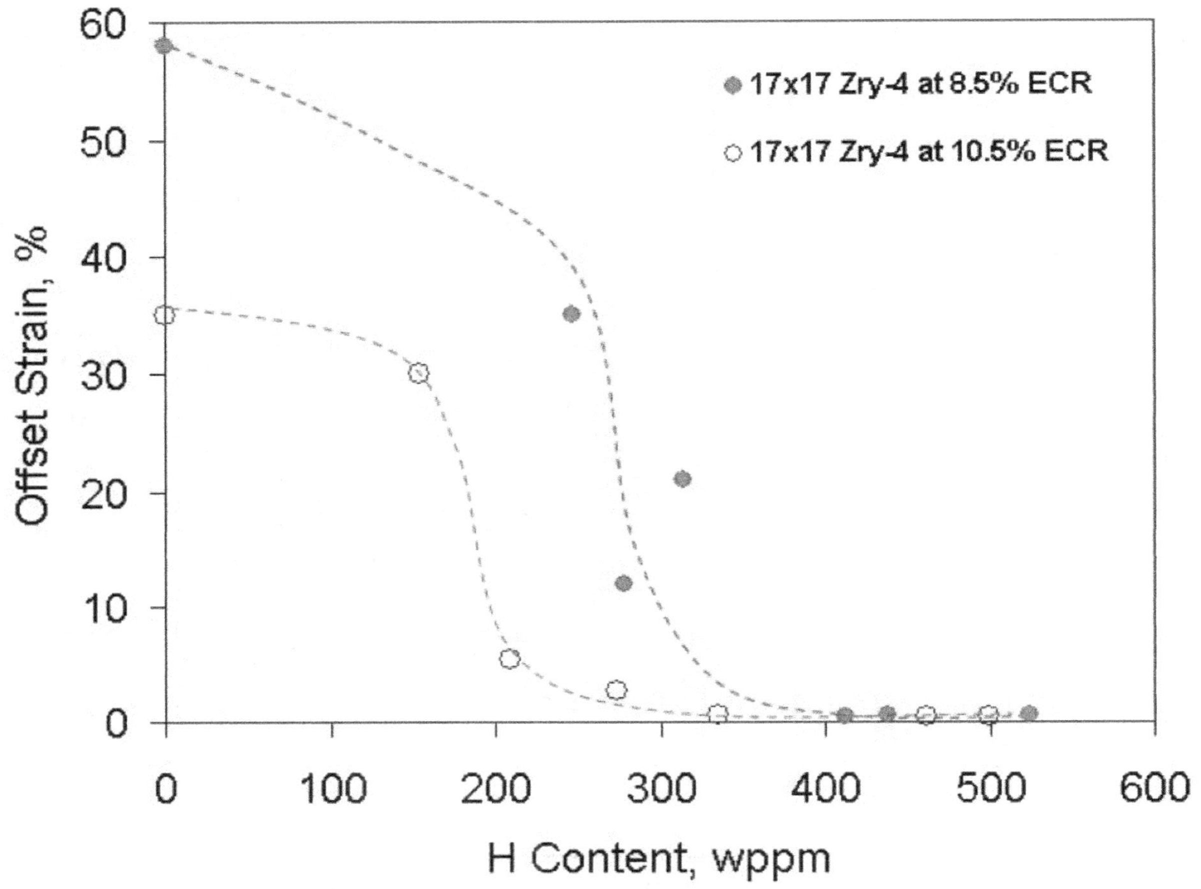

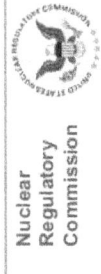

# Summary of Advanced Alloy PQD Results

- **RT Results for 1000-1100°C Oxidation & Quench at 800°C**

  - Zry-4, ZIRLO and M5 are ductile for CP-ECR $\leq$ 20%

  - E110 tubing oxidized at 1000°C is brittle for CP-ECR $\geq$ 7%

  - Mach.-and-pol. E110 oxidized at 1100°C is ductile for CP-ECR $\leq$ 19%

- **Results for 1200°C Oxidation & Quench at 800°C**

  - RT tests: significant embrittlement for CP-ECR > 9-11%

  - 135°C: 17x17 Zry-4, ZIRLO, and M5 are ductile for CP-ECR $\leq$ 17%

  - 100°C: 15x15 Zry-4 is ductile for CP-ECR $\leq$ 13% (see Yan)

  - 135°C: machined-and-polished E110 is ductile for CP-ECR $\leq$ 13%

- **135°C Results for Prehydrided Zry-4 Oxidized at 1200°C**

  - Measured ECR = 8.5%: embrittlement threshold $\approx$ 350-400 wppm

  - Measured ECR = 10.5%: embrittlement threshold $\approx$ 300 wppm

- **Embrittlement ECR Depends on Oxidation-T vs. t History**

# Overview of the CEA data on the influence of hydrogen on the metallurgical and thermal-mechanical behavior of Zircaloy-4 and M5™ alloys under LOCA conditions.

*J.C. Brachet(\*)*[(1)]*, L.. Portier*[(1)]*, V. Maillot*[(1)]*, T. Forgeron*[(2)]*, J.P. Mardon*[(3)]*, P. Jacques*[(4)]*, A. Lesbros*[(4)]

(1) CEA-DEN, DMN/SRMA, CEA/Saclay, 91191 Gif-sur-Yvette Cedex, France
(2) CEA-DSNI-Comb., CEA/Saclay, 91191 Gif-sur-Yvette Cedex, France
(3) FRAMATOME ANP Nuclear Fuel, 10 rue Juliette Récamier, 69456 Lyon Cedex 06, France
(4) EDF, SEPTEN, 12-14 avenue Dutrievoz, 69628 Villeurbanne Cedex, France
*(\*) corresponding author : jean-christophe.brachet@cea.fr*

**Abstract :**

A few years ago, within the framework of the CEA/EDF/Framatome-ANP R&D cooperative program, we made the assumption that the burn-up influence on the thermal-mechanical behavior of the fuel cladding tubes under LOCA conditions should be strongly linked to the hydrogen up-take due to the in-service oxidation (see for example discussion pp. 276-277 in [1]). Thus, since that time, an extensive experimental program has been conducted in CEA labs on as-received and pre-hydrided Zy-4 and M5™ advanced alloys of Framatome-ANP to get a better insight into the influence of the hydrogen on the thermal-mechanical cladding behavior during the first phase of the LOCA transient (ballooning and rupture) and for post-quenched conditions (residual ductility/toughness…) [2] [3] [4].

On the one hand, one of the main assumptions here was that the microstructural defects, and the resultant hardening produced under heavy neutron irradiation within the zirconium matrix, are annealed early upon the first phase of the LOCA transient (i.e. first thermal ramp) and thus, that the main effects of high burn-up should come from the hydrogen uptake. To assess this hypothesis, specific thermal-mechanical tests have been performed on virgin, pre-hydrided and irradiated cladding tubes. This confirmed that the effect of hydrogen uptake dominates over that of irradiation on the thermal-mechanical response of the materials.
So, in a first part of the presentation, we will briefly summarize the main results obtained here and, from the metallurgical point of view, we will illustrate the strong influence of hydrogen on the decrease of the alpha-to-beta phase transformation temperatures of the zirconium alloys studied.

On the other hand, studies have been performed on the post-quench mechanical behavior of as-received and pre-hydrided cladding tubes after single-face oxidation at 1000-1200°C and quenching. In parallel with these mechanical tests, in-depth metallurgical investigations have been developed [5], to be able to quantify the resultant phase thickness (that is, $ZrO_2$, Alpha(O) and Ex-Beta phase layers) and their specific chemical composition - especially their oxygen content which is known to influence strongly the residual mechanical properties. Also, fractograph analysis has been applied on failed samples to get a better knowledge of the failure mechanism as a function of the materials and of the hydrogen concentration, for different oxidation conditions.
So, in the second and major part of the presentation, we will focus on LOCA post-quenched behavior of as-received and prehydrided Zy-4 and M5™ cladding tubes for typical hydrogen contents ranging from ~200 up to ~600 wt-ppm depending on the alloy. Ring compression, impact, and bending tests at Room Temperature have been performed for different oxidation conditions. The mechanical results will be presented and briefly discussed, taking into account the metallurgical analysis (resultant phase morphology and thickness, chemical composition – oxygen contents, failure mode,…).

# References

[1] T. FORGERON, J.C. BRACHET, et al., « **Experiment and modelling of advanced fuel rod behaviour under LOCA conditions : $\alpha \Leftrightarrow \beta$ phase transformation kinetics and EDGAR methodology** », *Zirconium in the Nuclear Industry: 12$^{th}$. Int. Symposium, June 1998, Toronto, Canada*, ASTM STP 1354, (2000), pp. 256-278

[2] J.C. BRACHET, L. PORTIER, et al., "**Influence of hydrogen content on the $\alpha \Leftrightarrow \beta$ phase transformation temperatures and on the thermal-mechanical behavior of Zy-4, M4 (ZrSnFeV) and M5™ (ZrNbO) alloys during the first phase of LOCA transient,**" *Zirconium in the Nuclear Industry: 13$^{th}$. Int. Symposium, June 10-14 2001, Annecy, France,* ASTM STP 1423, (2002), pp. 673-701

[3] L. PORTIER, T. BREDEL, J.C. BRACHET, V. MAILLOT, J.P. MARDON, A. LESBROS, "**Influence of Long Service Exposures on the Thermal-Mechanical Behavior of Zy-4 and M5™ Alloys in LOCA Conditions**", *Zirconium in the Nuclear Industry: 14$^{th}$. Int. Symposium, June 13-17 2004, Stockholm, Sweden*

[4] J.-C. BRACHET, J. PELCHAT, D. HAMON, R. MAURY, P. JACQUES, J.-P. MARDON, « **Mechanical behavior at Room Temperature and Metallurgical study of Low-Tin Zy-4 and M5™ (Zr-NbO) alloys after oxidation at 1100°C and quenching**", *Proceeding of TCM on "Fuel behavior under transient and LOCA conditions", Sept. 10-14, 2001, organised by IAEA, Halden, Norway*

[5] J.C. BRACHET et al., "**Quantification of the $\alpha$(O) and Prior-$\beta$ phase fractions and their oxygen contents in high temperature (HT) oxidised Zr alloys (Zy-4, M5™)**", *Proceeding of "SEGFSM Topical Meeting on LOCA", 25-27$^{th}$. of May 2004, ANL, Chicago, USA*

Nuclear Energy Division

# Overview of the CEA data on the influence of hydrogen on the metallurgical and thermal-mechanical behavior of Zircaloy-4 and M5™ alloys under LOCA conditions.

JC. Brachet[1], L. Portier[1], V. Maillot[1], T. Forgeron[2],
JP. Mardon [3], P. Jacques[4], A. Lesbros[4]

(1) CEA-DEN, DMN/SRMA, CEA/Saclay, 91191 Gif-sur-Yvette Cedex, France
(2) CEA-DSNI-Comb., CEA/Saclay, 91191 Gif-sur-Yvette Cedex, France
(3) FRAMATOME ANP Nuclear Fuel, 10 rue Juliette Récamier, 69456 Lyon Cedex 06, France
(4) EDF, SEPTEN, 12-14 avenue Dutrievoz, 69628 Villeurbanne Cedex, France

Main objective : study of the influence of the hydrogen concentration within the fuel cladding tubes on their behavior in LOCA conditions => Hydrogen (due to in-service oxidation) is the 1st order parameter able to simulate the burn-up influence on the cladding tubes thermal-metallurgical-mechanical behavior in LOCA conditions

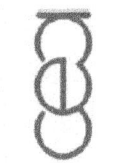

305

Nuclear Energy Division

# Hydrogen influence (~burn-up) on :

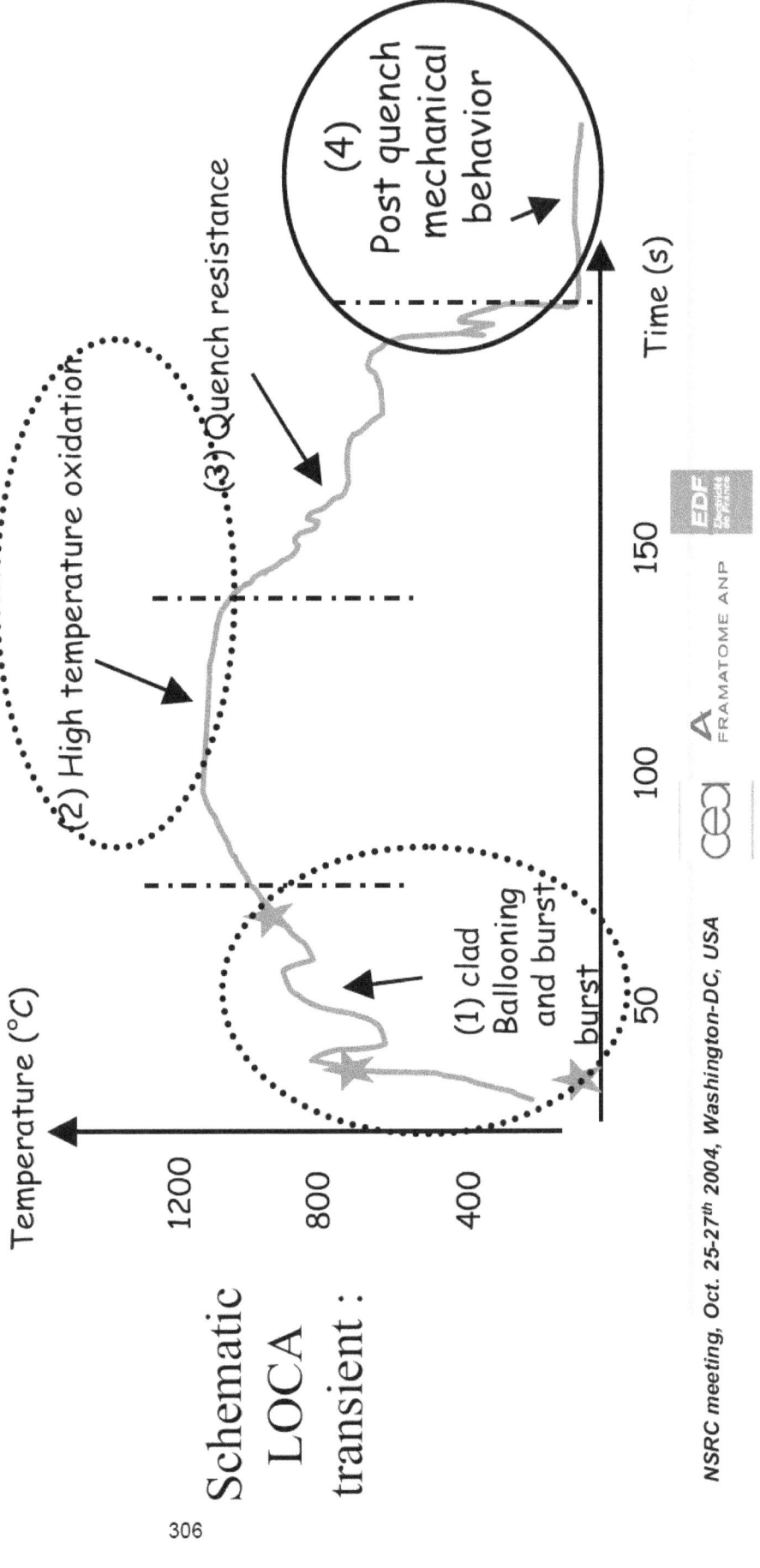

Schematic
LOCA
transient :

NSRC meeting, Oct. 25-27th 2004, Washington-DC, USA

*Nuclear Energy Division*

(1)  *[H] is 1st order parameter simulating the burn-up influence on cladding behavior during the 1st. phase of LOCA*

(2)  *Irradiation defects are annealed early upon the fast heating*

**Thermal ramps at constant axial load performed on irradiated Zy-4 and M5™ in hot lab:**

- Axial Load : 40MPa – 80MPa

- Direct Joule heating effect
=> T(°C) rate : 25 – 100°C/s

100°C/s  80MPa

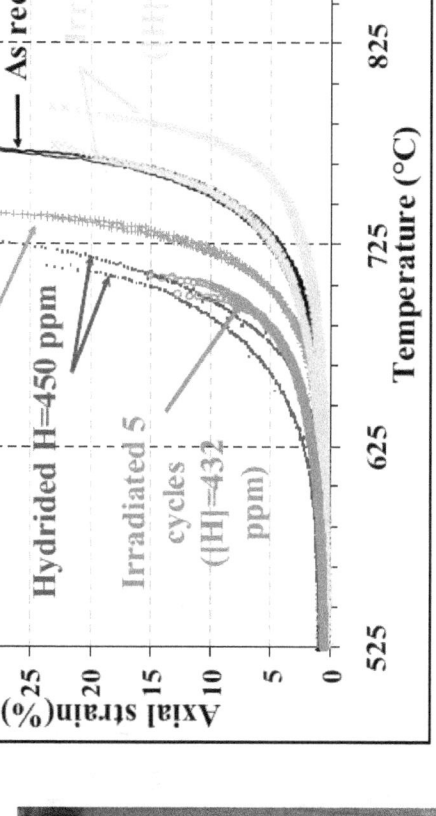

Axial strain(%)

Temperature (°C)

As received

Irradiated 3 cycles ([H]=50 ppm)

Hydrided H=150 ppm

Hydrided H=450 ppm

Irradiated 5 cycles ([H]=432 ppm)

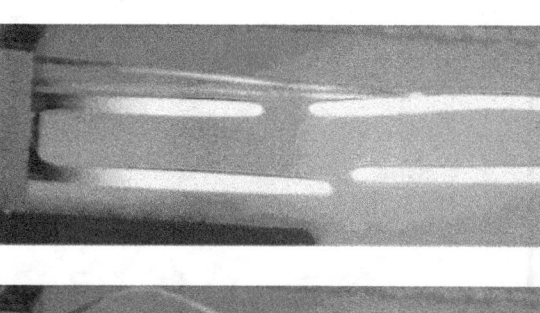

*Illustration of one of the results obtained on as-received, pre-hydrided and irradiated Zy-4*

*NSRC meeting, Oct. 25-27th 2004, Washington-DC, USA*

EDF
FRAMATOME ANP

# On pre-hydrided & irradiated materials [H] decreases the burst temperature :

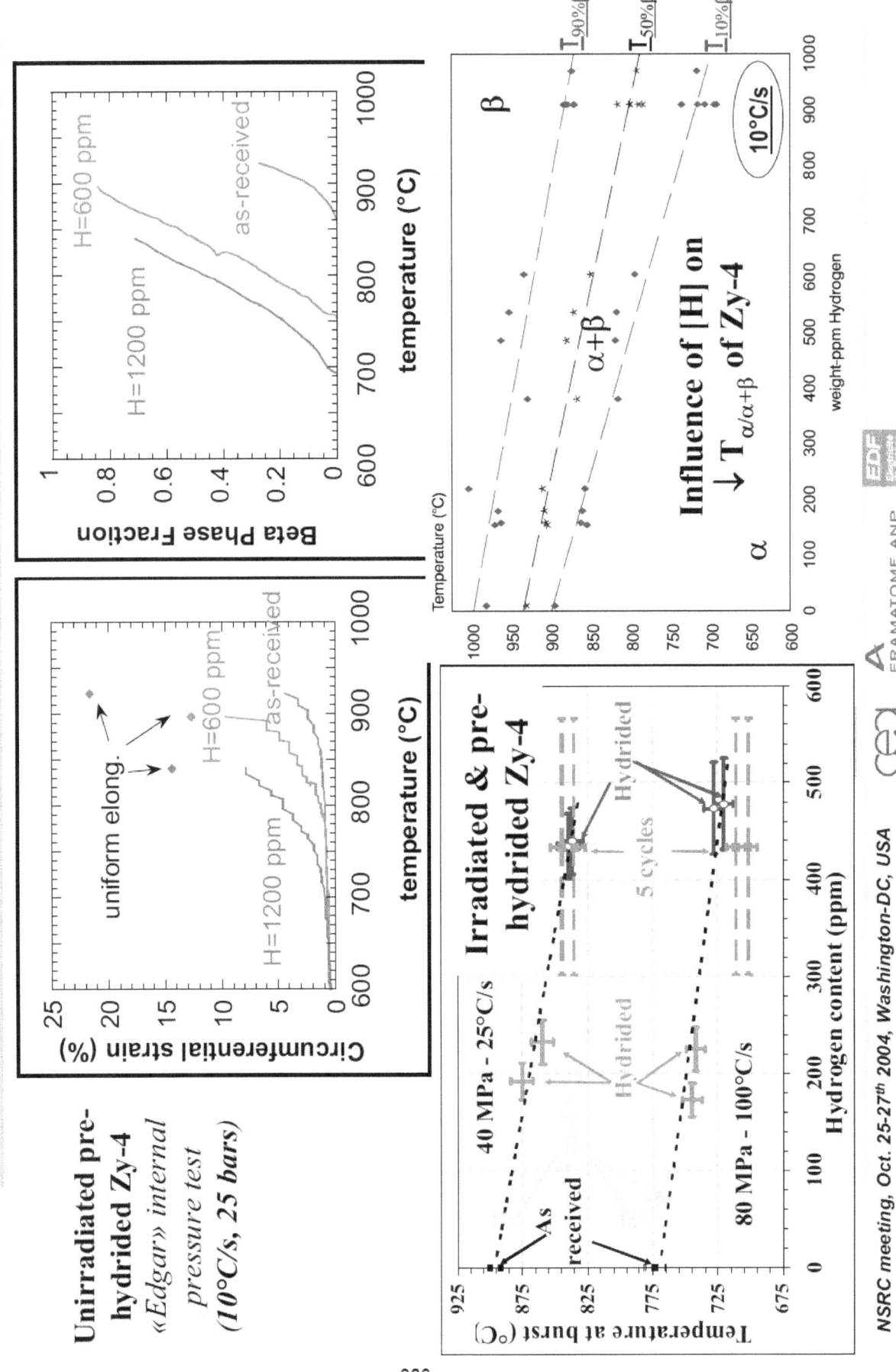

**Unirradiated pre-hydrided Zy-4**
*«Edgar» internal pressure test (10°C/s, 25 bars)*

*NSRC meeting, Oct. 25-27th 2004, Washington-DC, USA*

## *Partial conclusions*

*for cladding behavior upon fast LOCA representative thermal ramps*

- **On unirradiated+prehydrided materials** (from internal pressure EDGAR creep tests) : **[H] decreases both the creep strength and the ductility**

- **On unirradiated+pre-hydrided <u>and</u> irradiated (in-service hydrided) materials** (uniaxial loading, 25 and 100°C/s thermal ramps):

  **[H] decreases the burst temperature** (for irradiated materials, the defects produced by neutron irradiation are annealed early upon the first heating and do not play significant effects on the thermal-mechanical behavior of the cladding tubes at high temperatures (>700°C))

$\Rightarrow$ This can be related to : **(1)** $T_{\alpha/\beta}$ **shift** (H has a strong β-stabilizing effect)
  **(2) intrinsic effect of [H] in solid solution** within the Zr matrix, as observed for example in the α phase temperature range (i.e. : EDGAR creep tests at 600-700°C)

$\Rightarrow$ *Due to its lower in-service uptake of hydrogen, irradiated M5™ shows nearly the same behavior as that of as-received material*

*NSRC meeting, Oct. 25-27th 2004, Washington-DC, USA*

CEA    FRAMATOME ANP

309

# [H] influence on residual post-oxidation + (rapid)quench mechanical behavior of Zy-4 and M5™

## Dezirox device:

Single side oxidation in steam
/ sample length = 150mm

T : 1000 , 1100 , 1200°C

Weight gain : ~4 => ~38 mg/cm$^2$

ECR(Baker-Just) : ~5.5 => ~40.5%

[H]=600 ppm for Zy-4
~ "End-Of-Life" level (*)

and 200 ppm for M5™
>> "EOL" (< 100 ppm) (*)

(*) from in reactor feed-back experience

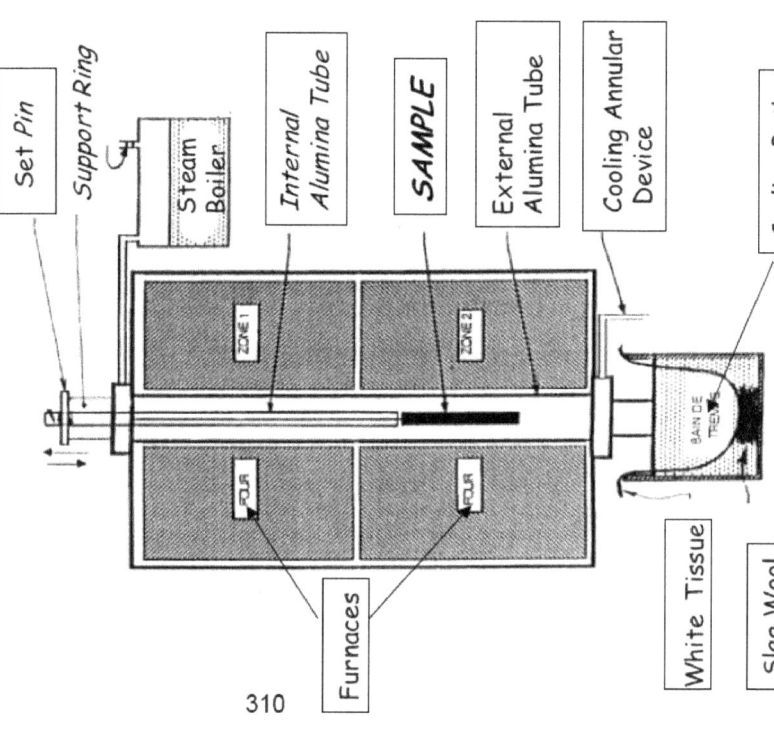

Set Pin

Support Ring

Steam Boiler

Internal Alumina Tube

SAMPLE

External Alumina Tube

Cooling Annular Device

Cooling Bath

ZONE 1

ZONE 2

FOUR

FOUR

Furnaces

White Tissue

Slag Wool

310

NSRC meeting, Oct. 25-27th 2004, Washington-DC, USA

FRAMATOME ANP

EDF

Nuclear Energy Division

$$T_{oxyd.} = 1200°C$$

*=> no significant [H] influence on the oxidation kinetics*

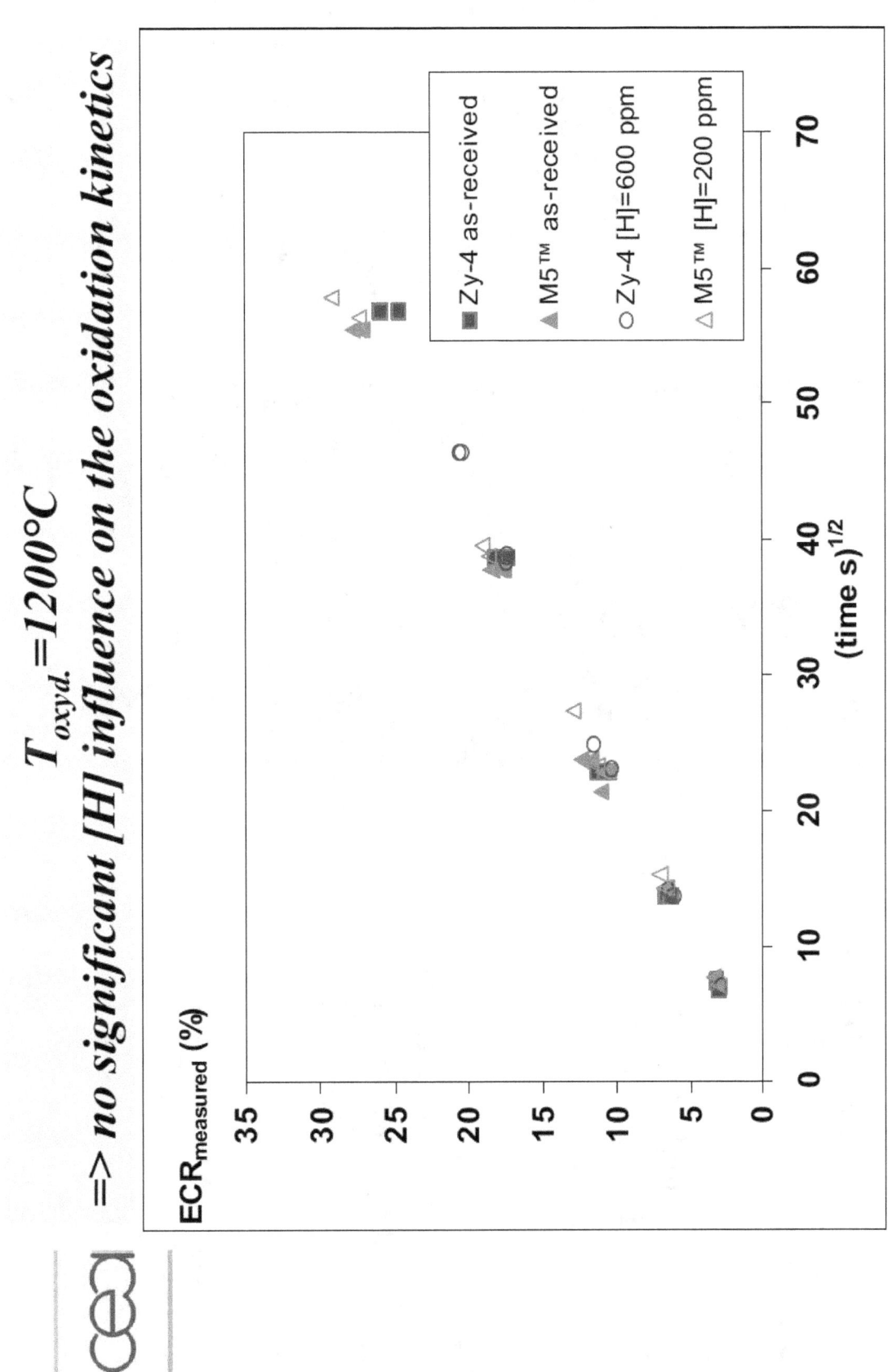

NSRC meeting, Oct. 25-27th 2004, Washington-DC, USA

*Nuclear Energy Division*

# *Post-quench mechanical tests (room temp.) – experimental*

**Ring Compression Tests**
samples of 10 mm in length,
displacement rate 0.5mm/mn

**3 Points Bend Tests:**
samples of 80 mm
in gauge length,
max. displacement ~7.5 mm

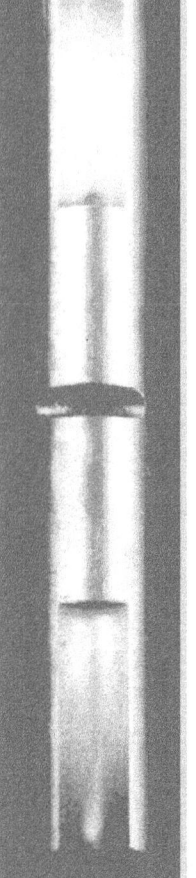

55 mm. length

*(Pre-notch : radius=1mm, depth = 8mm.)*

Impact pre-notched sample

**«Charpy» type
Impact Tests :**

*NSRC me*

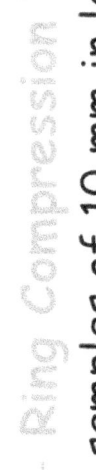

Determination of maximum load & displacement to rupture :
**3PBT tests give more accurate mechanical characteristics values than RCT** ; also, **allow a fractographic analysis** with no risk of degradation of the rupture surface.

3 points bend test:

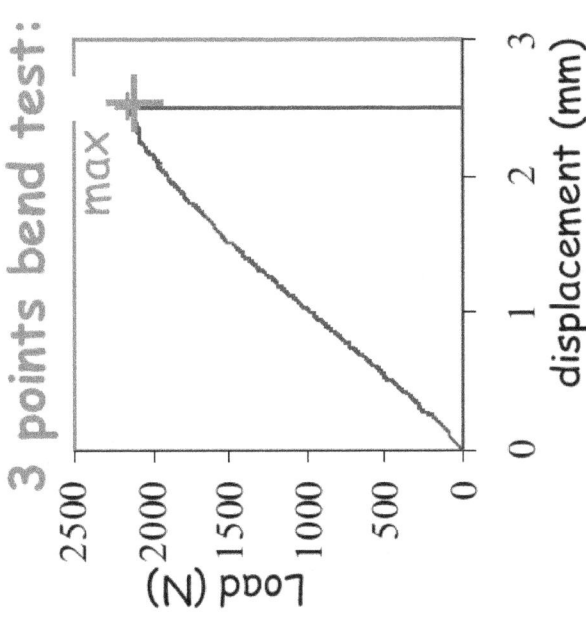

ring compression test:

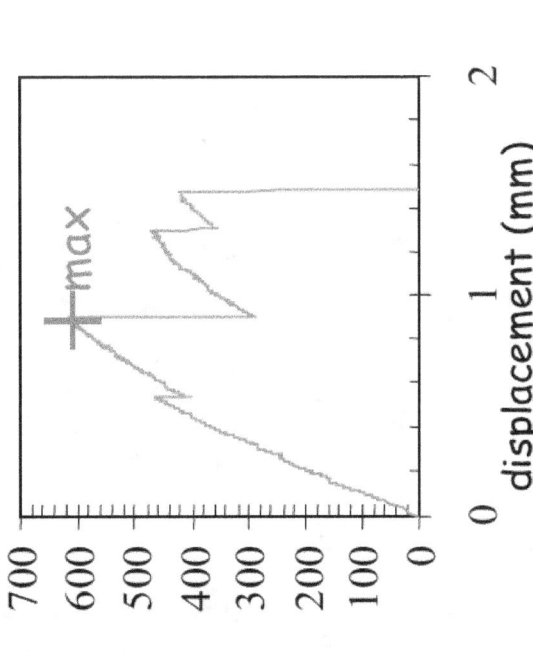

*The maximum load and displacement to rupture are determined at the first significant load drop indicating a brittle rupture.*
*For ring compression test it is difficult to determine it for intermediate oxidation times whereas it is always very easy for 3 points bend tests because there is only one load drop before rupture.*

*NSRC meeting, Oct. 25-27th 2004, Washington-DC, USA*

cea    FRAMATOME ANP

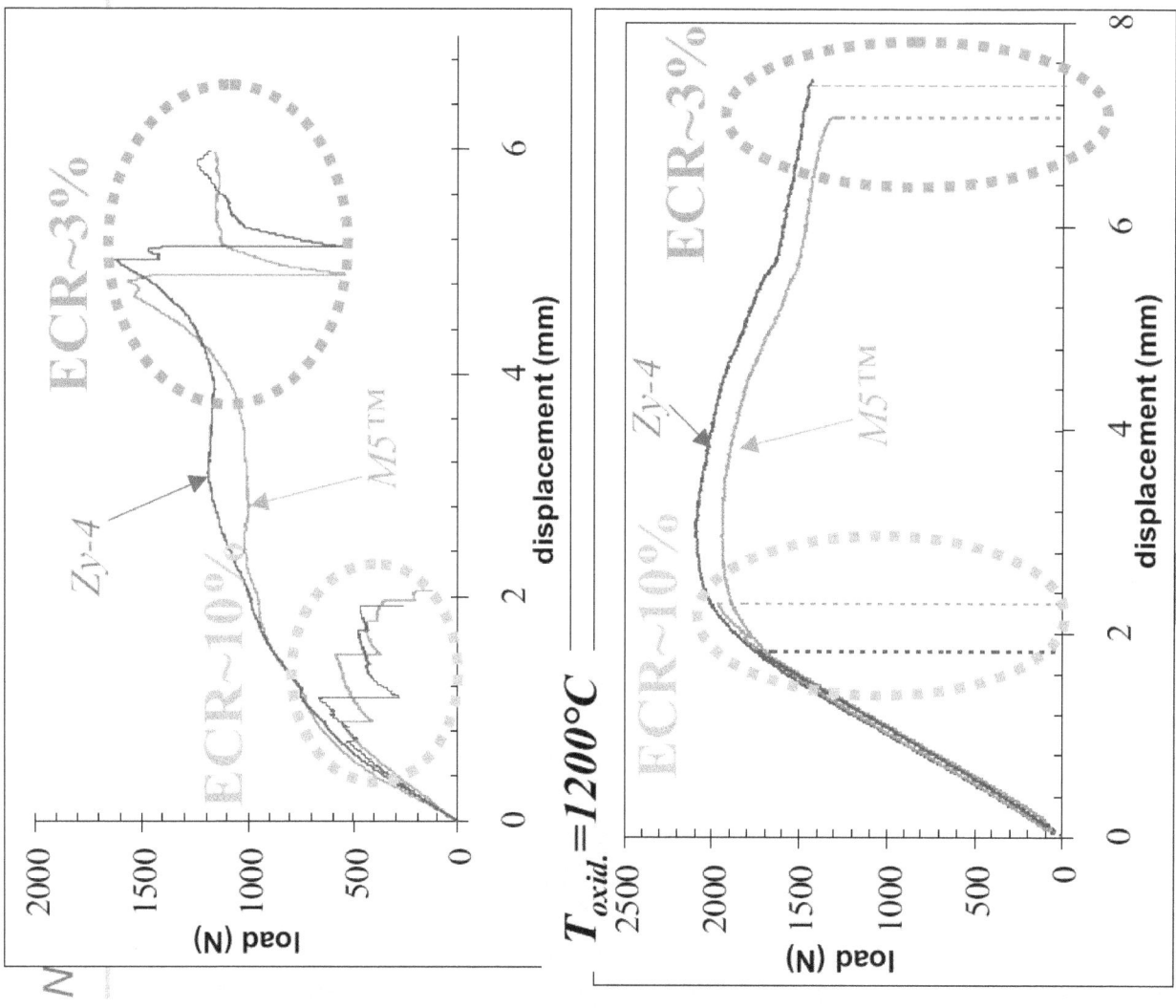

**As-received materials**
**(for 2 typical intermediate ECR$_{measured}$ values)**

$\Rightarrow$ 3 points bend test gives more «physical» and more accurate values for evaluation of residual strain-strength parameters in all the investigated ECR range

$\Rightarrow$ the failure seems to occur within the apparent elastic part of engineering curves for ECR$_{measured}$ > 10 % i.e. ECR$_{Baker-Just}$ > 15.5%

$\Rightarrow$ as-received Zy-4 and M5™ behave similarly

314

# Pre-hydrided materials :
## Zy-4 + 600 ppm [H]
## & M5™ + 200 ppm [H]
### (for 2 typical ECR$_{measured}$ values)

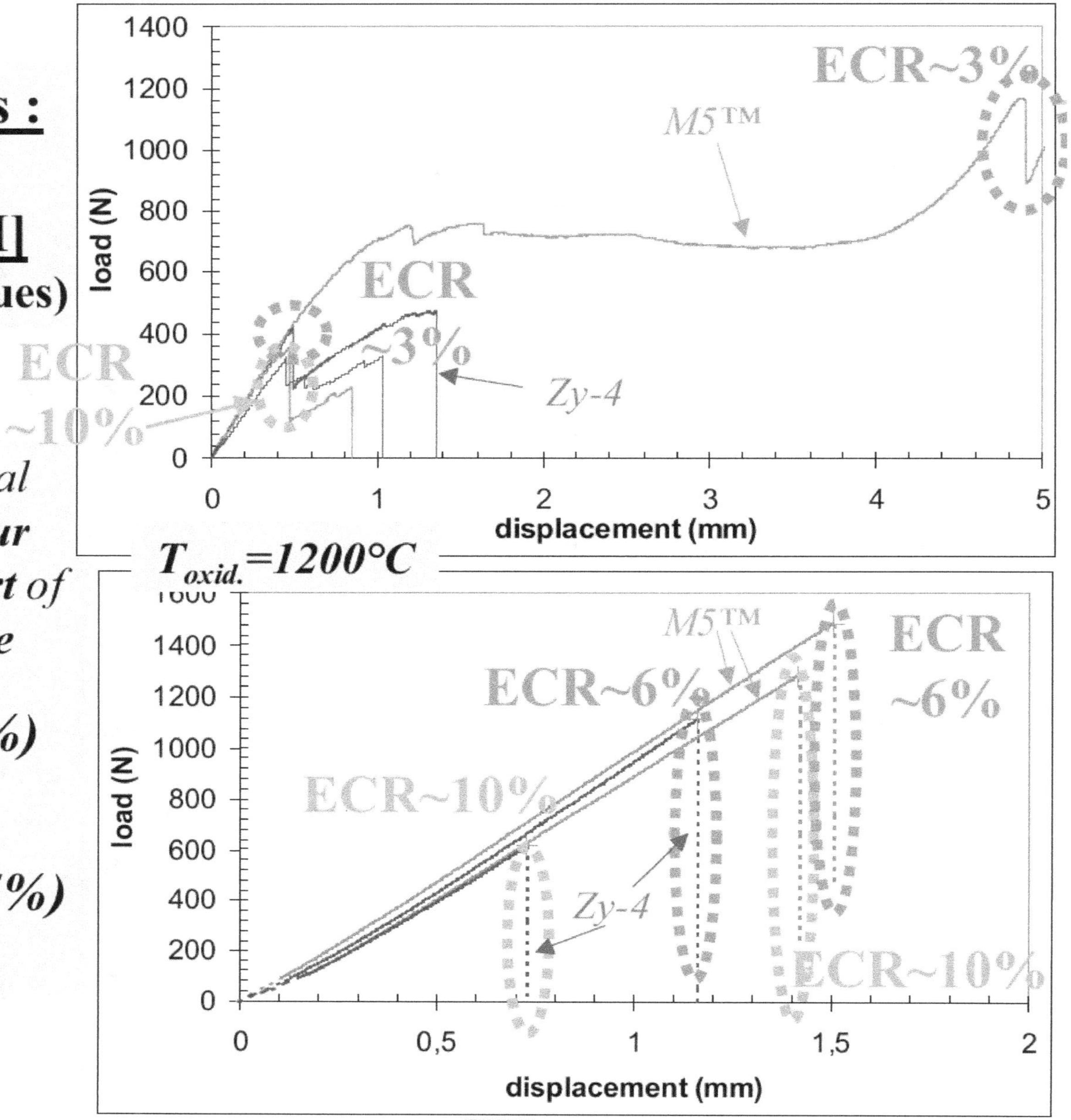

*From both types of mechanical test, the failure seems to occur within the apparent elastic part of the engineering curves above*

- ECR$_{measured}$ ~3% (BJ~5%) for Zy-4 + 600 ppm H

- ECR$_{measured}$ ~6%(BJ~9.5%) for M5™ + 200 ppm H

$T_{oxid.} = 1200°C$

315

EDF

CEQ

FRAMATOME ANP

# Ring Compression tests

*(room temperature)*

## as-received vs. hydrided materials, $T_{oxidation} = 1200°C$

*A marked decrease of the post-quench ductility is observed on hydrided materials,*

*especially on Zy-4 + 600 ppm H even for the lower ECR value tested ( i.e. for $ECR_{meas} \sim 3\% \Leftrightarrow BJ \sim 5\%$) while M5™ + 200 ppm H remains ductile for such an ECR*

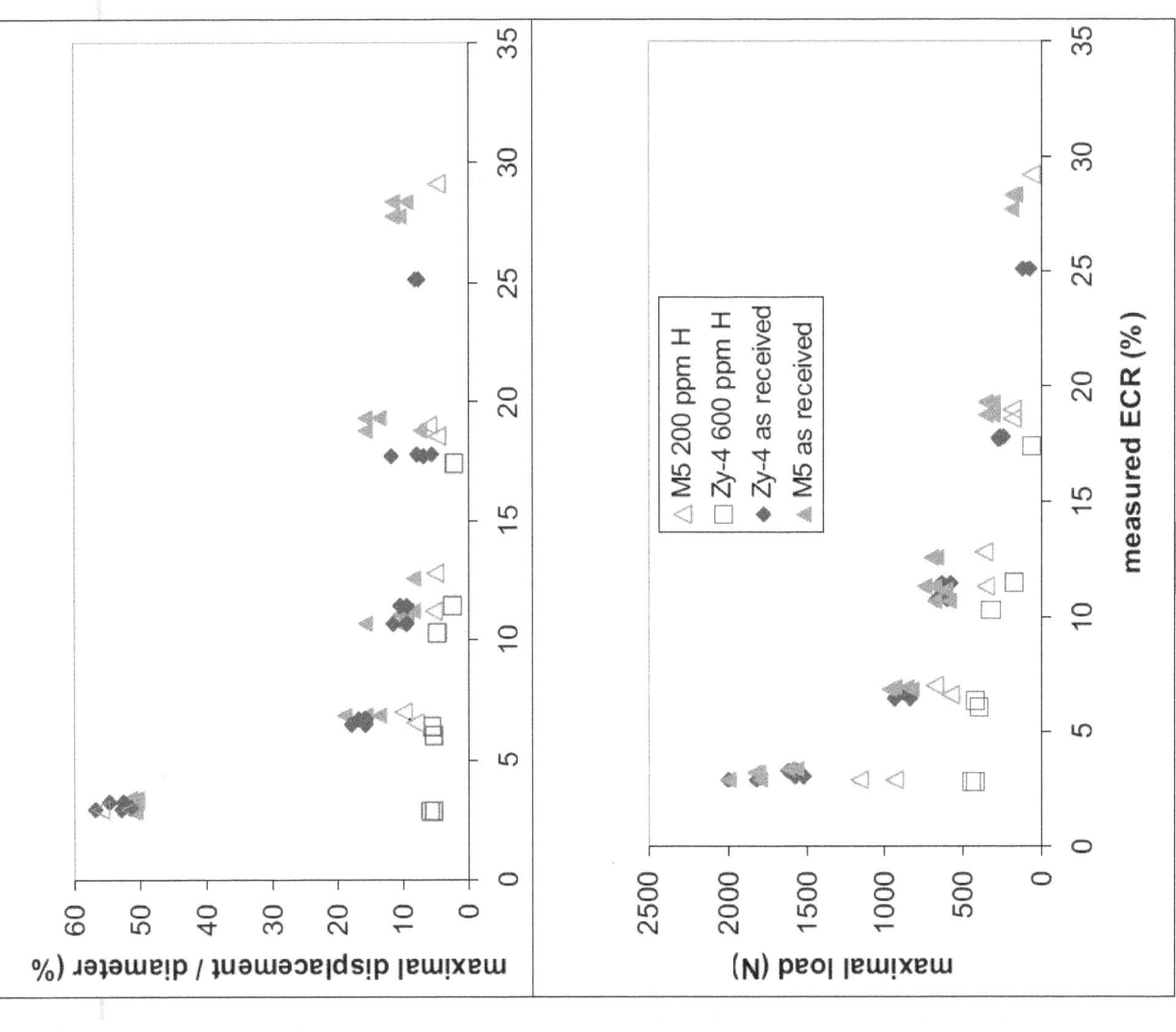

*NSRC meeting, Oct. 25-27th 2004, Washington-DC, USA*

EDF

FRAMATOME ANP

cea

# 3 Points Bend Test

*(room temperature)* **as-received vs. hydrided materials,** $T_{oxidation} = 1200°C$

*For both hydrided materials : a ductility ($\varepsilon_{total}$) tends to a (minimal) asymptotic value for a measured ECR ~ 10 => 17% (Baker-Just ~ 15.5 => 26.5 %)*

*But, considering the max load : hydrided M5™ ([H]=200 ppm) exhibits a more progressive decrease of this parameter compared to hydrided Zy-4 ([H]=600 ppm)*

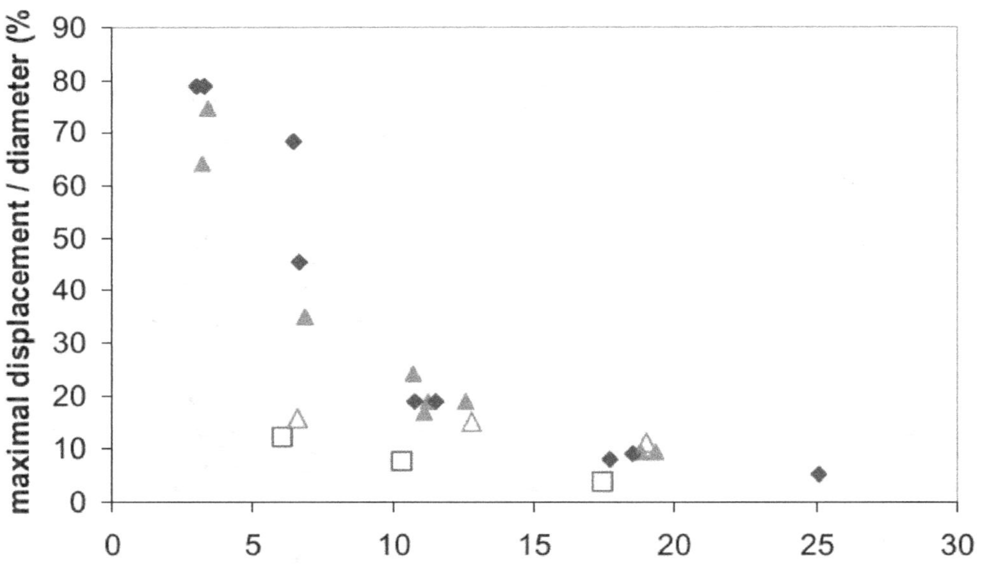

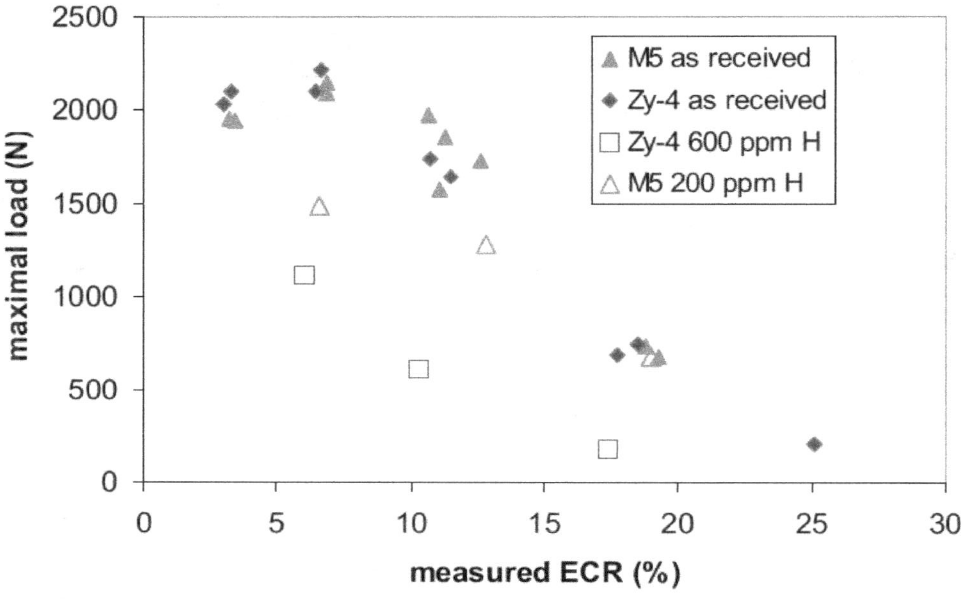

Legend (lower chart):
- ▲ M5 as received
- ◆ Zy-4 as received
- □ Zy-4 600 ppm H
- △ M5 200 ppm H

Upper chart: maximal displacement / diameter (%) vs measured ECR (%)

Lower chart: maximal load (N) vs measured ECR (%)

# Charpy impact tests *(room temperature)* / $T_{oxidation} = 1200°C$
## *As-received materials :*

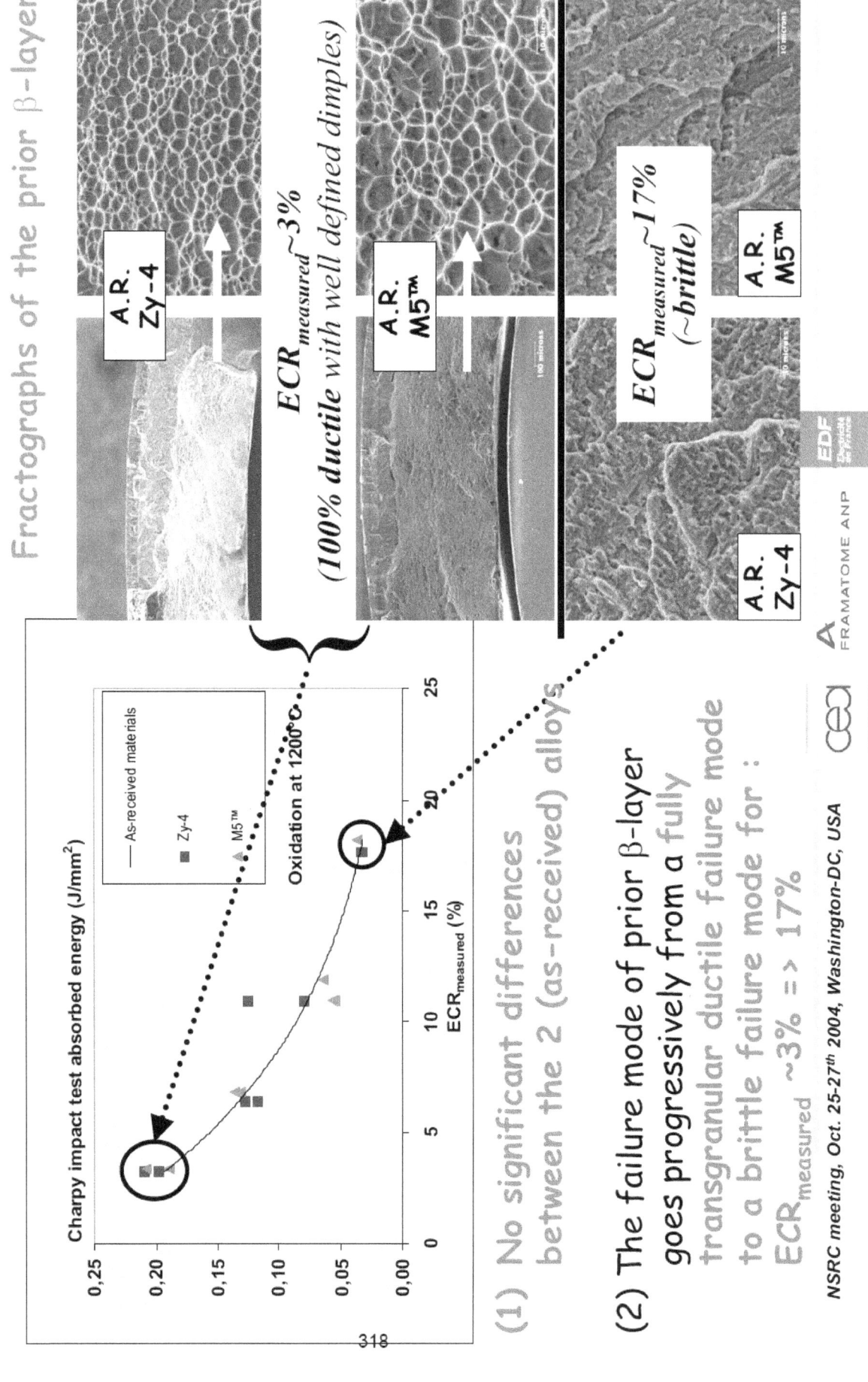

Fractographs of the prior β-layer

$ECR_{measured} \sim 3\%$
*(100% ductile with well defined dimples)*

A.R. Zy-4

A.R. M5™

$ECR_{measured} \sim 17\%$
*(~brittle)*

A.R. M5™

A.R. Zy-4

Charpy impact test absorbed energy (J/mm²)

As-received materials
Zy-4
M5™

Oxidation at 1200°C

$ECR_{measured}$ (%)

(1) No significant differences between the 2 (as-received) alloys

(2) The failure mode of prior β-layer goes progressively from a fully transgranular ductile failure mode to a brittle failure mode for : $ECR_{measured} \sim 3\%$ => 17%

*NSRC meeting, Oct. 25-27th 2004, Washington-DC, USA*

FRAMATOME ANP

EDF

cea

318

# *Charpy impact tests (room temperature) / $T_{oxidation} = 1200°C$*
## *Pre-hydrided materials :*

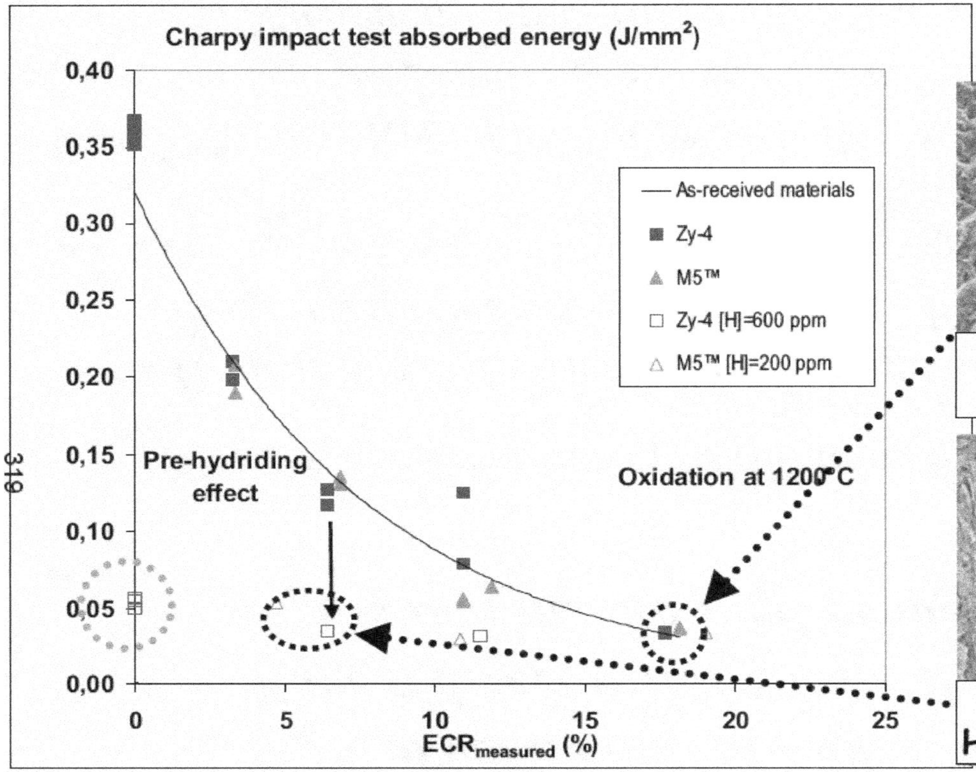

Fractographs of the prior β-layer

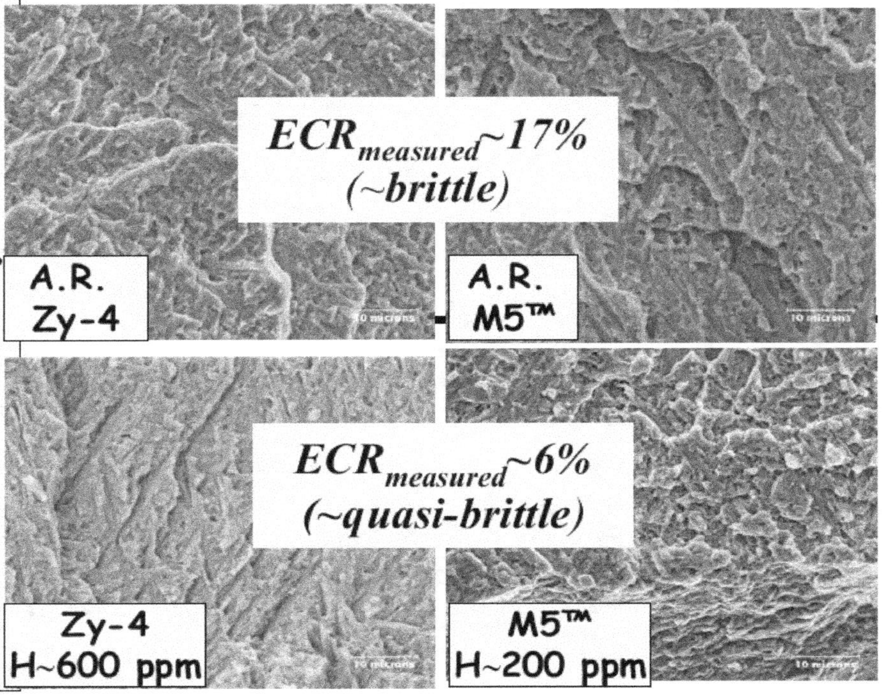

A.R. Zy-4

A.R. M5™

$ECR_{measured} \sim 17\%$ (~brittle)

Zy-4 H~600 ppm

M5™ H~200 ppm

$ECR_{measured} \sim 6\%$ (~quasi-brittle)

Charpy impact test absorbed energy (J/mm²)

— As-received materials
■ Zy-4
▲ M5™
□ Zy-4 [H]=600 ppm
△ M5™ [H]=200 ppm

Oxidation at 1200°C

Pre-hydriding effect

$ECR_{measured}$ (%)

(1) Hydrogen content induces a significant decrease in residual toughness even in the absence of oxidation (ex: Zy-4+600ppmH after a β thermal treatment at ~1050°C)

=> *What is the [H] embrittlement mechanism ?*

(2) *Consistent with other mechanical tests*

*NSRC meeting, Oct. 25-27th 2004, Washington-DC, USA*

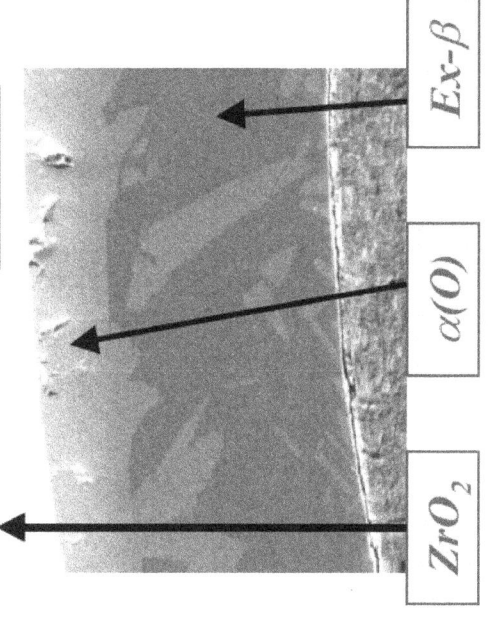

Zy-4, 1100°C
$ECR_{meas.} = 17\%$

O-K$\alpha$

ZrO$_2$

$\alpha$(O)

Ex-$\beta$

**ZrO$_2$, $\alpha$(O), <u>prior-$\beta$ phase layer</u>**
**thickness and their specific**
**chemical compositions – especially**
**their oxygen content – are known to**
**have a direct influence on the post-**
**quench mechanical properties**

$\Rightarrow$ *Systematic post-oxidation+quench metallurgical studies :*

- **Optical & SEM observation + image analysis**
  (*see for example proceeding of SEG-FSM meeting held in ANL, May 2004*)

- **EPMA** (*Cameca microprobe facility*)

- **Nuclear Reaction Analysis** (*collab. with CEA/DSM/DRECAM*)
  *(1) Van de Graaff accelerator – Nucl. react. : $^{16}O(d,p_1)^{17}O$ (Ph. Troulard)*
  *(2) Nuclear µprobe - Pierre Süe Lab. (P. Berger)*

- *µhardness measurements $\Leftrightarrow$ local oxygen hardening*

FRAMATOME ANP

320

*There is a systematic overestimation of the oxygen content by EPMA due to surface contamination => need of complementary NRA to quantify this phenomena*

Comparison between oxygen content measurements performed by EPMA and NRA on the internal Prior-β phase layer

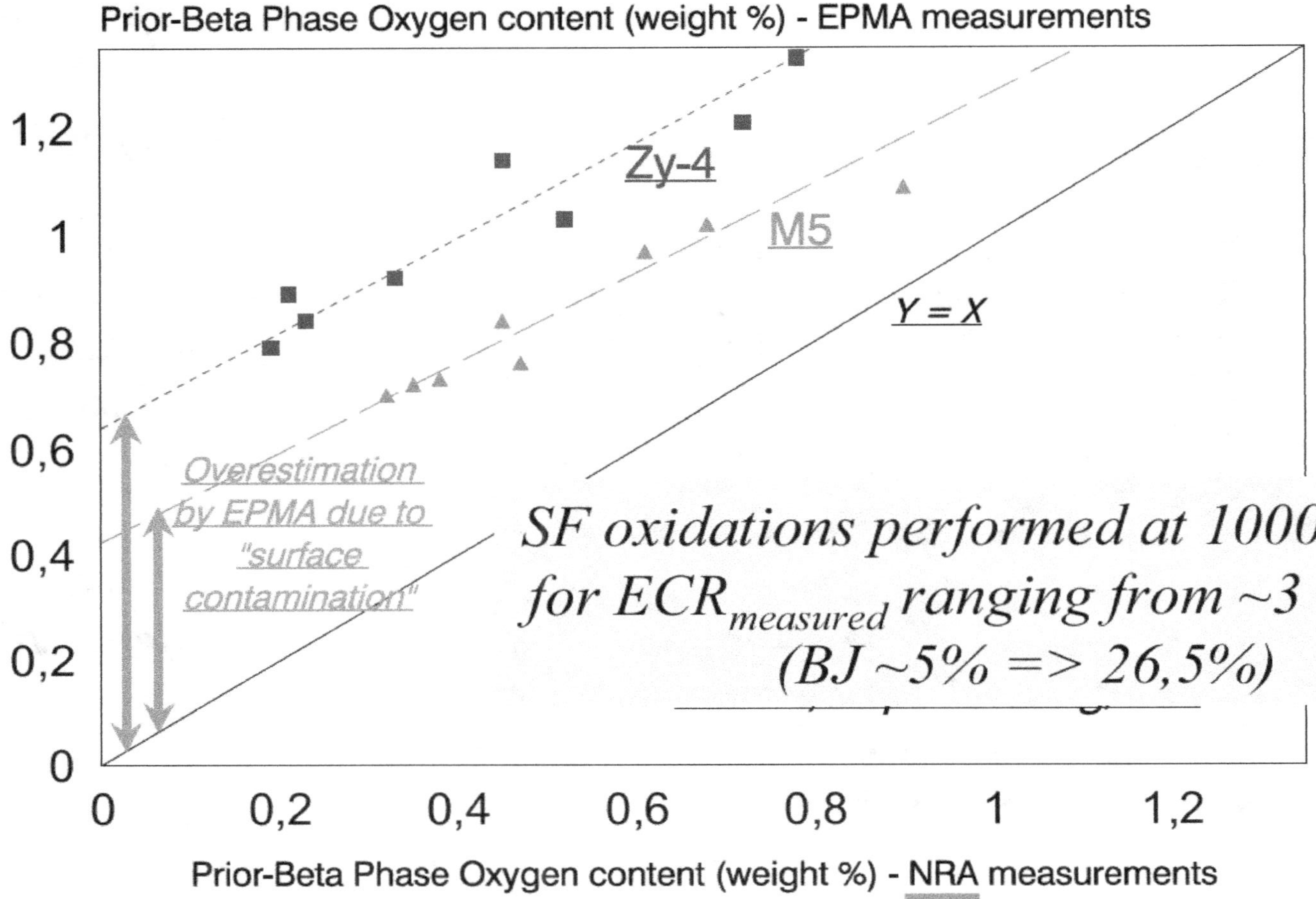

Prior-Beta Phase Oxygen content (weight %) - EPMA measurements

Y = X

M5

Zy-4

*Overestimation by EPMA due to "surface contamination"*

*SF oxidations performed at 1000-1200°C for $ECR_{measured}$ ranging from ~3 to ~17% (BJ ~5% => 26,5%)*

Prior-Beta Phase Oxygen content (weight %) - NRA measurements

*NSRC meeting, Oct. 25-27th 2004, Washington-DC, USA*

FRAMATOME ANP

EDF
Électricité de France

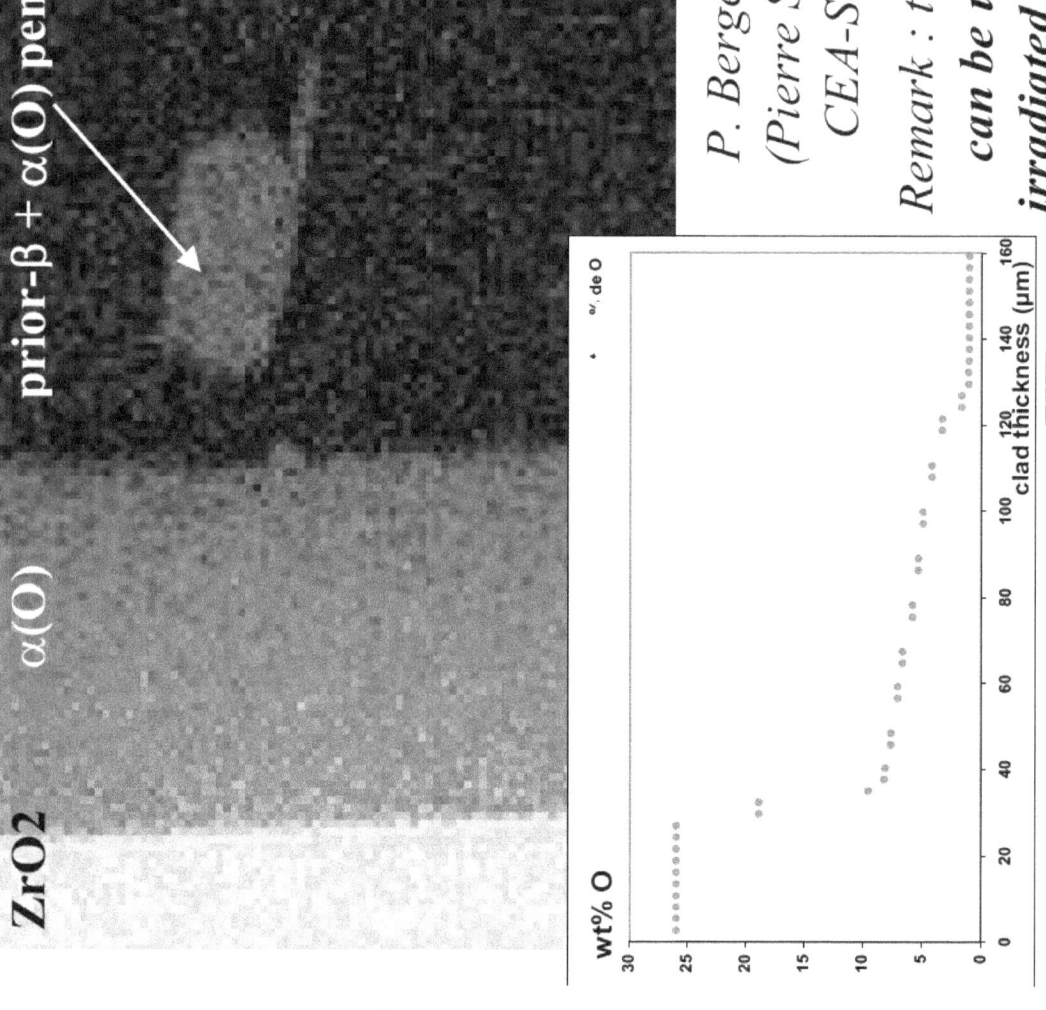

ZrO2

α(O)

prior-β + α(O) penetration

wt% O

% de O

clad thickness (µm)

*Nuclear μprobe application :*
*typical oxygen X-ray map* **and profile** *obtained in the thickness of a* **pre-hydrided** **Zy-4** *cladding tube after* **single-face oxidation** **(1200°C , ~600 s)**

*As for Van de Graaff accelerator, deconvolution of the contribution of oxygen atoms near the sample surface is possible but with a* **better spatial** **resolution (~1µm³)**

*P. Berger et al.*
*(Pierre Süe lab.*
*CEA-Saclay)*

*Remark : the facility* **can be used on** **irradiated materials**

cea        FRAMATOME ANP

322

# Correlation between prior-Beta phase oxygen content and μhardness : the measurements performed allow to quantify the hardening-embrittlement limit corresponding to ~0.7% of oxygen into the prior-β phase internal layer

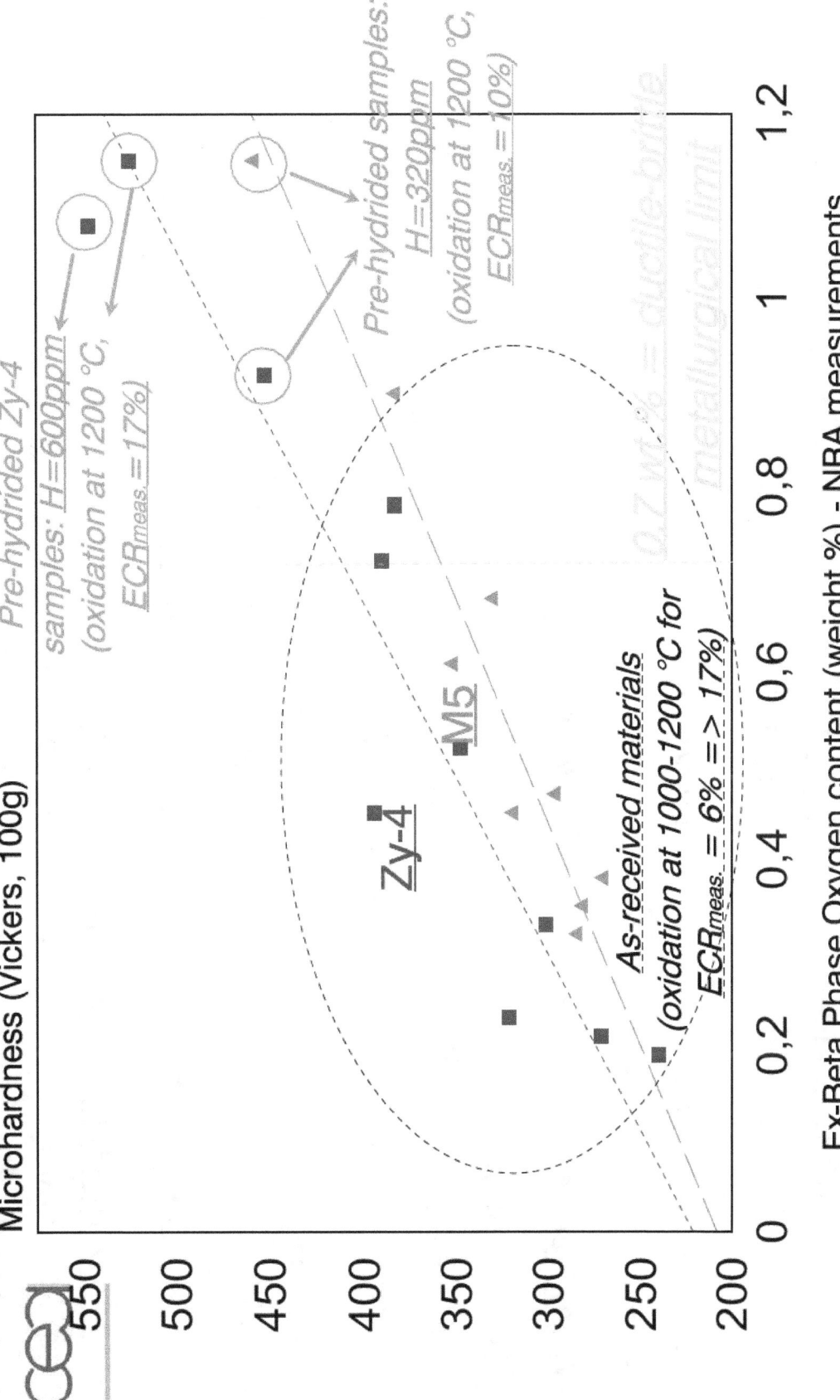

Ex-Beta Phase Oxygen content (weight %) - NRA measurements

NSRC meeting, Oct. 25-27th 2004, Washington-DC, USA

# Partial conclusions on the embrittlement phenomena after oxidation at 1200°C and quenching due to pre-hydriding (~ which simulates high burn-up).

*From NRA measurements of oxygen content within the internal prior-β phase layer :*

**(1) influence of [H] => oxygen solubility in β↑**

[H]=300-600ppm

=> $\Delta|O|^{\beta}_{1200°C} \sim$ 2000-4000 wt-ppm

=> $[O]_{Prior-\beta}$ *reaches concentration > 0.7wt-%*

=> *confirmed by thermodynamic calculation*
(« Thermocalc » + « Zircobase » (JNM, Vol.75, 1999))

$\Delta|O|_{\beta}\uparrow$ **above 0.7%** => $\Delta Hv\uparrow$ => **promotes brittle failure mode**

*Schematic phase diagram*

[H] effect

β

α + β

α

T(°C)

1200

1000

[O]

**(2) Intrinsic [H] embrittlement influence**
*as observed on «Beta» treated Zy-4 / 600 wt.ppm without oxidation*
*this phenomena is the 1st parameter responsible for the post-oxidation-quenching embrittlement of prehydrided materials for low ECR values (i.e. < 6%)*

# Final conclusions

- **As-received Zy-4 and M5™ achieve comparable post-quench mechanical behavior** in all the investigated range of ECR and oxidizing temperatures (i.e. **1000-1200°C**).

- For both Zy-4 and M5™, and for a given ECR, **the higher the hydrogen content, the lower the residual ductility and toughness after oxidation at 1200°C and quench.** Embrittlement due to :

    (1) At high ECR : a higher solubility of oxygen at 1200°C within the β-phase in the presence of hydrogen

    (2) At low ECR : + an intrinsic H embrittlement effect

* *Further work* - *especially on M5™ - study the influence of [H] for intermediate concentrations typical of MOL and EOL ones, that is : [H] ranging between ~40 and ~100 wt.ppm for M5™*

*=> aim : be able to quantify accuratly the post-LOCA(oxidation-quench) clad mechanical behavior as a function of the Burn-Up.*

CEA    FRAMATOME ANP     EDF

325

# *References*

[1] T. FORGERON, J.C. BRACHET, et al., « **Experiment and modelling of advanced fuel rod behaviour under LOCA conditions : α⇔β phase transformation kinetics and EDGAR methodology** », *Zirconium in the Nuclear Industry: 12th Int. Symposium, June 1998, Toronto, Canada, ASTM STP 1354, (2000), pp. 256-278*

[2] J.C. BRACHET, L. PORTIER, et al., **"Influence of hydrogen content on the α⇔β phase transformation temperatures and on the thermal-mechanical behavior of Zy-4, M4 (ZrSnFeV) and M5™ (ZrNbO) alloys during the first phase of LOCA transient,"** *Zirconium in the Nuclear Industry: 13th. Int. Symposium, June 10-14 2001, Annecy, France, ASTM STP 1423, (2002), pp. 673-701*

[3] L. PORTIER, T. BREDEL, J.C. BRACHET, V. MAILLOT, J.P. MARDON, A. LESBROS, **"Influence of Long Service Exposures on the Thermal-Mechanical Behavior of Zy-4 and M5™ Alloys in LOCA Conditions"**, *Zirconium in the Nuclear Industry: 14th. Int. Symposium, June 13-17 2004, Stockholm, Sweden., to be published in ASTM-STP*

[4] J.-C. BRACHET, J. PELCHAT, D. HAMON, R. MAURY, P. JACQUES, J.-P. MARDON, « **Mechanical behavior at Room Temperature and Metallurgical study of Low-Tin Zy-4 and M5™ (Zr-NbO) alloys after oxidation at 1100°C and quenching**", *Proceeding of TCM on "Fuel behavior under transient and LOCA conditions", Sept. 10-14, 2001, organised by IAEA, Halden, Norway*

[5] J.C. BRACHET et al., **"Quantification of the α(O) and Prior-β phase fractions and their oxygen contents in high temperature (HT) oxidised Zr alloys (Zy-4, M5™)"**, *Proceeding of "SEGFSM Topical Meeting on LOCA", 25-27th. of May 2004, ANL, Chicago, USA*

| NRC FORM 335<br>(9-2004)<br>NRCMD 3.7 | U.S. NUCLEAR REGULATORY COMMISSION | 1. REPORT NUMBER<br>(Assigned by NRC, Add Vol., Supp., Rev.,<br>and Addendum Numbers, if any.) |
|---|---|---|
| | **BIBLIOGRAPHIC DATA SHEET**<br>*(See instructions on the reverse)* | NUREG/CP-0192 |

| 2. TITLE AND SUBTITLE | 3. DATE REPORT PUBLISHED | |
|---|---|---|
| Proceedings of the Nuclear Fuels Sessions of the 2004 Nuclear Safety Research Conference | MONTH | YEAR |
| | October | 2005 |
| | 4. FIN OR GRANT NUMBER | |

| 5. AUTHOR(S) | 6. TYPE OF REPORT |
|---|---|
| Summaries of Conference Papers and Slides by various authors;<br>Compiled by Michelle Snell, RES/DSARE/SMSAB | Conference Summaries |
| | 7. PERIOD COVERED *(Inclusive Dates)*<br><br>October 25-27, 2004 |

8. PERFORMING ORGANIZATION  - NAME AND ADDRESS *(If NRC, provide Division, Office or Region, U.S. Nuclear Regulatory Commission, and mailing address; if contractor, provide name and mailing address.)*

Division of Systems Analysis and Regulatory Effectiveness

Office of Nuclear Regulatory Research
U.S. Nuclear Regulatory Commission
Washington, DC  20555-0001

9. SPONSORING ORGANIZATION - NAME AND ADDRESS *(If NRC, type "Same as above"; if contractor, provide NRC Division, Office or Region, U.S. Nuclear Regulatory Commission, and mailing address.)*

Same as Item 8 above

10. SUPPLEMENTARY NOTES

11. ABSTRACT *(200 words or less)*

This report contains papers from the nuclear fuels sessions of the 2004 Nuclear Safety Research Conference held at the Marriott Hotel at Metro Center in Washington, DC, October 25-27, 2004.

This information describes programs and results of work sponsored by the U.S. Nuclear Regulatory Commission's Office of Nuclear Regulatory Research. Also included are invited papers from others involved in nuclear fuels research.

The summaries, presentation slides, and full papers have been compiled here to provide a basis of information that was exchanged during the course of the meeting, and are provided in the order they were presented.

| 12. KEY WORDS/DESCRIPTORS *(List words or phrases that will assist researchers in locating the report.)* | 13. AVAILABILITY STATEMENT |
|---|---|
| nuclear fuels<br>nuclear safety research<br>reactor safety research | unlimited |
| | 14. SECURITY CLASS FICATION |
| | *(This Page)*<br>unclassified |
| | *(This Report)*<br>unclassified |
| | 15. NUMBER OF PAGES |
| | 16. PRICE |

www.ingramcontent.com/pod-product-compliance
Lightning Source LLC
Chambersburg PA
CBHW081431170526
45166CB00008B/2172